T0337979

FLUID FLOW FOR THE PRACTICING CHEMICAL ENGINEER

FLUID FLOW FOR THE PRACTICING CHEMICAL ENGINEER

J. Patrick Abulencia
Louis Theodore

A JOHN WILEY & SONS, INC., PUBLICATION

Published by John Wiley & Sons, Inc., Hoboken, New Jersey
Published simultaneously in Canada

For general information on our other products and services or for technical support, please contact our Customer Care Department within the United States at (800) 762-2974, outside the United States at (317) 572-3993 or fax (317) 572-4002.

Wiley also publishes its books in a variety of electronic formats. Some content that appears in print may not be available in electronic formats. For more information about Wiley products, visit our web site at www.wiley.com.

Library of Congress Cataloging-in-Publication Data:

Abulencia, James P.
    Fluid flow for the practicing chemical engineer / James P. Abulencia, Louis Theodore.
        p. cm.
    Includes index.
    ISBN 978-0-470-31763-1 (cloth)
    1. Fluid mechanics. 2. Fluidization. I. Title.
    TP156.F6T44 2009
    660.01′532051--dc22

                                                                        2008048336

Printed in the United States of America

10 9 8 7 6 5 4 3 2 1

To
my mother and father
who have unconditionally loved and supported me throughout my life
(J.P.A.)

To
Cecil K. Watson
a friend who has contributed mightily
to basketball and the youth of America
(L.T.)

CONTENTS

IV FLUID FLOW TRANSPORT AND APPLICATIONS   195

17 Prime Movers   197

18 Valves and Fittings   219

19 Flow Measurement   243

20 Ventilation   263

NOTE

Additional Problems for each chapter are available for all readers at www.wiley.com. Follow links for this title.

The above Problems may be used for training and/or homework purposes. Solutions to these Problems plus 10 exams with solutions (5 for each year or semester) are available to those who adopt the text for instructional purposes. A PowerPoint presentation covering all chapters is also available. Visit www.wiley.com for details; follow links for this title.

PREFACE

Persons attempting to find a motive in this narrative will be prosecuted;
Persons attempting to find a moral in it will be banished;
Persons attempting to find a plot in it will be shot
By order of the Author, Mark Twain (Samuel Langhorne Clemens, 1835–1910),
Adventures of Huckleberry Finn

It is becoming more and more apparent that engineering education must provide courses that will include material the engineering student will need and use both professionally and socially later in life. It is no secret that the teaching of Unit Operations—fluid flow, heat transfer, and mass transfer—is now required in any chemical engineering curriculum and is generally accepted as one of the key courses in applied engineering. In addition, this course, or its equivalent, is now slowly and justifiably finding its way into other engineering curricula.

Chemical engineering has traditionally been defined as a synthesis of chemistry, physics and mathematics, tempered with a concern for the dollar sign and applied in the service of humanity. During the 120 years (since 1888) that the profession has been in existence as a separate branch of engineering, humanity's needs have changed tremendously and so has chemical engineering. Thus it is that today, this changing profession faces a challenge and an opportunity to put to better use the advances that have occurred since its birth.

The teaching of Unit Operations at the undergraduate level has remained relatively static since the publication of several early to mid-1900 texts. At this time, however, these and some of the more recent texts in this field are considered by many to be too advanced and of questionable value for the undergraduate engineering student. The present text is the first of three texts to treat the three aforementioned unit operations—fluid flow, heat transfer, and mass transfer. This initial treatise has been written in order to offer the reader the fundamentals of fluid flow with appropriate practical applications, and to possibly serve as an introduction to the specialized and more sophisticated texts in this area.

It is no secret that the teaching of both stoichiometry (material and energy balances) and the three unit operations, including fluid flow, has been a major factor in the success of chemical engineers and chemical engineering since the early 1900s. The authors believe that the approach presented here is a logical step in the continual evolution of this subject that has come to be defined as a unit operation. This "new" treatment of fluid flow is offered in the belief that it will be more effective in training engineers for successful careers in and/or out of the chemical process industry.

The present book has primarily evolved from notes, illustrative examples, problems and exams prepared by the authors for a required three semester fluid flow course given to chemical engineering students at Manhattan College. The course is also offered as an elective to other engineering disciplines in the school and has occasionally been attended by students outside the Department. It is assumed the student has already taken basic college physics and chemistry, and should have as a minimum background in mathematics courses through differential equations.

The course at Manhattan roughly places equal emphasis on principles and applications. However, depending on the needs and desires of the lecturer, either area may be emphasized, and the material in this text is presented in a manner to permit this. Further, no engineering tool is complete without information on how to use it. By the same token, no engineering text is complete without illustrative examples that serve the important purpose of demonstrating the use of the procedures, equations, tables, graphs, etc., presented in the text. There are many such examples. There are also practice problems (available at a website) at the end of each chapter. It is believed that most, if not all, of the illustrative examples and practice problems are "original"; some have been drawn from National Science Foundation (NSF) workshops/seminars conducted at Manhattan College, and some have been employed for over such a long period of time that the original authors can no longer be identified and properly recognized. If that be the case, please accept the authors' apologies and be assured that appropriate credit (where applicable) will be given in the next printing.

In constructing this text, topics of interest to all practicing engineers have been included. The organization and contents of the text can be found in the table of contents. The table consists of six main parts—Introduction to Fluid Flow, Basic Laws, Fluid Transport Classification, Fluid Flow Applications, Fluid-Particle Applications, and Special Topics.

It is hoped that this writing will place in the hands of teachers and students of engineering, plus practicing engineers, a text covering the fundamental principles and applications of fluid flow in a thorough and clear manner. Upon completion of the course, the reader should have acquired not only a working knowledge of the principles of fluid flow, but also experience in their application; and, readers should find themselves approaching advanced texts and the engineering literature with more confidence.

Finally, the authors are particularly indebted to Shannon O'Brien for her extra set of eyes when it came time to proofreading the manuscript.

J. Patrick Abulencia
Louis Theodore

March, 2009

INTRODUCTION

No one means all he says, and yet very few say all they mean, for words are slippery and thought is viscous.

—Henry Brooks Adams (1837–1918)
The Education of Henry Adams

The history of unit operations is interesting. Chemical engineering courses were originally (late 1800 and early 1900s) based on the study of unit processes and/or industrial technologies. However, it soon became apparent that the changes produced in equipment from different industries where similar in nature, i.e., there was a commonality in the fluid flow, heat transfer, and mass transfer operations in the petroleum industry as with the utility industry. These similar operations became known as unit operations.

This book—"Fluid Flow"—was prepared as both a professional book and as an undergraduate text for the study of the principles and fundamentals of the first of the three aforementioned unit operations. Some of the introductory material is presented in the first two parts of the book. Understandably, more extensive coverage is given in the remainder of the book to applications and design. Furthermore, seven additional topics were included in the last part of the book—special topics. These topics are now all required by ABET (Accreditation Board for Engineering and Technology) to be emphasized in course offerings: each of these seven topics is briefly discussed below.

The first chapter in Part VI addresses environmental concerns; nearly one third of undergraduates chose environmental careers. The second topic is health, safety, and accident prevention; new and existing processes today require ongoing analyze in these areas. To better acquaint the student with human relations, engineering and environmental ethics is the third topic. Numerical methods are the next topic encountered since computers are not only used to design multi-component distillation columns but also routinely used in the work force. The success or failure of any business related activity is tied to economics and finance, and this too receives treatment. The "hot" topic—Biomedical Applications—receives treatment in Chapter 33. Finally, open-ended problems (problems that can have more than one solution), are treated in the last chapter. This final chapter requires the reader to ask questions, not always accept things at face value, and select a methodology that will yield the most effective and efficient solution. Illustrative examples on each of these topics are included within each chapter.

Although not a complete treatment of the subject, the text has attempted to present theory, principles, and applications of unit operation in a manner that will benefit the reader and/or prospective engineer in their career as a practicing engineer. Those desiring more information on these topics should proceed to specialized texts in these areas.

This book is the result of several years of effort by the Chemical Engineering Department at Manhattan College. The first rough draft was prepared during the 2001–2002 academic year and underwent peripheral classroom testing during the ensuing years; the manuscript underwent significant revisions during this past year, some of it based on the experiences gained from class testing.

In the final analysis, the problem of what to include and what to omit was particularly difficult. However, every attempt was made to offer engineering course material to individuals at a level that should enable them to better cope with some of the problems they will later encounter in practice. As such, the book was not written for the student planning to pursue advanced degrees; rather, it was primarily written for those individuals who are currently working as practicing engineers or plan to work as engineers in the future solving real world problems.

The entire book can be covered in a three-credit course. At Manhattan, Fluid Flow is taught in the second semester of the sophomore year (Heat and Mass Transfer are taught in the junior year). Finally, it should be again noted that the Manhattan approach is to place more emphasis on the macroscopic approach; however, some microscopic material is included.

I

INTRODUCTION TO
FLUID FLOW

This first part of the book provides an introduction to fluid flow. It contains six chapters and each serves a unique purpose in an attempt to treat important introductory aspects of fluid flow. From a practical point-of-view, systems and plants move liquids and gases from one point to another; hence, the student and/or practicing engineer is concerned with several key topics in this area. These receive some measure of treatment in the six chapters contained in this part. A brief discussion of each chapter follows.

Chapter 1 provides an overview of the History of Chemical Engineering—Fluid Flow. Chapter 2 is concerned with Units and Dimensional Analysis. Chapter 3 introduces Key Terms and Definitions. Chapter 4 provides a discussion of Transport Phenomena versus Unit Operations. The final two chapters introduce the reader to Newtonian Fluids (Chapter 5) and Non-Newtonian Flow (Chapter 6). These subjects are important in developing an understanding of the various fluid flow equipment and operations plus their design, which is discussed later in the text.

Fluid Flow for the Practicing Chemical Engineer. By J. Patrick Abulencia and Louis Theodore
Copyright © 2009 John Wiley & Sons, Inc.

1

HISTORY OF CHEMICAL ENGINEERING — FLUID FLOW

1.1 INTRODUCTION

Although the chemical engineering profession is usually thought to have originated shortly before 1900, many of the processes associated with this discipline were developed in antiquity. For example, filtration operations (see Chapter 27) were carried out 5000 years ago by the Egyptians. During this period, chemical engineering evolved from a mixture of craft, mysticism, incorrect theories, and empirical guesses.

In a very real sense, the chemical industry dates back to prehistoric times when people first attempted to control and modify their environment. The chemical industry developed as any other trade or craft. With little knowledge of chemical science and no means of chemical analysis, the earliest "chemical engineers" had to rely on previous art and superstition. As one would imagine, progress was slow. This changed with time. The chemical industry in the world today is a sprawling complex of raw-material sources, manufacturing plants, and distribution facilities which supplies society with thousands of chemical products, most of which were unknown over a century ago. In the latter half of the nineteenth century, an increased demand arose for engineers trained in the fundamentals of chemical processes. This demand was ultimately met by chemical engineers.

Fluid Flow for the Practicing Chemical Engineer. By J. Patrick Abulencia and Louis Theodore
Copyright © 2009 John Wiley & Sons, Inc.

1.2 FLUID FLOW

With respect to fluid flow, the history of pipes and fittings dates back to the Roman Empire. The ingenious "engineers" of that time came up with a solution for supplying the never-ending demand for fresh water to a city and then disposing of the waste-water produced by the Romans. Their system was based on pipes made out of wood and stone and the driving force of the water was gravity.[1] Over time, many improvements have been made to the piping system. These improvements have included the material choice, shape and size of the pipes; pipes are now made from different metals, plastic, and even glass, with different diameters and wall thicknesses. The next challenge was the connection of the pipes and that was accomplished with fittings. Changes in piping design ultimately resulted from the evolving industrial demands for specific requirements and the properties of fluids that needed to be transported.

The first pump can be traced back to 3000 B.C. in Mesopotamia. It was used to supply water to the crops in the Nile River valley.[2] The pump was a long lever with a weight on one side and a bucket on the other. The use of this first pump became popular in the Middle East and this technology was used for the next 2000 years. Sometimes, a series of pumps would be put in place to provide a constant flow of water to the crops far from the source. Another ancient pump was the bucket chain, a continuous loop of buckets that passed over a pulley-wheel; it is believed that this pump was used to irrigate the Hanging Gardens of Babylon around 600 B.C.[2] The most famous of these early pumps is the Archimedean screw. The pump was invented by the famous Greek mathematician and inventor Archimedes (287–212 B.C.). The pump was made of a metal pipe in which a helix-shaped screw was used to draw water upward as the screw turned. Modern force pumps were adapted from an ancient pump that featured a cylinder with a piston "at the top that create[d] a vacuum and [drew] water upward."[2] The first force pump was designed by Ctesibus of Alexandria, Egypt. Leonardo Da Vinci (1452–1519) was the first to come up with the idea of lifting water by means of centrifugal force; however, the operation of the centrifugal pump was first described scientifically by the French physicist Denis Papin (1647–1714) in 1687.[3] In 1754, Leonhard Euler further developed the principles on which centrifugal pumps operate and today the ideal pump performance term, "Euler head," is named after him.[4] In the United States, the first centrifugal pump to be manufactured was by the Massachusetts Pump Factory. James Stuart built the first multi-stage centrifugal pump in 1849.[3]

1.3 CHEMICAL ENGINEERING

The first attempt to organize the principles of chemical processing and to clarify the professional area of chemical engineering was made in England by George E. Davis. In 1880, he organized a Society of Chemical Engineers and gave a series of lectures in 1887, which were later expanded and published in 1901 as "A Handbook of Chemical Engineering." In 1888, the first course in chemical engineering in the

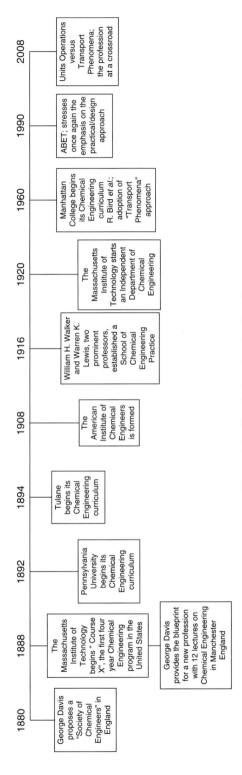

Figure 1.1 Chemical engineering time-line.

United States was organized at the Massachusetts Institute of Technology by Lewis M. Norton, a professor of industrial chemistry. The course applied aspects of chemistry and mechanical engineering to chemical processes.[5]

Chemical engineering began to gain professional acceptance in the early years of the twentieth century. The American Chemical Society was founded in 1876 and, in 1908, it organized a Division of Industrial Chemists and Chemical Engineers while authorizing the publication of the *Journal of Industrial and Engineering Chemistry*. Also in 1908, a group of prominent chemical engineers met in Philadelphia and founded the American Institute of Chemical Engineers.[5]

The mold for what is now called chemical engineering was fashioned at the 1922 meeting of the American Institute of Chemical Engineers when A. D. Little's committee presented its report on chemical engineering education. The 1922 meeting marked the official endorsement of the unit operations concept and saw the approval of a "declaration of independence" for the profession.[5] A key component of this report included the following:

> "Any chemical process, on whatever scale conducted, may be resolved into a coordinated series of what may be termed 'unit operations,' as pulverizing, mixing, heating, roasting, absorbing, precipitation, crystallizing, filtering, dissolving, and so on. The number of these basic unit operations is not very large and relatively few of them are involved in any particular process ... An ability to cope broadly and adequately with the demands of this (the chemical engineer's) profession can be attained only through the analysis of processes into the unit actions as they are carried out on the commercial scale under the conditions imposed by practice."

The key unit operations were ultimately reduced to three: Fluid Flow (the subject title of this text), Heat Transfer, and Mass Transfer. The Little report also went on to state that:

> "Chemical Engineering, as distinguished from the aggregate number of subjects comprised in courses of that name, is not a composite of chemistry and mechanical and civil engineering, but is itself a branch of engineering, ..."

A time line diagram of the history of chemical engineering between the profession's founding to the present day is shown in Fig. 1.1.[5] As can be seen from the time line, the profession has reached a crossroads regarding the future education/curriculum for chemical engineers. This is highlighted by the differences of Transport Phenomena and Unit Operations, a topic that is discussed in Chapter 4.

REFERENCES

1. *http://www.unrv.com/culture/roman-aqueducts.php*, 2004.
2. *http://www.bookrags.com/sciences/sciencehistory/water-pump-woi.html.*

3. A. H. Church and J. Lal, "Centrifugal Pumps and Blowers," John Wiley & Sons Inc., Hoboken, NJ, 1973.

4. R. D. Flack, "Fundamentals of Jet Propulsion with Applications," Cambridge University Press, New York, 2005.

5. N. Serino, "2005 Chemical Engineering 125th Year Anniversary Calendar," term project, submitted to L. Theodore, 2004.

2

UNITS AND DIMENSIONAL ANALYSIS

2.1 INTRODUCTION

This chapter is primarily concerned with units. The units used in the text are consistent with those adopted by the engineering profession in the United States. One usually refers to them as the English or engineering units. Since engineers are often concerned with units and conversion of units, both the English and SI system of units are used throughout the book. All the quantities and the physical and chemical properties are expressed using these two systems.

2.1.1 Units and Dimensional Consistency

Equations are generally dimensional and involve several terms. For the equality to hold, each term in the equation must have the same dimensions (i.e., the equation must be dimensionally homogeneous or consistent). This condition can be easily proved. Throughout the text, great care is exercised in maintaining the dimensional formulas of all terms and the dimensional consistency of each equation. The approach employed will often develop equations and terms in equations by first examining each in specific units (feet rather than length), primarily for the English system. Hopefully, this approach will aid the reader and will attach more physical significance to each term and equation.

Consider now the example of calculating the perimeter, P, of a rectangle with length, L, and height, H. Mathematically, this may be expressed as $P = 2L + 2H$.

Fluid Flow for the Practicing Chemical Engineer. By J. Patrick Abulencia and Louis Theodore
Copyright © 2009 John Wiley & Sons, Inc.

This is about as simple as a mathematical equation can be. However, it only applies when P, L, and H are expressed in the same units.

A conversion constant/factor is a term that is used to obtain units in a more convenient form. All conversion constants have magnitude and units in the term, but can also be shown to be equal to 1.0 (unity) with *no* units. An often used conversion constant is

$$12 \text{ inches/foot}$$

This term is obtained from the following defining equation:

$$12 \text{ in} = 1 \text{ ft}$$

If both sides of this equation are divided by 1 ft one obtains

$$12 \text{ in/ft} = 1.0$$

Note that this conversion constant, like all others, is also equal to unity without any units. Another defining equation is

$$1 \text{ lb}_f = 32.2 \frac{\text{lb} \cdot \text{ft}}{\text{s}^2}$$

If this equation is divided by lb_f, one obtains

$$1.0 = 32.2 \frac{\text{lb} \cdot \text{ft}}{\text{lb}_f \cdot \text{s}^2}$$

This serves to define the conversion constant g_c. Other conversion constants are given in Table A.1 of the Appendix.

Illustrative Example 2.1   Convert the following:

1. 8.03 yr to seconds (s)
2. 150 mile/h to yard/h
3. 100.0 m/s^2 to ft/min^2
4. 0.03 g/cm^3 to lb/ft^3

Solution

1. The following conversion factors are needed:
   365 day/yr
   24 h/day
   60 min/h
   60 s/min

The following is obtained by arranging the conversion factors so that units cancel to leave only the desired units.

$$(8.03 \text{ yr}) \left(\frac{365 \text{ day}}{\text{yr}} \right) \left(\frac{24 \text{ h}}{\text{day}} \right) \left(\frac{60 \text{ min}}{\text{h}} \right) \left(\frac{60 \text{ s}}{\text{min}} \right) = 2.53 \times 10^8 \text{ s}$$

2. In a similar fashion,

$$\left(\frac{150 \text{ mile}}{\text{h}} \right) \left(\frac{5280 \text{ ft}}{\text{mile}} \right) \left(\frac{\text{yd}}{3 \text{ ft}} \right) = 2.6 \times 10^5 \text{ yd/h}$$

3. $(100.0 \text{ m/s}^2) \left(\frac{100 \text{ cm}}{\text{m}} \right) \left(\frac{\text{ft}}{30.48 \text{ cm}} \right) \left(\frac{60 \text{ s}}{\text{min}} \right)^2 = 1.181 \times 10^6 \text{ ft/ min}^2$

4. $(0.03 \text{ g/cm}^3) \left(\frac{\text{lb}}{454 \text{ g}} \right) \left(\frac{30.48 \text{ cm}}{\text{ft}} \right)^3 = 2.0 \text{ lb/ft}^3$

Terms in equations must also be constructed from a "magnitude" viewpoint. Differential terms cannot be equated with finite or integral terms. Care should also be exercised in solving differential equations. In order to solve differential equations to obtain a description of the pressure, temperature, composition, etc., of a system, it is necessary to specify boundary and/or initial conditions for the system. This information arises from a description of the problem or the physical situation. The number of boundary conditions (BC) that must be specified is the sum of the highest-order derivative for each independent differential term. A value of the solution on the boundary of the system is one type of boundary condition. The number of initial conditions (IC) that must be specified is the highest-order time derivative appearing in the differential equation. The value for the solution at time equal to zero constitutes an initial condition. For example, the equation

$$\frac{d^2 v_y}{dz^2} = 0 \tag{2.1}$$

requires 2 BCs (in terms of z). The equation

$$\frac{dT}{dt} = 0; \quad t = \text{time} \tag{2.2}$$

requires 1 IC. And finally, the equation

$$\frac{\partial c_A}{\partial t} = D \frac{\partial^2 c_A}{\partial y^2}; D = \text{diffusivity} \tag{2.3}$$

requires 1 IC and 2 BCs (in terms of y).

2.2 DIMENSIONAL ANALYSIS

Problems are frequently encountered in fluid flow and other engineering work that involve several variables. Engineers are generally interested in developing functional relationships (equations) between these variables. When these variables can be grouped together in such a manner that they can be used to predict the performance of similar pieces of equipment, independent of the scale or size of the operations, something very valuable has been accomplished.

Consider, for example, the problem of establishing a method of calculating the power requirements for mixing liquids in open tanks. The obvious variables would be the depth of liquid in the tank, the density and viscosity of the liquid, the speed of the agitator, the geometry of the agitator, and the diameter of the tank. There are therefore six variables that affect the power, or a total of seven terms that must be considered. To generate a general equation to describe power variation with these variables, a series of tanks having different diameters would have to be set up in order to gather data for various values of each variable. Assuming that ten different values of each of six variables were imposed on the process, 10^6 runs would be required. Obviously, a mathematical method for handling several variables that requires considerably less than one million runs to establish a *design method* must be available. In fact, such a method is available and it is defined as *dimensional analysis.*[1]

Dimensional analysis is a powerful tool that is employed in planning experiments, presenting data compactly, and making practical predictions from models without detailed mathematical analysis. The first step in an analysis of this nature is to write down the units of each variable. The end result of a dimensional analysis is a list of pertinent dimensionless numbers. A partial list of common dimensionless numbers used in fluid flow analyses is given in Table 2.1.

Dimensional analysis is a relatively "compact" technique for reducing the number and the complexity of the variables affecting a given phenomenon, process or calculation. It can help obtain not only the most out of experimental data but also scale-up data from a model to a prototype. To do this, one must achieve similarity between the prototype and the model. This similarity may be achieved through dimensional analysis by determining the important dimensionless numbers, and then designing the model and prototype such that the important dimensionless numbers are the same in both.

There are three steps in dimensional analysis. These are:

1. List all parameters and their primary units.
2. Formulate dimensionless numbers (or ratios).
3. Develop the relation between the dimensionless numbers experimentally.

Further details on this approach are provided in the next section.

Table 2.1   Dimensionless numbers

Parameter	Definition	Importance	Qualitative Ratio
Cavitation number	$Ca = \dfrac{P - p'}{\rho v^2/2}$	Cavitation	$\dfrac{\text{Pressure}}{\text{Inertia}}$
Eckert number	$Ec = \dfrac{v^2}{C_p \Delta T}$	Dissipation	$\dfrac{\text{Kinetic energy}}{\text{Inertia}}$
Euler number	$Eu = \dfrac{\Delta P}{\rho v^2/2}$	Pressure drop	$\dfrac{\text{Pressure}}{\text{Inertia}}$
Froude number	$Fr = \dfrac{v^2}{gL}$	Free surface flow	$\dfrac{\text{Inertia}}{\text{Gravity}}$
Mach number	$Ma = \dfrac{v}{c}$	Compressible flow	$\dfrac{\text{Flow speed}}{\text{Sound speed}}$
Poiseuille number	$P_0 = \dfrac{D^2 \Delta P}{\mu L v}$	Laminar flow in pipes	$\dfrac{\text{Pressure}}{\text{Viscous forces}}$
Relative roughness	$\dfrac{k}{D}$	Turbulent flow, rough walls	$\dfrac{\text{Wall roughness}}{\text{Body length}}$
Reynolds number	$Re = \dfrac{\rho v D}{\mu} = \dfrac{vD}{v}$	Various uses	$\dfrac{\text{Inertia forces}}{\text{Viscous forces}}$
Strouhal number	$St = \dfrac{\varpi L}{v}$	Oscillating flow	$\dfrac{\text{Oscillation speed}}{\text{Mean speed}}$
Weber number	$We = \dfrac{\rho v^2 L}{\sigma}$	Surface forces effect	$\dfrac{\text{Inertia}}{\text{Surface tension}}$

Note: p' = vapor pressure, C_p = heat capacity.

2.3   BUCKINGHAM Pi (π) THEOREM

This theorem provides a simple method to obtain dimensionless numbers (or ratios) termed π parameters. The steps employed in obtaining the dimensionless π parameters are given below[2]:

1. List all parameters. Define the number of parameters as n.
2. Select a set of primary dimensions, e.g., kg, m, s, K (English units may also be employed). Let $r =$ the number of primary dimensions.
3. List the units of all parameters in terms of the primary dimensions, e.g., L [=] m, where "[=]" means "has the units of." This is a critical step and often requires some creativity and ingenuity on the part of the individual performing the analysis.

4. Select a number of variables from the list of parameters (equal to r). These are called repeating variables. The selected repeating parameters must include all r independent primary dimensions. The remaining parameters are called "non-repeating" variables.
5. Set up dimensional equations by combing the repeating parameters with each of the other non-repeating parameters in turn to form the dimensionless parameters, π. There will be $(n-r)$ dimensionless groups of (πs).
6. Check that each resulting π group is in fact dimensionless.

Note that it is permissible to form a different π group from the product or division of other πs, e.g.,

$$\pi_5 = \frac{2\pi_1 \pi_2}{\pi_3^2} \quad \text{or} \quad \pi_6 = \frac{1}{\pi_4} \tag{2.4}$$

Note, however, that a dimensional analysis approach will fail if the fundamental variables are not correctly chosen. The Buckingham Pi theorem approach to dimensionless numbers is given in the Illustrative Example that follows.

Illustrative Example 2.2 When a fluid flows through a horizontal circular pipe, it undergoes a pressure drop, $\Delta P = (P_2 - P_1)$. For a rough pipe, ΔP will be higher than a smooth pipe. The extent of non-smoothness of a material is expressed in terms of the roughness, k. For steady state incompressible Newtonian (see Chapter 5) fluid flow, the pressure drop is believed to be a function of the fluid average velocity v, viscosity μ, density ρ, pipe diameter D, length L, and roughness k (discussed in more detail in Chapter 14), and the speed of sound in fluid (an important variable if the flow is compressible) c, i.e.,

$$\Delta P = f(v, \ \mu, \ \rho, \ D, \ L, \ k, \ c)$$

Determine the dimensionless numbers of importance for this flow system.

Solution A pictorial representation of the system in question is provided in Fig. 2.1.

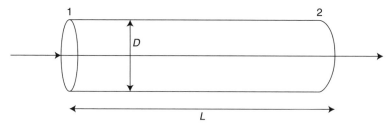

Figure 2.1 Pipe.

List all parameters and find the value of n:

$$\Delta P, \ v, \ \mu, \ \rho, \ D, \ L, \ k, \ c$$

Therefore $n = 8$.
   Choose primary units (employ SI)

$$\text{m, s, kg, K}$$

List the primary units of each parameter:

$$\Delta P \ [=] \ \text{Pa} = \text{kg m}^{-1}\text{s}^{-2}$$

$$v \ [=] \ \text{m s}^{-1}$$

$$\mu \ [=] \ \text{kg m}^{-1}\text{s}^{-1}$$

$$D \ [=] \ \text{m}$$

$$L \ [=] \ \text{m}$$

$$\rho \ [=] \ \text{kg m}^{-3}$$

$$k \ [=] \ \text{m}$$

$$c \ [=] \ \text{m s}^{-1}$$

Therefore $r = 3$ with primary units m, s, kg.
   Select three parameters from the list of eight parameters. These are the repeating variables:

$$D \ [=] \ \text{m}$$

$$\rho \ [=] \ \text{kg m}^{-3}$$

$$v \ [=] \ \text{m s}^{-1}$$

The non-repeating parameters are then ΔP, μ, k, c, and L.
   Determine the number of πs:

$$n - r = 8 - 3 = 5$$

Formulate the first π, π_1, employing ΔP as the non-repeating parameter

$$\pi_1 = \Delta P v^a \rho^b D^f$$

Determine a, b, and f by comparing the units on both sides of the following equation:

$$0 \ [=] \ (\text{kg m}^{-1}\text{s}^{-2})(\text{m s}^{-1})^a(\text{kg m}^{-3})^b(\text{m})^f$$

Compare kg:

$$0 = 1 + b. \text{ Therefore } b = -1$$

Compare s:

$$0 = -2 - a. \text{ Therefore } a = -2$$

Compare m:

$$0 = -1 + a - 3b + f. \text{ Therefore } f = 0$$

Substituting back into π_1 leads to:

$$\pi_1 = \Delta P v^{-2} \rho^{-1} = \frac{\Delta P}{\rho v^2}$$

This represents the Euler number (see Table 2.1). Formulate the second π, π_2 as

$$\pi_2 = \mu v^a \rho^b D^f$$

Determine a, b, and f by comparing the units on both sides:

$$0 \,[=]\, (\text{kg m}^{-1}\text{s}^{-1})(\text{m s}^{-1})^a(\text{kg m}^{-3})^b(\text{m})^f$$

Compare kg:

$$0 = 1 + b. \text{ Therefore } b = -1$$

Compare s:

$$0 = -1 - a. \text{ Therefore } a = -1$$

Compare m:

$$0 = -1 + a - 3b + f. \text{ Therefore } f = -1$$

Substituting back into π_2 yields:

$$\pi_2 = \mu v^{-1} \rho^{-1} D^{-1} = \frac{\mu}{v \rho D}$$

Replace π_2 by its reciprocal:

$$\pi_2 = \frac{v \rho D}{\mu} = \text{Re}$$

where Re = Reynolds number (see Chapter 12).
    Similarly, the remaining non-repeating variables lead to

$$\pi_3 = k v^a \rho^b D^f \longrightarrow \frac{k}{D}$$

and

$$\pi_4 = cv^a \rho^b D^f \longrightarrow \frac{c}{v} \text{ take inverse}$$

$$\pi_4 = \frac{v}{c} = \text{the Mach number (see Chapter 15)}$$

Similarly,

$$\pi_5 = \frac{L}{D}$$

Combine the πs into an equation, expressing π_1 as a function of π_2, π_3, π_4, and π_5:

$$Eu = \frac{\Delta P}{\rho v^2/2} = f\left(\text{Re}, \frac{k}{D}, Ma, \frac{L}{D}\right) = \text{the Euler number}$$

Consider the case of incompressible flow

$$Eu = \frac{\Delta P}{\rho v^2/2} = f\left(\text{Re}, \frac{k}{D}, \frac{L}{D}\right)$$

The result indicates that to achieve similarity between a model (m) and a prototype (p), one must have the following:

$$\text{Re}_m = \text{Re}_p,$$

$$(k/D)_m = (k/D)_p, \text{ and}$$

$$(L/D)_m = (L/D)_p$$

Since $Eu = f(\text{Re}, k/D, L/D)$, then it follows that $Eu_m = Eu_p$ (see Table 2.1).

2.4 SCALE-UP AND SIMILARITY

To scale-up (or scale-down) a process, it is necessary to establish geometric and dynamic similarities between the model and the prototype. These two similarities are discussed below.

Geometric similarity implies using the same geometry of equipment. A circular pipe prototype should be modeled by a tube in the model. Geometric similarity establishes the scale of the model/prototype design. A 1/10th scale model means that the characteristic dimension of the model is 1/10th that of the prototype.

Dynamic similarity implies that the important dimensionless numbers must be the same in the model and the prototype. For a particle settling in a fluid, it has been shown (see Chapter 23) that the drag coefficient, C_D, is a function of the dimensionless Reynolds number, Re, i.e.:

$$C_D = f(\text{Re}) \tag{2.5}$$

By selecting the operating conditions such that Re in the model equals the Re in the prototype, then the drag coefficient (or *friction factor*) in the prototype equals the friction factor in the model.

REFERENCES

1. I. Farag and J. Reynolds, "Fluid Flow", A Theodore Tutorial, East Williston, NY, 1995.
2. W. Badger and J. Banchero, "Introduction to Chemical Engineering", McGraw-Hill, New York, 1955.

NOTE: Additional problems are available for all readers at www.wiley.com. Follow links for this title.

3

KEY TERMS AND DEFINITIONS

3.1 INTRODUCTION

This chapter is concerned with key terms and definitions in fluid flow. Since fluid flow is an important subject that finds wide application in engineering, the understanding of "fluid" flow jargon is therefore important to the practicing engineer. The handling and flow of either gases or liquids is much simpler, cheaper, and less troublesome than solids. Consequently, the engineer attempts to transport most quantities in the form of gases or liquids whenever possible. It is important to note that throughout this book, the word "fluid" will always be used to include both liquids and gases.

The mechanics of fluids are treated in most physics courses and form the basis of the subject of fluid flow and hydraulics. Key terms in these two topics that are of special interest to engineers are covered in this chapter. Fluid mechanics includes two topics: statics and dynamics. Fluid statics treats fluids at rest while fluid dynamics treats fluids in motion. The definition of key terms in this subject area is presented in Section 3.2.

3.1.1 Fluids

For the purpose of this text, a fluid may be defined as a substance that does not permanently resist distortion. An attempt to change the shape of a mass of fluid will result in layers of fluid sliding over one another until a new shape is attained. During the change in shape, shear stresses (forces parallel to a surface) will result,

Fluid Flow for the Practicing Chemical Engineer. By J. Patrick Abulencia and Louis Theodore
Copyright © 2009 John Wiley & Sons, Inc.

the magnitude of which depends upon the viscosity (to be discussed shortly) of the fluid and the rate of sliding. However, when a final shape is reached, all shear stresses will have disappeared. Thus, a fluid at equilibrium is free from shear stresses. This definition applies for both liquids and gases.

3.2 DEFINITIONS

Standard key definitions, particularly as they apply to fluid flow, follow.

3.2.1 Temperature

Whether in a gaseous, liquid, or solid state, all molecules possess some degree of kinetic energy; that is, they are in constant motion—vibrating, rotating, or translating. The kinetic energies of individual molecules cannot be measured, but the combined effect of these energies in a very large number of molecules can. This measurable quantity is known as *temperature*; it is a macroscopic concept only and as such does not exist on the molecular level.

Temperature can be measured in many ways; the most common method makes use of the expansion of mercury (usually encased inside a glass capillary tube) with increasing temperature. (However, thermocouples or thermistors are more commonly employed in industry.) The two most commonly used temperature scales are the Celsius (or Centigrade) and Fahrenheit scales. The Celsius scale is based on the boiling and freezing points of water at 1-atm pressure; to the former, a value of 100°C is assigned, and to the latter, a value of 0°C. On the older Fahrenheit scale, these temperatures correspond to 212°F and 32°F, respectively. Equations (3.1) and (3.2) show the conversion from one scale to the other:

$$°F = 1.8(°C) + 32 \qquad (3.1)$$

$$°C = (°F - 32)/1.8 \qquad (3.2)$$

where °F = a temperature on the Fahrenheit scale and °C = a temperature on the Celsius scale.

Experiments with gases at low-to-moderate pressures (up to a few atmospheres) have shown that, if the pressure is kept constant, the volume of a gas and its temperature are linearly related (see Chapter 11—Charles' law) and that a decrease of 0.3663% or (1/273) of the initial volume is experienced for every temperature drop of 1°C. These experiments were not extended to very low temperatures, but if the linear relationship were extrapolated, the volume of the gas would *theoretically* be zero at a temperature of approximately −273°C or −460°F. This temperature has become known as *absolute zero* and is the basis for the definition of two *absolute* temperature scales. (An *absolute* scale is one that does not allow negative quantities.) These absolute temperature scales are the Kelvin (K) and Rankine (°R) scales; the former is defined by shifting the Celsius scale by 273°C so that 0 K is equal to −273°C. The Rankine scale is defined by shifting the Fahrenheit scale by 460°.

Equation (3.3) shows this relationship for both absolute temperatures:

$$K = {}^{\circ}C + 273$$
$$^{\circ}R = {}^{\circ}F + 460$$

(3.3)

3.2.2 Pressure

There are a number of different methods used to express a pressure term or measurement. Some of them are based on a force per unit area (e.g., pound-force per square inch, dyne, and so on) and others are based on fluid height (e.g., inches of water, millimeters of mercury, etc.). Pressure units based on fluid height are convenient when the pressure is indicated by a difference between two levels of a liquid. Standard barometric (or atmospheric) pressure is 1 atm and is equivalent to 14.7 psi, 33.91 ft of water, and 29.92 inches of mercury.

Gauge pressure is the pressure relative to the surrounding (or atmospheric) pressure and it is related to the absolute pressure by the following equation:

$$P = P_a + P_g$$

(3.4)

where P is the absolute pressure (psia), P_a is the atmospheric pressure (psi) and P_g is the gauge pressure. The absolute pressure scale is absolute in the same sense that the absolute temperature scale is absolute; i.e., a pressure of zero psia is the lowest possible pressure theoretically achievable—a perfect vacuum.

In stationary fluids subjected to a gravitational field, the *hydrostatic pressure difference* between two locations A and B is defined as

$$P_A - P_B = -\int_{z_A}^{z_B} \rho g \, dz$$

(3.5)

where z is a vertical upwards direction, g is the gravitational acceleration, and ρ is the fluid density. This equation will be revisited in Chapter 10.

Expressed in various units, the standard atmosphere is equal to 1.00 atmosphere (atm), 33.91 feet of water (ft H_2O), 14.7 pound-force per square inch absolute (psia), 2116 pound-force per square foot (psfa), 29.92 inches of mercury (in Hg), 760.0 millimeters of mercury (mm Hg), and 1.013×10^5 Newtons per square meter (N/m^2). The pressure term will be reviewed again in several later chapters.

Vapor pressure, usually denoted p', is an important property of liquids and, to a much lesser extent, of solids. If a liquid is allowed to evaporate in a confined space, the pressure in the vapor space increases as the amount of vapor increases. If there is sufficient liquid present, a point is eventually reached at which the pressure in the vapor space is exactly equal to the pressure exerted by the liquid at its own surface. At this point, a dynamic equilibrium exists in which vaporization and condensation take place at equal rates and the pressure in the vapor space remains constant. The pressure exerted at equilibrium is called the vapor pressure of the liquid. The magnitude of this pressure for a given liquid depends on the temperature, but not on the amount of liquid present. Solids, like liquids, also exert a vapor pressure. Evaporation of solids (called *sublimation*) is noticeable only for those with appreciable vapor pressures.

3.2.3   Density

At a given temperature and pressure, a fluid possesses density, ρ, which is measured as mass per unit volume. The density of a fluid depends on both temperature and pressure; if a fluid is not affected by changes in pressure, it is said to be incompressible, and most liquids are incompressible. The density of a liquid can, however, change if there are extreme changes in temperature, and not appreciably affected by moderate changes in pressure. In the case of gases, the density may be affected appreciably by both temperature and pressure. Gases subjected to small changes in pressure and temperature vary so little in density that they can be considered incompressible and the change in density can be neglected without serious error. Density, specific gravity, and other similar properties have the same significance for fluids as for solids.

3.2.4   Viscosity

Viscosity, μ, is an important fluid property that provides a measure of the resistance to flow. The viscosity is frequently referred to as the *absolute* or *dynamic* viscosity. The principal reason for the difference in the flow characteristics of water and of molasses is that molasses has a much higher viscosity than water. Note also that the viscosity of a liquid decreases with increasing temperature, while the viscosity of a gas increases with increasing temperature.

One set of units of viscosity in SI units is $g/(cm \cdot s)$, which is defined as a poise (P). Since this numerical unit is somewhat high for many engineering applications, viscosities are frequently reported in centipoises (cP) where one poise is 100 centipoises. In English or engineering units, the dimensions of viscosity are in lb/ft $\cdot$ s. To convert from poises to this unit, one may simply multiply by (30.48/453.6) or (0.0672); to convert from centipoises, multiply by 6.72×10^{-4}. To convert centipoises to lb/ft $\cdot$ hr, multiply by 2.42.

Kinematic viscosity, ν, is the absolute viscosity divided by the density (μ/ρ) and has the dimensions of (volume)/length $\cdot$ time. The corresponding unit to the poise is the stoke, having the SI dimensions of cm^2/s. The specific viscosity is the ratio of the viscosity to the viscosity of a standard fluid expressed in the same units and measured at the same temperature and pressure. Although all real fluids possess viscosity, an ideal fluid is a hypothetical fluid that has a viscosity of zero and possesses no resistance to shear.

The viscosity is a fluid property listed in many engineering books, including Perry's Handbook.[1] Data are given as tables, charts, or nomographs. Figures B.1 and B.2 (see Appendix) are two nomographs that can be used to obtain the absolute (or dynamic) viscosity of liquids and gases, respectively.[2,3] In addition, the kinematic viscosities of some common liquids and gases at a temperature of 20°C are listed[2,3] in Tables A.2 and A.3, respectively (see Appendix).

Illustrative Example 3.1   To illustrate the use of nomograph, calculate the dynamic viscosity of a 98% sulfuric acid solution at 45°C.

Solution   From Fig. B.1 in the Appendix, the coordinates of 98% H_2SO_4 are given as X = 7.0 and Y = 24.8 (number 97). Locate these coordinates on the grid and call it point A. From 45°C, draw a straight line through point A and extend it to cut the viscosity axis. The intersection occurs at approximately 12 centipoise (cP). Therefore,

$$\mu = 12\,cP = 0.12\,P = 0.12\,g/cm \cdot s$$

3.2.5   Surface Tension: Capillary Rise

A liquid forms an interface with another fluid. At the surface, the molecules are more densely packed than those within the fluid. This results in surface tension effects and interfacial phenomena. The surface tension coefficient, σ, is the force per unit length of the circumference of the interface, or the energy per unit area of the interface area. The surface tension for water is listed in Table A.4 (see Appendix).

Surface tension causes a *contact angle* to appear when a liquid interface is in contact with a solid surface, as shown in Fig. 3.1. If the contact angle θ is <90°, the liquid is termed *wetting*. If $\theta > 90°$, it is a *nonwetting* liquid. Surface tension causes a fluid interface to rise (or fall) in a capillary tube. The capillary rise is obtained by equating the vertical component of the surface tension force, F_σ, to the weight of the liquid of height h, F_g (see Fig. 3.2). These two forces are shown

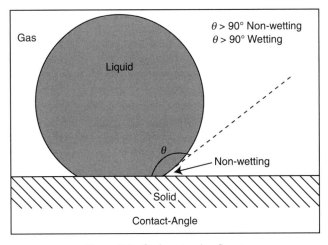

Figure 3.1   Surface tension figure.

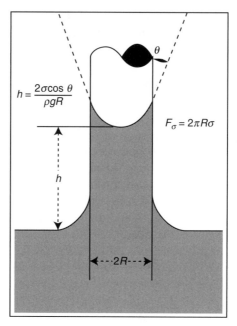

Figure 3.2 Capillary rise in a circular tube.

in the following equations:

$$F_\sigma = 2\pi R\sigma \cos\theta \qquad (3.6)$$

$$F_g = \rho g \pi R^2 h \qquad (3.7)$$

Equating the above two forces gives:

$$2\pi r\sigma \cos\theta = \rho g \pi R^2 h$$

$$h = \frac{2\sigma \cos\theta}{\rho g R} \qquad (3.8)$$

where σ is the surface tension (N/m), θ the contact angle, ρ the liquid density (kg/m³), g is the acceleration due to gravity (9.807 m/s²), and R is the tube radius (m).

For a droplet, the pressure is higher on the inside than on the outside. The pressure increase in the interior of the liquid droplet is balanced by the surface tension force. By applying a force balance on the interior of a spherical droplet, see Fig. 3.3, one can obtain the force due to the pressure increase, F_p, which equals the surface tension force on the ring, F_σ (see Eqs. 3.9 and 3.10). This force balance neglects the weight of the liquid in the droplet

$$F_p = \pi r^2 \Delta P \qquad (3.9)$$

$$F_\sigma = 2\pi r\sigma \qquad (3.10)$$

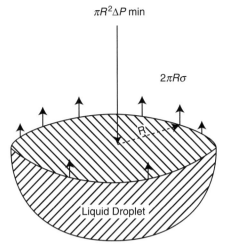

$\pi R^2 \Delta P$ min

$2\pi R\sigma$

Liquid Droplet

Figure 3.3 Surface tension in a spherical droplet.

Equating the two forces gives,

$$\pi r^2 \Delta P = 2\pi r \sigma \tag{3.11}$$

The pressure increase is therefore,

$$\Delta P = \frac{2\sigma}{r} \tag{3.12}$$

where ΔP is the pressure increase (Pa or psi) and r is the droplet radius (m or ft).

Illustrative Example 3.2   A capillary tube is inserted into a liquid. Determine the rise, h, of the liquid interface inside the capillary tube. Data are provided below.

Liquid-gas system is water-air
Temperature is 30°C and pressure is 1 atm
Capillary tube diameter = 8 mm = 0.008 m
Water density = 1000 kg/m^3
Contact angle, $\theta = 0°$

Solution   The height equation is first written

$$h = \frac{2\sigma \cos\theta}{\rho g R} \tag{3.8}$$

The surface tension of water (see Table A.4 in the Appendix) at 30°C is

$$\sigma = 0.0712\,\text{N/m} = 0.0712\,\text{kg/s}^2$$

The height is therefore

$$h = \frac{(2)(0.0712)\cos 0°}{(1000)(9.807)(0.004)}$$

$$= 0.00363 \text{ m} = 3.63 \text{ mm}$$

Note that for most industrial applications involving pipes, the diameters are large enough that any capillary rise may be neglected.

Illustrative Example 3.3   At 30°C, what diameter glass tube is necessary to keep the capillary height change of water less than one millimeter? Assume negligible angle of contact.

Solution   For air-water-glass, assume the contact angle $\theta = 0$, noting that $\cos(0°) = 1$. Obtain the properties of water from Table A.2 in the Appendix.

$$\rho = 996 \text{ kg/m}^3$$

$$\sigma = 0.071 \text{ N/m} \text{ (surface tension)}$$

Use the capillary rise Equation (3.8) to calculate the tube radius

$$h = \frac{2\sigma\cos\theta}{\rho g R}$$

$$R = \frac{2\sigma\cos\theta}{\rho g h} = \frac{2(0.071)(1)}{(996)(9.807)(0.001)} = 0.0145 \text{ m} = 14.5 \text{ mm}$$

$$D = 29 \text{ mm}$$

If the tube diameter is greater than 29 mm, then the capillary rise will be less than 1 mm.

3.2.6   Newton's Law

The relationship between force mass, velocity, and acceleration may be expressed by Newton's second law with force equaling the time rate of change of momentum, $\dot{M}$.

$$F = \frac{1}{g_c}\frac{\mathrm{d}(mv)}{\mathrm{d}t} = \frac{\mathrm{d}M}{\mathrm{d}t} = \dot{M} \tag{3.13}$$

If the mass is constant,

$$F = \frac{ma}{g_c} \tag{3.14}$$

where a = acceleration or $\mathrm{d}v/\mathrm{d}t$.

In the English engineering system of units, the pound-force (lb_f) is defined as that force which accelerates 1 pound-mass (lb) 32.174 ft/s. Newton's law must therefore include a dimensional conversion constant for consistency. This constant, g_c, is 32.174 $(lb/lb_f)(ft/s^2)$. When employing SI units, the value of g_c becomes unity and has no dimensions associated with it, i.e., $g_c = 1.0$ (see previous chapter for more details). Thus, the g_c term is normally retained in equations involving force where English units are employed. The SI unit of force is the Newton (N), which simply expresses force F as the product of mass m and acceleration a (see Equation 3.14 once again). The Newton is defined as the force, when applied to a mass of 1 kg, produces an acceleration of 1 m/s^2; the term g_c is not retained in this (and similar) equations when SI units are employed.

The term g_c is carried in most of the force and force-related terms and equations presented in this and the following chapters. Although both sets of units are employed in the Illustrative Examples and Problems, the reader should note that despite statements to the contrary by academics and theorists, English units are almost exclusively employed by industry in the US.

As described earlier, pressure is a force per unit area. The conversion of force per unit area (S) to a height of fluid follows from Newton's law, i.e.,

$$P = \frac{F}{S} = m\frac{g}{g_c} \bigg/ S \tag{3.15}$$

and

$$m = \rho S h \tag{3.16}$$

Thus, a vertical column of a given fluid under the influence of gravity exerts a pressure at its base that is directly proportional to its height so that pressure may also be expressed as the equivalent height of a fluid column. The pressure to which a fluid height corresponds may be determined from the density of the fluid and the local acceleration of gravity.

Forces that act on a fluid can be classified as either *body forces* or *surface forces*. Body forces are distributed throughout the material, e.g., gravitational, centrifugal, and electromagnetic forces. *Body forces* therefore act on the bulk of the object from a distance and are proportional to its mass; the most common examples are the aforementioned gravitational and electromagnetic forces. *Surface forces* are forces that act on the surface of a material. Surface forces are exerted on the surface of the object by other objects in contact with it; they generally increase with increasing contact area. *Stress* is a force per unit area. If the force is parallel to the surface, the force per unit area is called *shear stress*. When the force is perpendicular (normal) to a surface, the force per unit area is called *normal stress* or *pressure*.

For a stationary (static, non-moving) fluid, the sum of all forces acting on the fluid ($\sum F$) is zero. Newton's second law simplifies to

$$\sum F = 0 \tag{3.17}$$

When there are two opposing forces, for example, a gravity force and a pressure force, P, (acting on a surface) is then

$$F_{pres} = F_{grav}$$

$$F_{pres} = P(S)$$

$$F_{grav} = m(g/g_c)$$

Equating the two forces gives the result described in Equation (3.15)

$$m(g/g_c) = PS \tag{3.18}$$

Illustrative Example 3.4   Given a force $F = 10\,lb_f$, acting on a surface of area $S = 2\,ft^2$, at an angle $\theta = 30°$ to the normal of the surface. Determine the magnitude of the normal and parallel force components, the shear stress, and the pressure.

Solution   When a force acts at an angle to a surface, the component of the force parallel to that surface is $F\cos\theta$. Noting that $\cos(30°) = 0.866$.

$$F_{para} = F\cos\theta = 10\cos(30°)$$
$$= 8.66\,lb_f$$

The normal (perpendicular) component of the force is $F\sin\theta$, noting that $\sin(30°) = 0.500$.

$$F_{norm} = F\sin\theta = 10\sin(30°)$$
$$= 5\,lb_f$$

The shear stress, τ, is defined as

$$\tau = \frac{F_{para}}{S} = \frac{8.66}{2}$$
$$= 4.33\,psf$$

Likewise, the pressure, P, is defined as

$$P = \frac{F_{norm}}{S} = \frac{5}{2}$$
$$= 2.50\,psf$$

3.2.7   Kinetic Energy

Consider a body of mass, m, that is acted upon by a force, F. If the mass is displaced a distance, dL, during a differential interval of time, dt, the energy expended is given by

$$dE_k = m\frac{a}{g_c}dL \tag{3.19}$$

Since the acceleration is given by $a = dv/dt$,

$$dE_k = \frac{m}{g_c}\frac{dv}{dt}\,dL = \frac{m}{g_c}\frac{dL}{dt}\,dv \tag{3.20}$$

Noting that $v = dL/dt$, the above expression becomes:

$$dE_k = m\frac{v}{g_c}\,dv \tag{3.21}$$

If this equation is integrated from v_1 to v_2, the change in energy is

$$\Delta E_k = \frac{m}{g_c}\int_{v_1}^{v_2} v\,dv = \frac{m}{g_c}\left(\frac{v_2^2}{2} - \frac{v_1^2}{2}\right) \tag{3.22}$$

or

$$\Delta E_k = \left(\frac{mv_2^2}{2g_c} - \frac{mv_1^2}{2g_c}\right) = \Delta\left(\frac{mv^2}{2g_c}\right) \tag{3.23}$$

The term above is defined as the change in kinetic energy.

The reader should note that for flow through conduits, the above kinetic energy term can be retained as written if the velocity profile is uniform; that is, the local velocities at all points in the cross-section are the same. Ordinarily, there is a velocity gradient across the passage; this introduces an error, the magnitude of which depends on the nature of the velocity profile and the shape of the cross section. For the usual case where the velocity is approximately uniform (e.g., turbulent flow) (see Chapter 14), the error is not serious, and since the error tends to cancel because of the appearance of kinetic terms on each side of any energy balance equation, it is customary to ignore the effect of velocity gradients. When the error cannot be ignored, the introduction of a correction factor, that is used to multiply the v^2/g_c term, is needed. This is quantitatively treated in Chapter 8.

3.2.8 Potential Energy

A body of mass m is raised vertically from an initial position z_1 to z_2. For this condition, an upward force at least equal to the weight of the body must be exerted on it, and this force must move through the distance $z_2 - z_1$. Since the weight of the body is the force of gravity on it, the minimum force required is again given by Newton's law:

$$F = \frac{ma}{g_c} = m\frac{g}{g_c} \tag{3.24}$$

where g is the local acceleration of gravity. The minimum work required to raise the body is the product of this force and the change in vertical displacement, that is,

$$\Delta E_{PE} = F(z_2 - z_1) = m\frac{g}{g_c}(z_2 - z_1) = \Delta\left(m\frac{g}{g_c}z\right) \tag{3.25}$$

The term above is defined as the potential energy of the mass.

Illustrative Example 3.5 As part of a fluid flow course, a young environmental engineering major has been requested to determine the potential energy of water before it flows over a waterfall 10 meters in height above ground level conditions.

Solution The potential energy of water depends on two considerations:

1. the quantity of water, and
2. a reference height.

For the problem at hand, take as a basis 1 kilogram of water and assume the potential energy to be zero at ground level conditions. Apply Equation (3.25) based on the problem statement, set $z_1 = 0$ m and $z_2 = 10$ m, so that

$$\Delta z = 10\,\text{m}$$

At ground level conditions,

$$PE_1 = 0$$

Therefore

$$\Delta(PE) = PE_2 - PE_1 = PE_2$$
$$PE_2 = m(g/g_c)z_2$$
$$= (1\,\text{kg})(9.8\,\text{m/s}^2)(10\,\text{m})$$
$$= 98\,\text{kg} \cdot \text{m}^2/\text{s}^2$$
$$= 98\,\text{J}$$

REFERENCES

1. A. Foust, L. Wenzel, C. Clump, L. Maus, and L. Andrews, "Principles of Unit Operations", John Wiley & Sons, Hoboken, NJ, 1950.
2. J. Santoleri, J. Reynolds, and L. Theodore, "Introduction to Hazardous Waste Incineration", 2nd edition, John Wiley and Sons, Hoboken, NJ, 2000.
3. D. Green and R. Perry, "Perry's Chemical Engineers' Handbook", 8th edition, McGraw-Hill, New York, 2008.
4. C. Lapple, "Fluid and Particle Mechanics", University of Delaware, Newark, Delaware, 1951.

NOTE: Additional problems are available for all readers at www.wiley.com. Follow links for this title.

4

TRANSPORT PHENOMENA VERSUS UNIT OPERATIONS

4.1 INTRODUCTION

As indicated in Chapter 1, chemical engineering courses were originally based on the study of unit processes and/or industrial technologies. It soon became apparent that the changes produced in equipment from different industries were similar in nature; i.e., there was a commonality in the fluid flow operations in the petroleum industry as with the utility industry. These similar operations became known as Unit Operations. This approach to chemical engineering was promulgated in the Little report as discussed earlier in Chapter 1 and to varying degrees and emphasis, has dominated the profession to this day.

The Unit Operations approach was adopted by the profession soon after its inception. During many years (since 1880) that the profession has been in existence as a branch of engineering, society's needs have changed tremendously and, in turn, so has chemical engineering.

The teaching of Unit Operations at the undergraduate level remained relatively static since the publication of several early-to-mid 1900 texts. However, by the middle of the 20th century, there was a slow movement from the unit operation concept to a more theoretical treatment called transport phenomena. The focal point of this science was the rigorous mathematical description of all physical rate processes in terms of mass, heat, or momentum crossing boundaries. This approach took hold of the education/curriculum of the profession with the publication of the

Fluid Flow for the Practicing Chemical Engineer. By J. Patrick Abulencia and Louis Theodore
Copyright © 2009 John Wiley & Sons, Inc.

first edition of the Bird *et al.* book.[1] Some, including both authors of this text, feel that this concept set the profession back several decades since graduating chemical engineers, in terms of training, were more applied physicists that traditional chemical engineers.

There has fortunately been a return to the traditional approach of chemical engineering in recent years, primarily due to the efforts of the Accreditation Board for Engineering Technology (ABET). Detractors to this approach argue that this type of practical education experience provides the answers to 'what' and 'how', but not 'why' (i.e., a greater understanding of both physical and chemical processes). However, the reality is that nearly all practicing engineers (including chemical engineers) are in no way presently involved with the 'why' questions; material normally covered here has been replaced, in part, with a new emphasis on solving design and open-ended problems. This approach is emphasized in this text.

4.2   THE DIFFERENCES

This section attempts to qualitatively describe the differences between the two approaches discussed above. Both deal with the transfer of certain quantities (momentum, energy, and mass) from one point in a system to another. Three basic transport mechanisms are involved in a process. They are:

1. Radiation.
2. Convection.
3. Molecular diffusion.

The first mechanism, radiative transfer, arises due to wave motion and is not considered since it may be justifiably neglected in most engineering applications. Convective transfer occurs simply due to bulk motion. One may define molecular diffusion as the transport mechanism arising due to gradients. For example, momentum is transferred in the presence of a velocity gradient; energy in the form of heat is transferred due to a temperature gradient; mass is transferred in the presence of a concentration gradient. These molecular diffusion effects are described by phenomenological laws.

Momentum, energy, and mass are all conserved. As such, each quantity obeys the conservation law within a system:

$$
\left\{ \begin{array}{c} \text{quantity} \\ \text{into} \\ \text{system} \end{array} \right\} - \left\{ \begin{array}{c} \text{quantity} \\ \text{out of} \\ \text{system} \end{array} \right\} + \left\{ \begin{array}{c} \text{quantity} \\ \text{generated in} \\ \text{system} \end{array} \right\} = \left\{ \begin{array}{c} \text{quantity} \\ \text{accumulated} \\ \text{in system} \end{array} \right\} \tag{4.1}
$$

This equation may also be written on a time rate basis:

$$\left\{ \begin{array}{c} \text{rate} \\ \text{into} \\ \text{system} \end{array} \right\} - \left\{ \begin{array}{c} \text{rate} \\ \text{out of} \\ \text{system} \end{array} \right\} + \left\{ \begin{array}{c} \text{rate} \\ \text{generated in} \\ \text{system} \end{array} \right\} = \left\{ \begin{array}{c} \text{rate} \\ \text{accumulated} \\ \text{in system} \end{array} \right\} \qquad (4.2)$$

The conservation law may be applied at the macroscopic, microscopic or molecular level. One can best illustrate the differences in these methods with an example. Consider a system in which a fluid is flowing through a cylindrical tube (see Fig. 4.1) and define the system as the fluid contained with the tube between points 1 and 2 at any time.

If one is interested in determining changes occurring at the inlet and outlet of the system, the conservation law is applied on a "macroscopic" level to the entire system. The resultant equation describes the overall changes occurring *to* the system (or equipment). This approach is usually applied in the Unit Operation (or its equivalent) courses, an approach which is highlighted in this text. Resulting equations are almost always algebraic.

In the microscopic approach, detailed information concerning the behavior within a system is required and this is occasionally requested of and by the engineer. The conservation law is then applied to a differential element within the system which is large compared to an individual molecule, but small compared to the entire system. The resulting equation is then expanded via an integration to describe the behavior of the entire system. This has been defined as the transport phenomena approach.

The molecular approach involves the application of the conservation laws to individual molecules. This leads to a study of statistical and quantum mechanics—both of which are beyond the scope of this text. In any case, the description of individual particles at the molecular level is of little value to the practicing engineer. However, the statistical averaging of molecular quantities in either a differential or finite element within a system can lead to a more meaningful description of the behavior of a system.

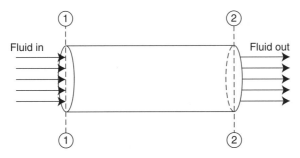

Figure 4.1 Flow through cylinder.

Both the microscopic and molecular approaches shed light on the physical reasons for the observed macroscopic phenomena. Ultimately, however, for the practicing engineer, these approaches may be valid but are akin to killing a fly with a machine gun. Developing and solving these equations (in spite of the advent of computer software packages) is typically not worth the trouble.

Traditionally, the applied mathematician has developed the differential equations describing the detailed behavior of systems by applying the appropriate conservation law to a differential element or shell within the system. Equations were derived with each new application. The engineer later removed the need for these tedious and error-prone derivations by developing a general set of equations that could be used to describe systems. These are referred to as the transport equations. In recent years, the trend toward expressing these equations in vector form has also gained momentum (no pun intended). However, the shell-balance approach has been retained in most texts, where the equations are presented in componential form—in three particular coordinate systems—rectangular, cylindrical and spherical. The componential terms can be "lumped" together to produce a more concise equation in vector form. The vector equation can be in turn be re-expanded into other coordinate systems. This information is available in the literature.[2,3]

As noted above, the microscopic approach receives limited treatment in this text. It is introduced in the next chapter and again in Chapter 9 (Conservation Law for Momentum) and Chapter 14 (Turbulent Flow).

4.3 WHAT IS ENGINEERING?

A discussion on chemical engineering is again warranted before proceeding to the fluid flow material presented in this text. A reasonable question to ask is: What is Chemical Engineering? An outdated but once official definition provided by the American Institute of Chemical Engineers is:

> "Chemical Engineering is that branch of engineering concerned with the development and application of manufacturing processes in which chemical or certain physical changes are involved. These processes may usually be resolved into a coordinated series of unit physical operation (hence the name of the topic and book) and chemical processes. The work of the chemical engineer is concerned primarily with the design, construction, and operation of equipment and plants in which these unit operations and processes are applied. Chemistry, physics, and mathematics are the underlying sciences of chemical engineering, and economics is its guide in practice."

The above definition has been appropriate up until a few decades ago since the profession grew out of the chemical industry (see Chapter 1). Today, that definition has changed. Although it is still based on chemical fundamentals and physical principles, these principles have been deemphasized in order to allow the expansion of the profession to other areas (biotechnology, semiconductors, fuel cells, environmental, etc.). These areas include environmental management, health and safety, computer applications, and economics and finance. This has led to many new definitions of chemical engineering, several of which are either too specific or too vague. A definition proposed by one of the authors is simply, "chemical engineers solve problems." This definition can be extended to all engineers and state "engineers solve problems."

REFERENCES

1. R. N. Bird, W. Stewart, and E. Lightfoot, "Transport Phenomena", John Wiley & Sons, Hoboken, NJ, 1960.
2. L. Theodore, "Transport Phenomena for Engineers", International Textbook Company, Scranton, PA, 1971 (with permission).
3. R. N. Bird, W. Stewart, and E. Lightfoot, "Transport Phenomena", 2nd edition, John Wiley & Sons, Hoboken, NJ, 2002.

NOTE: Additional problems are available for all readers at www.wiley.com. Follow links for this title.

5

NEWTONIAN FLUIDS

5.1 INTRODUCTION

This chapter is introduced by examining the units of some of the pertinent quantities that will be encountered below. The momentum of a system is defined as the product of the mass and velocity of the system.

$$\text{Momentum} = (\text{Mass})(\text{Velocity}) \tag{5.1}$$

One set of units for momentum are, therefore, $\text{lb} \cdot \text{ft/s}$. The units of time rate of change of momentum (hereafter referred to as rate of momentum) are simply the units of momentum divided by time, i.e.,

$$\text{Rate of momentum} \equiv \frac{\text{lb} \cdot \text{ft}}{\text{s}^2} \tag{5.2}$$

The above units can be converted to lb_f if multiplied by an appropriate constant. The conversion constant in this case is a term that was developed in Chapter 2.

$$g_c = 32.2 \frac{(\text{lb} \cdot \text{ft})}{(\text{lb}_f \cdot \text{s}^2)} \tag{5.3}$$

Fluid Flow for the Practicing Chemical Engineer. By J. Patrick Abulencia and Louis Theodore
Copyright © 2009 John Wiley & Sons, Inc.

This serves to define the conversion constant g_c. If the rate of momentum is divided by g_c as 32.2 $(\text{lb} \cdot \text{ft})/(\text{lb}_f \cdot \text{s}^2)$—the following units result:

$$\text{Rate of momentum} \equiv \left(\frac{\text{lb} \cdot \text{ft}}{\text{s}^2}\right)\left(\frac{\text{lb}_f \cdot \text{s}^2}{\text{lb} \cdot \text{ft}}\right)$$

$$\equiv \text{lb}_f \tag{5.4}$$

One may conclude from the above dimensional analysis that a force is equivalent to a rate of momentum.

5.2 NEWTON'S LAW OF VISCOSITY

The above development is now extended to Newton's law of viscosity. Consider a fluid flowing between the region bounded by two infinite parallel horizontal plates separated by a distance h. The flow is steady and only in the y-direction. Part of the system is represented in Fig. 5.1. A sufficient force F is being applied to the upper plate at $z = h$ to maintain the upper plate in motion with a velocity $v_y = V_h$. If the fluid density is constant and the flow is everywhere isothermal and laminar, the linear velocity gradient in the two-dimensional representation in Fig. 5.2 will result.

It has been shown by experiment that the applied force per unit area F/A required to maintain the upper plate in motion with velocity V_h is proportional to the velocity gradient, i.e.,

$$\frac{F}{A} \propto \frac{V_h}{h}$$

For a slightly more general form, one may write

$$\frac{F}{A} \propto \frac{\Delta v_y}{\Delta z} \tag{5.5}$$

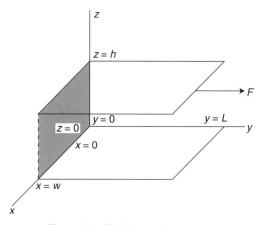

Figure 5.1 Fluid/two-plate system.

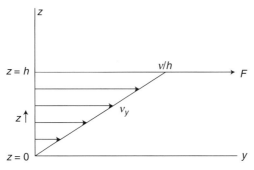

Figure 5.2   Velocity profile.

The difference term Δ can be removed by applying Equation (5.5) to a differential width dz:

$$\frac{F}{A} \propto \frac{dv_y}{dz} \tag{5.6}$$

Equation (5.6) may be written in equation form by replacing the proportionality sign with a proportionality constant, $-\mu$:

$$\frac{F}{A} = -\mu \frac{dv_y}{dz} \tag{5.7}$$

The term μ is defined as the coefficient of viscosity, or simply the aforementioned viscosity of the fluid. The term F/A is a shear stress since F is exerted parallel to the direction of motion. This applied force per unit area is now designated by τ_{zy},

$$\tau_{zy} = -\mu \frac{dv_y}{dz} \tag{5.8}$$

A fluid whose shear stress is described by Equation (5.8) is defined as a Newtonian fluid.

A word of interpretation is in order for Equation (5.8). The applied force at $z = h$ has resulted in a velocity V_h at $z = h$. The fluid at this point possesses momentum due to this velocity. As z decreases the momentum of the fluid decreases since the velocity decreases in this direction. We have already shown that the force applied to a fluid is equivalent to the fluid receiving a rate of momentum. Part of the momentum imparted to the fluid at $z = h$ is transferred at the specified rate to the slower-moving fluid immediately below it. This momentum maintains the velocity of the fluid at that point, and is, in turn, transported to the slower fluid below it, and so on. This momentum transfer process is occurring in the z-direction throughout the fluid. One may therefore conclude the applied force in the positive y-direction has resulted in the transfer of momentum in the negative z-direction. The first subscript in τ_{zy} is retained as a reminder of this fact. The subscript y indicates the direction of motion. The negative sign in Equation (5.7) was introduced since momentum is transferred in the negative z-direction due to a positive velocity gradient.

The force per unit area term τ is equivalent to a rate of momentum per unit area. Therefore, the shear stress and its components are also defined as the momentum flux.

Referring once again to the shear stress component τ_{zy}, one may divide the RHS of Equation (5.8) by g_c,

$$\tau_{zy} = -\frac{\mu}{g_c}\frac{dv_y}{dz} \tag{5.8}$$

If τ_{zy} has the units lb_f/ft^2, the viscosity μ assumes the units $\text{lb}/\text{ft} \cdot \text{s}$.

A term that will frequently be employed in the text is the kinematic viscosity v (see Chapter 3). This is defined as the ratio of the viscosity to the density of the fluid.

$$v = \frac{\mu}{\rho} \tag{5.9}$$

The units of v can be shown to be ft^2/s.

All components of the shear stress for a Newtonian fluid can be expressed in terms of the viscosity of the fluid and a velocity gradient. These are presented, but not derived, in Table 5.1[1,2] for rectangular, cylindrical, and spherical coordinates. The equations are applicable to all Newtonian fluids provided:

1. The system is isothermal.
2. Flow is laminar.
3. The fluid density is constant.

Procedures for predicting viscosity values from theory are beyond the scope of this text, but available in the literature.[1]

Illustrative Example 5.1   A fluid of viscosity μ is flowing in the y-direction between two infinite horizontal parallel plates. The velocity profile of the fluid is given by

$$v_y = V\left(\frac{z}{h} - z^2\right)$$

where V and h are constants.

Calculate the shear stress at the surface $z = 0$ in terms of μ, V, and h.

Solution   This problem is solved using rectangular coordinates. First note that v_x and v_z equal zero and v_y is solely a function of z. From Table 5.1,

$$\tau_{zy} = -\frac{\mu}{g_c}\left(\frac{\partial v_y}{\partial z} + \cancel{\frac{\partial v_z}{\partial y}}^{\,0}\right)$$

$$= -\frac{\mu}{g_c}\frac{dv_y}{dz} \quad \text{(since } v_y \text{ is solely a function of } z)$$

Table 5.1 Shear–stress components

Component	Rectangular Coordinates	Cylindrical Coordinates	Spherical Coordinates
τ_{11}	$\tau_{xx} = -\dfrac{\mu}{g_c}\left[2\left(\dfrac{\partial v_x}{\partial x}\right)\right]$	$\tau_{rr} = -\dfrac{\mu}{g_c}\left[2\left(\dfrac{\partial v_r}{\partial r}\right)\right]$	$\tau_{rr} = -\dfrac{\mu}{g_c}\left[2\left(\dfrac{\partial v_r}{\partial r}\right)\right]$
τ_{12}	$\tau_{xy} = -\dfrac{\mu}{g_c}\left[\dfrac{\partial v_x}{\partial y} + \dfrac{\partial v_y}{\partial x}\right]$	$\tau_{r\phi} = -\dfrac{\mu}{g_c}\left[r\dfrac{\partial}{\partial r}\left(\dfrac{v_\phi}{r}\right) + \dfrac{1}{r}\left(\dfrac{\partial v_r}{\partial \phi}\right)\right]$	$\tau_{r\theta} = -\dfrac{\mu}{g_c}\left[r\dfrac{\partial}{\partial r}\left(\dfrac{v_\theta}{r}\right) + \dfrac{1}{r}\left(\dfrac{\partial v_r}{\partial \theta}\right)\right]$
τ_{13}	$\tau_{xz} = -\dfrac{\mu}{g_c}\left[\dfrac{\partial v_z}{\partial x} + \dfrac{\partial v_x}{\partial z}\right]$	$\tau_{rz} = -\dfrac{\mu}{g_c}\left[\dfrac{\partial v_z}{\partial r} + \dfrac{\partial v_r}{\partial z}\right]$	$\tau_{r\phi} = -\dfrac{\mu}{g_c}\left[\dfrac{1}{r\sin\theta}\left(\dfrac{\partial v_r}{\partial \phi}\right) + r\dfrac{\partial}{\partial r}\left(\dfrac{v_\phi}{r}\right)\right]$
τ_{21}	$\tau_{yx} = -\dfrac{\mu}{g_c}\left[\dfrac{\partial v_x}{\partial y} + \dfrac{\partial v_y}{\partial x}\right]$	$\tau_{\phi r} = -\dfrac{\mu}{g_c}\left[r\dfrac{\partial}{\partial r}\left(\dfrac{v_\phi}{r}\right) + \dfrac{1}{r}\left(\dfrac{\partial v_r}{\partial \phi}\right)\right]$	$\tau_{\theta r} = -\dfrac{\mu}{g_c}\left[r\dfrac{\partial}{\partial r}\left(\dfrac{v_\theta}{r}\right) + \dfrac{1}{r}\left(\dfrac{\partial v_r}{\partial \theta}\right)\right]$
τ_{22}	$\tau_{yy} = -\dfrac{\mu}{g_c}\left[2\left(\dfrac{\partial v_y}{\partial y}\right)\right]$	$\tau_{\phi\phi} = -\dfrac{\mu}{g_c}\left[2\left(\dfrac{1}{r}\dfrac{\partial v_\phi}{\partial \phi} + \dfrac{v_r}{r}\right)\right]$	$\tau_{\theta\theta} = -\dfrac{\mu}{g_c}\left[2\left(\dfrac{1}{r}\dfrac{\partial v_\theta}{\partial \theta} + \dfrac{v_r}{r}\right)\right]$
τ_{23}	$\tau_{yz} = -\dfrac{\mu}{g_c}\left[\dfrac{\partial v_y}{\partial z} + \dfrac{\partial v_z}{\partial y}\right]$	$\tau_{\phi z} = -\dfrac{\mu}{g_c}\left[\dfrac{\partial v_\phi}{\partial z} + \dfrac{1}{r}\left(\dfrac{\partial v_z}{\partial \phi}\right)\right]$	$\tau_{\theta\phi} = -\dfrac{\mu}{g_c}\left[\dfrac{\sin\theta}{r}\dfrac{\partial}{\partial \theta}\left(\dfrac{v_\phi}{\sin\theta}\right) + \dfrac{1}{r\sin\theta}\left(\dfrac{\partial v_\theta}{\partial \phi}\right)\right]$
τ_{31}	$\tau_{zx} = -\dfrac{\mu}{g_c}\left[\dfrac{\partial v_z}{\partial x} + \dfrac{\partial v_x}{\partial z}\right]$	$\tau_{zr} = -\dfrac{\mu}{g_c}\left[\dfrac{\partial v_z}{\partial r} + \dfrac{\partial v_r}{\partial z}\right]$	$\tau_{\phi r} = -\dfrac{\mu}{g_c}\left[\dfrac{1}{r\sin\theta}\left(\dfrac{\partial v_r}{\partial \phi}\right) + r\dfrac{\partial}{\partial r}\left(\dfrac{v_\phi}{r}\right)\right]$
τ_{32}	$\tau_{zy} = -\dfrac{\mu}{g_c}\left[\dfrac{\partial v_y}{\partial z} + \dfrac{\partial v_z}{\partial y}\right]$	$\tau_{z\phi} = -\dfrac{\mu}{g_c}\left[\dfrac{\partial v_\phi}{\partial z} + \dfrac{1}{r}\left(\dfrac{\partial v_z}{\partial \phi}\right)\right]$	$\tau_{\phi\theta} = -\dfrac{\mu}{g_c}\left[\dfrac{\sin\theta}{r}\dfrac{\partial}{\partial \theta}\left(\dfrac{v_\phi}{\sin\theta}\right) + \dfrac{1}{r\sin\theta}\left(\dfrac{\partial v_\theta}{\partial \phi}\right)\right]$
τ_{33}	$\tau_{zz} = -\dfrac{\mu}{g_c}\left[2\left(\dfrac{\partial v_z}{\partial z}\right)\right]$	$\tau_{zz} = -\dfrac{\mu}{g_c}\left[2\left(\dfrac{\partial v_z}{\partial z}\right)\right]$	$\tau_{\phi\phi} = -\dfrac{\mu}{g_c}\left[2\left(\dfrac{1}{r\sin\theta}\dfrac{\partial v_\phi}{\partial \phi} + \dfrac{v_r}{r} + \dfrac{v_\theta\cot\theta}{r}\right)\right]$

The velocity profile is given as:

$$v_y = V\left[\frac{z}{h} - z^2\right]$$

so that,

$$\frac{dv_y}{dz} = \frac{V}{h} - 2Vz$$

$$-\mu\frac{dv_y}{dz} = 2\mu Vz - \frac{V\mu}{h}$$

The shear stress at the surface $z = 0$ is denoted by $\tau_{zy}|_{z=0}$:

$$\tau_{zy}\bigg|_{z=0} = \frac{\mu}{g_c}\left[2V(0) - \frac{V}{h}\right]$$

$$= -\mu V/g_c h$$

Illustrative Example 5.2   Two vertical parallel plates are spaced 1 inch apart. The plate on the left side is moving at a velocity of 5 ft/min in the z-direction and the plate on the right side is stationary. The space between the plates contains a gas whose kinematic viscosity is 1.66 ft^2/hr and density is 0.08 lb/ft^3.

1. Calculate the force necessary to maintain the movement of the left plate.
2. Calculate the momentum flux at the surface of the left plate and at the surface of the right plate.

Solution   Note that based on no slip conditions, the velocity of the gas at the surface of the moving plate is equal to the velocity of the plate and the velocity of the gas at the surface of the stationary plate is zero.

1. Calculate the force per unit area of plate; this is the shear stress (τ_y) that can be evaluated from the appropriate equation in Table 5.1. For this application,

$$\tau_{xy} = -\frac{\mu}{g_c}\left(\frac{\partial v_y}{\partial x}\right)$$

Since $x_1 = 0$, $x_2 = 0.0833$ ft, $v_1 = (5)(60)$ ft/hr, $v_2 = 0$,

$$\tau_{xy} = -(1.66)(0.08)\left(\frac{0 - 300}{0.0833 - 0}\right)$$

$$= 478\frac{\text{lb} \cdot \text{ft/hr}}{\text{ft}^2 \cdot \text{hr}}$$

Since,

$$g_c = 32.2 \frac{\text{ft} \cdot \text{lb/s}}{\text{lb}_f \cdot \text{s}} = 4.17 \times 10^8 \frac{\text{ft} \cdot \text{lb/hr}}{\text{lb}_f \cdot \text{hr}}$$

$$\tau_{xy} = \frac{478}{4.17 \times 10^8}$$

$$= 1.15 \times 10^{-6} \, \text{lb}_f / \text{ft}^2$$

5.3 VISCOSITY MEASUREMENTS

One of the simplest methods to measure viscosity is to time the discharge of a known volume of fluid through a nozzle. A vessel with a short capillary tube is employed. This equipment is known as the *Saybolt viscometer*. It has been used to determine the viscosities of oils and paints. Another common technique is to measure the torque required to rotate a torque element in a liquid (e.g., *Brookfield viscometer* and *coaxial cylindrical viscometers*). On a *Couette–Hatschek viscometer* (or *MacMichael viscometer*), the outer member of a pair of closely fitting coaxial cylinders is rotated, while in the *Stormer viscometer*, the inner member of a pair of closely fitting cylinders is rotated (see Figs. 5.3 and 5.4). The clearance between the two

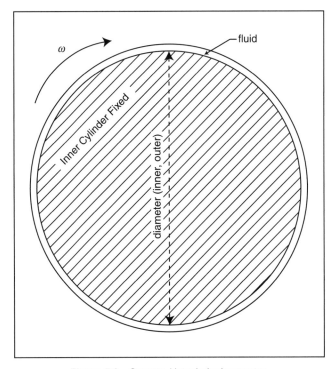

Figure 5.3   Couette-Hatschek viscometer.

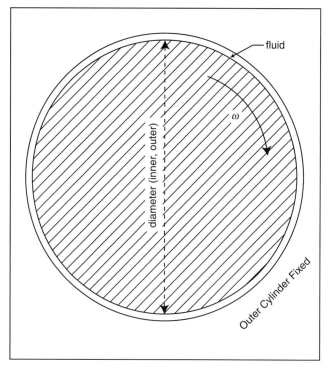

Figure 5.4   Stormer viscometer.

cylinders is so small (relative to the cylinder dimensions) that a linear velocity profile may be assumed in the fluid filling the gap. By measuring the torque, T, required to rotate the cylinder at a specified angular velocity, ω (rad/s), it is possible to calculate the fluid absolute viscosity and/or *fluidity*, where the *fluidity* is defined as the reciprocal of viscosity. In the SI system of units, the fluidity unit is known as the "rhe" (1 rhe = 1/poise = 1 s/g · cm).

The definitions and equations for the calculation of the fluid viscosity from these viscometers are given below:

$$\text{Torque, } T = (\text{force})(\text{cylinder radius}) = (\text{force})(\text{diameter}/2)$$
$$= (F)(D/2) \tag{5.10}$$

$$\text{Force, } F = (\text{shear stress})(\text{surface area of cylinder})$$
$$= (\tau)(\pi DL) \tag{5.11}$$

$$\text{Shear stress, } \tau = (\text{absolute viscosity})(\text{velocity gradient}) = (\mu)(dv/dy)$$
$$= (\text{viscosity})(\text{velocity at the rotating cylinder})/$$
$$(\text{gap separation}) = (\mu)(v/D) \tag{5.12}$$

$$\text{Velocity, } v = \text{velocity of the rotating cylinder}$$
$$= (\omega)(R) = (\omega)(D/2) \tag{5.13}$$

Radius, $R =$ radius of inside (or outside) cylinder

Diameter, $D =$ diameter of inside (or outside) cylinder

Height, $L =$ height of cylinder

Friction power loss, $W_L =$ (force)(velocity) $= (F)(v)$

$d =$ gap separation, clearance   (5.14)

Illustrative Example 5.3   A Couette–Hatschek viscometer is used to measure the viscosity of an oil (SG $= 0.97$). The viscometer used has a fixed inner cylinder of 3 inches diameter and 6 inches height, and a rotating outer cylinder of the same height. The clearance, d, between the two cylinders is 0.001 inch. The measured torque is 15.3 ft · lb$_f$ at an angular rotation speed of 250 rpm. Determine the shear stress in the oil. Assume the viscometer clearance gap is so small that the velocity distribution is assumed linear, that is, $dv/dy = \Delta v/\Delta y = v/d$.

Solution   Calculate the force, F, employing Equation (5.10). Since $D = 3$ in $= 0.25$ ft and $L = 6$ in $= 0.5$ ft.

$$F = \frac{2T}{D} = \frac{2(15.3)}{0.25} = 122.4\,\text{lb}_f = 544.5\,\text{N}$$

Calculate the shear stress, τ (force parallel to the surface), using Equation (5.11),

$$\tau = \frac{F}{\pi DL} = \frac{122.4}{\pi(0.25)(0.5)} = 311.7\,\text{psf} = 14.924\,\text{kPa}$$

Illustrative Example 5.4   Refer to Illustrative Example 5.3. Determine the dynamic and kinematic viscosities.

Solution   Calculate the linear velocity of the oil, v, from its angular velocity ω. See Equation (5.13).

$$\omega = 250\,\text{rpm} = \left(250\frac{\text{rev}}{\text{min}}\right)\left(2\pi\frac{\text{rad}}{\text{rev}}\right)\left(\frac{\text{min}}{60\,\text{sec}}\right) = 26.2\,\text{rad/s}$$

$$v = \frac{\omega D}{2} = \frac{26.2(0.25)}{2} = 3.27\,\text{ft/s}$$

Calculate the velocity gradient

$$\frac{dv}{dy} = \frac{v}{d} = \frac{3.27}{(0.001/12)} = 39{,}270\,\text{s}^{-1}$$

Assume Newton's law of viscosity to apply and calculate the viscosity, μ, noting that $\tau = (\mu/g_c)(dv/dy)$. Rearranging yields

$$\mu = \frac{g_c\tau}{dv/dy} = \frac{(32.174)(311.7)}{39{,}270} = 0.256\,\text{lb}/\text{ft}\cdot\text{s}$$

The kinematic viscosity, v, is

$$v = \frac{\mu}{\rho} = \frac{0.256}{(62.4)(0.97)} = 0.00423\ \text{ft}^2/\text{s}$$

5.4 MICROSCOPIC APPROACH

Consider the following application. A fluid is flowing through a long horizontal cylindrical duct of radius R under steady-state conditions (see Fig. 5.5). The general equation for the velocity profile in a pipe as a function of the pressure drop per unit length in the direction of motion has been shown to take the form[1,2]

$$v_z = -\frac{g_c\Delta P}{4\mu L}r^2 + A\ln r + B \tag{5.15}$$

where A and B are integration constants that are evaluated from the boundary and/or initial conditions (Ba/oICs) for the system in question. An equation describing the velocity profile in the tube can be generated. Referring to Fig. 5.5, one concludes

$$\text{BC(1):}\quad v_z = 0\quad\text{at }r = R$$

and

$$\text{BC(2):}\quad v_z = \text{finite}\quad\text{at }r = 0$$

or the equivalent:

$$\frac{dv_z}{dr} = 0\quad\text{at }r = 0$$

based on physical grounds

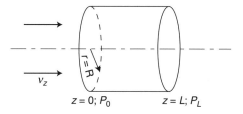

Figure 5.5 Horizontal flow in a tube.

Substituting BC(2) into Equation (5.15) yields

$$A = 0$$

BC(1) gives

$$0 = -\frac{g_c \Delta P}{4\mu L} R^2 + B$$

$$B = \frac{g_c \Delta P}{4\mu L} R^2$$

Equation (5.15) now becomes

$$v_z = \frac{g_c \Delta P}{4\mu L}(R^2 - r^2) \tag{5.16}$$

This equation will be derived and reviewed again in Chapters 9 and 13.

Another application involves fluid flowing between the region bounded by two infinite parallel horizontal plates separated by a distance h. The flow is steady and only in the y-direction. Part of the system is represented in Figs. 5.1 and 5.2. A sufficient force is applied to the upper plate to maintain a velocity V_h. The general equation for the velocity profile is given by[2]

$$v_y = \frac{g_c z^2}{2\mu} \frac{\Delta P}{\Delta y} + Bz + A \tag{5.17}$$

The boundary conditions (BC) are

$$\text{BC(1):} \quad v_y = 0 \quad \text{at } z = 0$$
$$\text{BC(2):} \quad v_y = V_h \quad \text{at } z = h$$

Substituting BC(1) into Equation (5.17) yields

$$0 = 0 + 0 + A$$
$$A = 0$$

Substituting BC(2) into Equation (5.17) gives

$$V_h = Bh + \frac{g_c h^2}{2\mu} \frac{\Delta P}{\Delta y}$$

$$B = \frac{V_h}{h} - \frac{g_c h}{2\mu} \frac{\Delta P}{\Delta y}$$

Therefore,

$$v_y = V_h \left(\frac{z}{h}\right) - \frac{g_c h z}{2\mu} \frac{\Delta P}{\Delta y} + \frac{g_c z^2}{2\mu} \frac{\Delta P}{\Delta y}$$

$$= V_h \left(\frac{z}{h}\right) - \frac{g_c z}{2\mu}(h - z)\frac{\Delta P}{\Delta y} \tag{5.18}$$

Since the fluid is not moving relative to fixed points on both plates, $\Delta P/\Delta y = 0$ and

$$v_y = V_h \left(\frac{z}{h}\right) \tag{5.19}$$

It would be wise at this point to verify that the above solution satisfies both the differential equation and BCs. We leave this exercise to the reader.

REFERENCES

1. R. Bird, W. Stewart, and E. Lightfoot, "Transport Phenomena", 2nd edition, John Wiley & Sons, Hoboken, NJ, 2002.
2. L. Theodore, "Transport Phenomena for Engineers", International Textbook Company, Scranton, PA, 1971.

NOTE: Additional problems are available for all readers at www.wiley.com. Follow links for this title.

6

NON-NEWTONIAN FLOW

6.1 INTRODUCTION

The study of the mechanics of the flow of liquids and suspensions comes under the science of *Rheology*. The name Rheology was chosen by Prof. John R. Crawford of Lafayette College, PA, and is defined as the study of the flow and deformation of matter. (The name is a combination of the Greek words "Rheo"-flow and "Logos"-theory.)

The shear-stress equations developed in the previous chapter were written for fluids with a viscosity that is constant at constant temperature and independent of the rate of shear and the time of application of shear. Fluids with this property were defined as Newtonian fluids. All gases and pure low-molecular-weight liquids are Newtonian. Miscible mixtures of low-molecular-weight liquids are also Newtonian. On the other hand, high-viscosity liquids as well as polymers, colloids, gels, concentrated slurries and solutions of macromolecules generally do not exhibit Newtonian properties; i.e., a strict proportionality between stress and strain rate. Interestingly, non-Newtonian properties are sometimes desirable. For example, non-Newtonian behavior is exhibited in many paints. During brush working, certain paints flow readily to cover the surface, but upon standing, the original highly viscous condition returns and the paint will not run.

The study of non-Newtonian fluids has not progressed far enough to develop many useful theoretical approaches. As noted in the previous chapter, if the liquid or suspension is found to be Newtonian, the pressure drop can be calculated from the "Poiseuille" equation for laminar flow (see Chapter 13) and the Fanning equation for turbulent flow (see Chapter 14), using the density and viscosity of the liquid or

Fluid Flow for the Practicing Chemical Engineer. By J. Patrick Abulencia and Louis Theodore
Copyright © 2009 John Wiley & Sons, Inc.

suspension. For non-Newtonian liquids and suspensions, the viscosity is a variable, and the procedure for computing the pressure drop is more involved.

The remainder of this section will discuss non-Newtonian liquids and suspensions. Useful engineering design procedures and prediction equations receive treatment that are limited to isothermal laminar (viscous) flow. The turbulent flow of non-Newtonian fluids (as with Newtonian ones) is characterized by the presence of random eddies and whirls of fluid that cause the instantaneous values of velocity and pressure at any point in the system to fluctuate wildly. Because of these fluctuations, flow problems cannot be easily solved. Since non-Newtonian turbulent flow rarely occurs, it has not received much attention.

6.2 CLASSIFICATION OF NON-NEWTONIAN FLUIDS

Fluids can be classified based on their viscosity. An imaginary fluid of zero viscosity is called a *Pascal fluid*. The flow of a Pascal fluid is termed *inviscid* (or non-viscous) flow. Viscous fluids are classified based on their rheological (viscous) properties. These are detailed below:

1. Newtonian fluids, as described in the previous chapter, obey Newton's law of viscosity (i.e., the fluid shear stress is linearly proportional to the velocity gradient). All gases are considered Newtonian fluids. Newtonian liquid examples are water, benzene, ethyl alcohol, hexane and sugar solutions. All liquids of a simple chemical formula are normally considered Newtonian fluids.

2. Non-Newtonian fluids do not obey Newton's law of viscosity. Generally they are complex mixtures (e.g., polymer solutions, slurries, and so on). Non-Newtonian fluids are classified into three types:

    a. Time-independent fluids are fluids in which the viscous properties do not vary with time.

    b. Time-dependent fluids are fluids in which the viscous properties vary with time.

    c. Visco-elastic or memory fluids are fluids with elastic properties that allow them to "spring back" after the release of a shear force. Examples include egg-white and rubber cement.

    Additional details on the first two classes of fluids follow.

3. Time-independent, non-Newtonian fluids are further classified into three types.

    a. *Pseudoplastic* or shear thinning fluids are characterized by a fluid resistance decrease with increasing stress (e.g., polymers).

    b. *Dilatant* or shear thickening fluids increase resistance with increasing velocity gradient or applied stress. These are uncommon, but an example is quicksand.

c. *Bingham plastics* are fluids that resist a small shearing stress. At low shear stress these fluids do not move. At high shear the fluids move. The fluid just starts moving when sufficient stress is applied. This stress is termed the *yield stress*. When the applied stress exceeds the yield stress, the Bingham plastic flows. Examples are toothpaste, jelly, and bread-dough.

4. Time-dependent, non-Newtonian fluids are further classified into two types.

a. *Rheopectic* fluids are characterized by an increasing viscosity with time. Rubber cement is an example.

b. *Thixotropic* fluids have a decreasing viscosity with time. Examples are slurries or solutions of polymers.

6.2.1 Non-Newtonian Fluids: Shear Stress[1]

There are the aforementioned class of fluids that do not obey Newton's law of viscosity. These were defined as *non-Newtonian* and several different types of these fluids exist. The shear stress equation equivalent to Equation (5.7) for one of the more common types of non-Newtonian fluids is given by the so-called "power law" equation:

$$\tau_{zy} = -\frac{K}{g_c}\left(\frac{dv_y}{dz}\right)^n \tag{6.1}$$

K is defined as the *consistency number* and may in special cases equal μ. The exponent n is defined as the *flow-behavior index* and is a real number that usually assumes a value other than unity. Although n is considered a physical property of the fluid, it is not necessarily a constant; rather, it may vary with the shear rate, dv_y/dz. Equation (6.1) may be written in terms of the apparent viscosity μ_a for non-Newtonian fluids (most non-Newtonian fluids have apparent viscosities that are relatively high compared with the viscosity of water)

$$\frac{\mu_a}{g_c} = -\frac{\tau_{zy}}{dv_y/dz}$$

or

$$\mu_a = K\left(\frac{dv_y}{dz}\right)^{n-1} \tag{6.2}$$

In order to remove the problem arising when the velocity gradient is a negative quantity, Equation (6.1) is rewritten as

$$\tau_{zy} = -\frac{K}{g_c}\frac{dv_y}{dz}\left|\frac{dv_y}{dz}\right|^{n-1} \tag{6.3}$$

A typical shear stress vs. shear rate (dv_y/dz) curve (often referred to as a rheogram), is shown for a non-Newtonian fluid in Fig. 6.1 on arithmetic coordinates. Newtonian

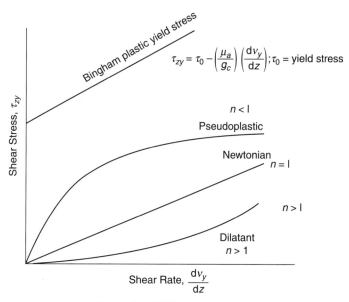

Figure 6.1   Fluid shear diagrams.

behavior is also depicted in the diagram. Due to the exponential nature of the shear rate of this type of non-Newtonian fluid, a straight line would be obtained on a log-log plot as demonstrated in Equations (6.4) and (6.5):

$$\tau_{zy} = -\frac{K}{g_c}\left(\frac{dv_y}{dz}\right)^n \tag{6.4}$$

$$\log \tau_{zy} = -\log\left(\frac{K}{g_c}\right) + n\log\left(\frac{dv_y}{dz}\right) \tag{6.5}$$

One notes that a Newtonian fluid yields a slope of 1.0 on log-log coordinates. The slope of a non-Newtonian fluid generally differs from unity. The slope, n, can be thought of as an index to the degree of non-Newtonian behavior in that the farther that n is from unity (above or below), the more pronounced is the non-Newtonian characteristics of the fluid.

Illustrative Example 6.1   For each of the following four classes of fluids, indicate the line type (straight or curved) on a logarithmic shear diagram, i.e., shear stress versus shear rate.

1. Newtonian
2. Pseudoplastic
3. Dilatant
4. Bingham plastic

Solution   Refer to Fig. 6.1.

1. Newtonian: Straight
2. Pseudoplastic: Straight
3. Dilatant: Straight
4. Bingham plastic: Curved

Illustrative Example 6.2   For each of the following four classes of fluids, indicate the line slope (>1, 1, or <1) on a logarithmic shear diagram, i.e., shear stress versus shear rate.

1. Newtonian
2. Pseudoplastic
3. Dilatant
4. Bingham plastic

Solution   Refer once again to Fig. 6.1.

1. Newtonian: 1
2. Pseudoplastic: <1
3. Dilatant: >1
4. Bingham plastic: not applicable

Illustrative Example 6.3   Classify the following substances according to their rheological behavior: paint, grease, toothpaste, tar, silly putty, and ordinary putty.

Solution   The classification is tabulated below:

SUBSTANCE	EXPLANATION
Paint	Shear-thinning (pseudo-plastic). Also rheopectic (it hardens with time)
Grease	Bingham plastic (needs a yield stress before flowing). Visco-elastic
Toothpaste	Ideal Bingham plastic
Tar	Pseudoplastic at high temperature
Silly putty	Dilatant (shear thickening). Visco-elastic
Ordinary putty	Visco-elastic

6.3 MICROSCOPIC APPROACH

Most non-Newtonian fluids either follow the power law relationship provided in Equation (6.4) or may be approximated by it for engineering purposes. The presentation below is therefore limited to power-law applications.

6.3.1 Flow in Tubes

The reader is referred to the Microscopic Approach section in the previous chapter on Newtonian flow. One may now re-examine the flow of a fluid through a horizontal tube under the condition that it follows the power law relationship

$$\tau_{rz} = K\left(\frac{dv_z}{dr}\right)^n \tag{6.6}$$

Wohl[2] has shown that the velocity profile for the above system is given by

$$v_z = \left(\frac{n}{n+1}\right)\left(\frac{\Delta P}{2LK}\right)^{1/n}[R^{(n+1)n} - r^{(n+1)/n}] \tag{6.7}$$

Illustrative Example 6.4   Verify that Equation (6.7) reduces to the velocity profile relationship provided in Equation (5.16) for a Newtonian fluid.

Solution   One notes that for $n = 1$, where the fluid is Newtonian, Equation (6.7) reduces to

$$v_z = \left(\frac{\Delta P}{4\mu L}\right)[R^2 - r^2]$$

This is, as one would expect, the same equation provided in the previous chapter for flow of a Newtonian fluid through a pipe (see Eq. (5.16)).

The above equation may also be written in terms of the maximum centerline velocity, v_{max}.

$$v_z = v_{max}\left[1 - \left(\frac{r}{R}\right)^{(n+1)/n}\right] \tag{6.8}$$

Alternately, the local velocity can be expressed in terms of the average velocity, v_{av}.

$$v_z = v_{av}\left(\frac{3n+1}{n+1}\right)\left[1 - \left(\frac{r}{R}\right)^{(n+1)n}\right] \tag{6.9}$$

For a Bingham plastic, the local velocity is given by

$$v_z = \frac{\Delta P}{4\mu L}(R^2 - r^2) - \frac{\tau_0}{\mu}(R - r) \tag{6.10}$$

For values of $r_p < r < R$ where

$$r_p = \frac{2L\tau_0}{\Delta P} \tag{6.11}$$

For $r_p > r > 0$, the describing equation

$$v_z = \frac{\Delta P}{2L\mu}(R - r)^2 \tag{6.12}$$

6.3.2 Flow Between Parallel Plates

For flow between parallel plates of height H, length L, and width W, Wohl[3] has shown that the local velocity is given by

$$v = \left(\frac{n}{n+1}\right)\left(\frac{\Delta P}{LK}\right)^{1/n}\left[\left(\frac{H}{2}\right)^{(n+1)/n} - (z)^{(n+1)/n}\right] \tag{6.13}$$

where z is the vertical Cartisian coordinate constrained by $z = \Delta \pm H/2$.

Illustrative Example 6.5   Refer to Equation (6.13). Generate an equation that describes the maximum velocity.

Solution   For the maximum velocity, set $z = 0$ in Equation (6.13).

$$v = \left(\frac{n}{n+1}\right)\left(\frac{\Delta P}{LK}\right)^{1/n}\left[\left(\frac{H}{2}\right)^{(n+1)/n} - (0)^{(n+1)/n}\right]$$

$$= \left(\frac{n}{n+1}\right)\left(\frac{\Delta P}{LK}\right)^{1/n}\left(\frac{H}{2}\right)^{(n+1)/n}$$

Illustrative Example 6.6   Starting with Equation (6.13), obtain the equation for the velocity profile if the fluid is Newtonian.

Solution   If the fluid is Newtonian, $n = 1$. Therefore, set $n = 1$ and $K = \mu$ in Equation (6.13).

$$v = \left(\frac{\Delta P}{L\mu}\right)\left[\left(\frac{H}{2}\right) - z\right]$$

$$= \frac{\Delta P}{2\mu L}(H - 2z)$$

Illustrative Example 6.7 Refer to Equation (6.13). Obtain an equation describing the volumetric flow rate q.

Solution By definition, the integral below

$$q_{1/2} = \int\limits_{z=0}^{z=H/2} W v \, dz$$

provides the volumetric flow rate passing the upper half of the system. Substituting for v,

$$q_{1/2} = W \int\limits_{z=0}^{z=H/2} \left(\frac{n}{n+1}\right)\left(\frac{\Delta P}{LK}\right)^{1/n}\left[\left(\frac{H}{2}\right)^{(n+1)/n} - (z)^{(n+1)/n}\right] dz$$

$$= W \left(\frac{n}{n+1}\right)\left(\frac{\Delta P}{LK}\right)^{1/n} \int\limits_{0}^{H/2}\left[\left(\frac{H}{2}\right)^{(n+1)/n} - (z)^{(n+1)/n}\right] dz$$

$$= W \left(\frac{n}{n+1}\right)\left(\frac{\Delta P}{LK}\right)^{1/n} \left[\left(\frac{H}{2}\right)^{(n+1)/n}(z) - \frac{(z)^{[(n+1)/n]+1}}{[(n+1)/n]+1}\right]_{0}^{H/2}$$

$$= W \left(\frac{n}{n+1}\right)\left(\frac{\Delta P}{LK}\right)^{1/n} \left[\left(\frac{H}{2}\right)^{(n+1)/n}\left(\frac{H}{2}\right) - \frac{(H/2)^{(2n+1)/n}}{2n+1}\right]$$

$$= W \left(\frac{n}{n+1}\right)\left(\frac{\Delta P}{LK}\right)^{1/n} \left[\left(\frac{H}{2}\right)^{(2n+1)/n} - \frac{(H/2)^{(2n+1)/n}}{(2n+1)/n}\right]$$

$$= W \left(\frac{n}{n+1}\right)\left(\frac{\Delta P}{LK}\right)^{1/n} \left(\frac{H}{2}\right)^{1/n}\left(\frac{H}{2}\right)^2 \left[1 - \frac{1}{(2n+1)/n}\right]$$

$$= W \left(\frac{n}{2n+1}\right)\left(\frac{H\Delta P}{2LK}\right)^{1/n} \left(\frac{H}{4}\right)^2 \left[\frac{2n+1}{n}\right]$$

$$= \frac{WH^2}{4} \left(\frac{n}{2n+1}\right)\left(\frac{H\Delta P}{2LK}\right)^{1/n}$$

$$= \left(\frac{n}{8n+4}\right) WH^2 \left(\frac{H\Delta P}{2LK}\right)^{1/n}$$

The total volumetric flow rate q is

$$q = 2q_{1/2}$$

$$= \left(\frac{n}{4n+2}\right) WH^2 \left(\frac{H\Delta P}{LK}\right)^{1/n}$$

For $n = 0.5$

$$q = \left(\frac{0.5}{4}\right) WH^2 \left(\frac{H\Delta P}{LK}\right)^2$$

$$= \frac{1}{8} WH^2 \left(\frac{H\Delta P}{LK}\right)^2$$

$$= \frac{1}{8} WH^2 \left(\frac{H\Delta P}{2LK}\right)$$

$$= \left(\frac{1}{2}\right) WH^2 \left(\frac{H\Delta P}{2LK}\right)^2$$

6.3.3 Other Flow Geometries

Kozicki[4] has developed simple and useful expressions for the flow of several time-independent non-Newtonian fluids in ducts of various shapes. The equations contain two shape factors, and a function of the stress, which characterize the fluid. Numerical values of the shape factors have been determined for circular, slit, concentric annular, rectangular, elliptical, and isosceles triangular ducts. The reader is referred to Kozicki's work[4] for the formulas by which these shape factors are calculated, and for a tabulated list of values to four significant figures. The derived equations are for the average and the maximum velocities as functions of the shape factors, hydraulic radius, parameters of the constitutive equations, and average shear stress at the duct wall. This average shear stress is defined by:

$$\tau_0 = r_H \Delta P / L \tag{6.14}$$

REFERENCES

1. L. Theodore, "Transport Phenomena for Engineers", International Textbook Company, Scranton, PA, 1971.
2. M. Wohl, "Isothermal Terminal Flow of Non-Newtonian Fluids in Pipes", Chemical Engineering, New York, April 8, 1968.

3. M. Wohl, "Dynamics of Flow Between Parallel Plates and in Noncircular Ducts", Chemical Engineering, New York, May 6, 1968.

4. W. Kozicki et al., "Chemical Engineering", New York, *21*, 665, 1966.

NOTE: Additional problems are available for all readers at www.wiley.com. Follow links for this title.

II

BASIC LAWS

This part of the book is concerned with Basic Laws. It contains five chapters and each serves a unique purpose in an attempt to treat several important aspects of fluid flow. This part provides an overview of the Basic Laws that will be employed in the fluid flow chapters of this book, and the motivation for their inclusion in this text. Topics covered include:

1. Chapter 7: Conservation Law for Mass
2. Chapter 8: Conservation Law for Energy
3. Chapter 9: Conservation Law for Momentum
4. Chapter 10: Law of Hydrostatics
5. Chapter 11: Ideal Gas Law

Finally, the reader should realize that this part and the contents of much of this book are geared towards engineering and science students and practitioners, not chemical engineering per se. Simply put, topics of interest to all practicing engineers have been included.

Fluid Flow for the Practicing Chemical Engineer. By J. Patrick Abulencia and Louis Theodore
Copyright © 2009 John Wiley & Sons, Inc.

7

CONSERVATION LAW
FOR MASS

7.1 INTRODUCTION

The three principles underlying the analysis of fluid flow are: the conservation law of mass, the conservation law of energy, and thirdly the conservation law of momentum. The first is expressed in the form of a material (mass) balance, the second in the form of an energy balance, and the third in the form of a momentum balance. These three topics are treated in this and the following two chapters.

7.2 CONSERVATION OF MASS

The conservation law for mass can be applied to any process or system. The general form of this law is given by Equation (7.1):

$$\{\text{mass in}\} - \{\text{mass out}\} + \{\text{mass generated}\} = \{\text{mass accumulated}\} \qquad (7.1)$$

or on a time rate basis

$$\{\text{rate of mass in}\} - \{\text{rate of mass out}\} + \{\text{rate of mass generated}\}$$
$$= \{\text{rate of mass accumulated}\} \qquad (7.2)$$

This has also come to be defined as the continuity equation, a topic which also receives treatment in this chapter.

Fluid Flow for the Practicing Chemical Engineer. By J. Patrick Abulencia and Louis Theodore
Copyright © 2009 John Wiley & Sons, Inc.

This equation may be applied either to the total mass involved or to a particular species, on either a mole or mass basis. In many processes, it is often necessary to obtain quantitative relationships by writing mass balances on the various elements in the system.

In order to isolate a system for study, it is separated from the surroundings by a boundary or envelope. This boundary may be real (e.g., the walls of an incinerator) or imaginary. Mass crossing the boundary and entering the system is part of the *mass in* term in Equation (7.2), while that crossing the boundary and leaving the system is part of the *mass out* term.

Equation (7.2) may be written for any compound, the quantity of which is not changed by chemical reaction, and for any chemical element whether or not it has participated in a chemical reaction. It may be written for one piece of equipment, around several pieces of equipment, or around an entire process. It may be used to calculate an unknown quantity directly, to check the validity of experimental data, or to express one or more of the independent relationships among the unknown quantities in a particular problem situation.

This law can be applied to steady-state or unsteady state (transient) processes and to batch or continuous systems. A steady-state process is one in which there is no change in conditions (pressure, temperature, composition, etc.) or rates of flow with time at any given point in the system. The accumulation term in Equation (7.2) is then zero. (If there is no chemical or nuclear reaction, the generation term is also zero.) All other processes are unsteady state. In a batch process, a given quantity of reactants is placed in a container, and by chemical and/or physical means, a change is made to occur. At the end of the process, the container (or adjacent containers to which material may have been transferred) holds the product or products. In a continuous process, reactants are continuously removed from one or more points. A continuous process may or may not be steady-state. A coal-fired power plant, for example, operates continuously. However, because of the wide variation in power demand between peak and slack periods, there is an equally wide variation in the rate at which the coal is fired. For this reason, power plant problems may require the use of average data over long periods of time. Most industrial operations are assumed to be steady-state and continuous.

As indicated previously, Equation (7.2) may be applied to the total mass of each stream (referred to as an *overall* or *total* material balance) or to the individual component(s) of the streams (referred to as a *componential* or *component* material balance). The primary task in preparing a material balance in engineering calculations is often to develop the quantitative relationships among the streams. The primary factors, therefore, are those that tie the streams together. An element, compound, or unreactive mass (ash, for example) that enters or leaves in a single stream or passes through a process unchanged is so convenient for this purpose that it may be considered a key to the calculations. If sufficient data is given about this component, it can be used in a component balance to determine the total masses of the entering and exiting streams. Such a component is sometimes referred to as a *key component*. Since a key component does not react in a process, it must retain its identity as it passes through the process. Obviously, except for nuclear reactions, elements may always be used as key components because they do not change identity even though they may be involved in a chemical reaction. Thus, CO (carbon monoxide) may be used as a key component only when it does not react, but C (carbon) may always be used as a key component. A component that enters the system in only one stream and leaves in only one stream is usually the most convenient choice for a key component.

Four important processing concepts are *bypass, recycle, purge*, and *makeup*. With *bypass*, part of the inlet stream is diverted around the equipment to rejoin the (main) stream after the unit (see Fig. 7.1). This stream effectively moves in parallel with the stream passing through the equipment. In *recycle*, part of the product stream is sent back to mix with the feed. If a small quantity of nonreactive material is present in the feed to a process that includes recycle, it may be necessary to remove the nonreactive material in a *purge* stream to prevent its building up above a maximum tolerable value. This can also occur in a process without recycle; if a nonreactive material is added in the feed and not totally removed in the products, it will accumulate until *purged*. The *purging* process is sometimes referred to as *blowdown*. *Makeup*, as its name implies, involves adding or making up part of a stream that has been removed from a process. *Makeup* may be thought of, in a final sense, as the opposite of *purge* and/or *blowdown*.[1,2]

The conservation law for mass may be applied to a flowing fluid in a process. Consider the equipment depicted in Fig. 7.2, focusing attention on the part between Sections 1 and 2. No fluid is entering or leaving the process between these sections. Under steady conditions, the mass flow rates at each section are identical since there would otherwise be a progressive accumulation or depletion of fluid within the unit between the two sections. The result can be expressed in the following form:

$$\dot{m} = \rho_1 v_1 S_1 = \rho_2 v_2 S_2 = G_1 S_1 = G_2 S_2 \tag{7.3}$$

where $\dot{m}$ represents the mass flow rate; S, the cross-section area; v, the average velocity (equal to the volumetric flow rate, q, divided by the cross-section area); ρ, the mass density (i.e., mass per unit volume); and, G, the mass velocity or mass

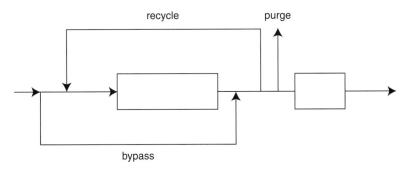

Figure 7.1 Recycle, bypass, and purge.

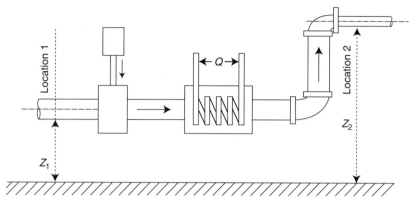

Figure 7.2 Process flow.

flux, equal to the mass flow rate divided by the cross-section, that is, the *mass flux, G*, is the mass flow rate of fluid, $\dot{m}$ per unit area of the duct, S:

$$G = \frac{\dot{m}}{S} \tag{7.4}$$

This is also equal to

$$G = \rho v \tag{7.5}$$

Illustrative Example 7.1   A gaseous waste is fed into an incinerator at a rate of 4000 kg/hr in the presence of 6000 kg/hr of air. Due to the low heating value of the waste, 550 kg/hr of methane is added to assist the combustion of the pollutants in the waste stream. Determine the rate of product gases exiting the incinerator in kg/hr. Assume steady-state operation.

Solution   Apply the conservation law for mass to the incinerator on a rate basis. See Equation (7.2)

{rate of mass in} − {rate of mass out} + {rate of mass generated}
= {rate of mass accumulated}

Rewrite the equation subject to the conditions in the example statement

{rate of mass in} = {rate of mass out}

or

$$\dot{m}_{in} = \dot{m}_{out}$$

Calculate $\dot{m}_{in}$

$$\dot{m}_{in} = 4000 + 8000 + 550 = 12{,}550 \, \text{kg/hr}$$

Also, determine $\dot{m}_{\text{out}}$. For steady-state flow conditions,

$$\dot{m}_{\text{in}} = \dot{m}_{\text{out}} = 12{,}550\,\text{kg/hr}$$

Illustrative Example 7.2 Water ($\rho = 1000\,\text{kg/m}^3$) flows in a converging circular pipe (see Fig. 7.3). It enters at Section 1 and leaves at Section 2. At Section 1, the inside diameter is 14 cm and the velocity is 2 m/s. At Section 2, the inside diameter is 7 cm. Determine the mass and volumetric flow rates, the mass flux of water, and the velocity at Section 2. Assume steady-state flow.

Solution Calculate the flow rates, q and $\dot{m}$, based on the information at Section (station) 1:

$$S_1 = \frac{\pi D_1^2}{4} = \frac{\pi (0.14)^2}{4} = 0.0154\,\text{m}^2$$

$$q_1 = S_1 v_1 = 0.0154(2) = 0.031\,\text{m}^3/\text{s}$$

$$\dot{m}_1 = \rho q_1 = 1000(0.031) = 31\,\text{kg/s}$$

Obtain the mass flux, G, with units of $\text{kg/m}^2 \cdot \text{s}$—see Equation (7.4):

$$G = \frac{\dot{m}_1}{S_1} = \frac{31}{0.0154} = 2013\,\text{kg/m}^2 \cdot \text{s}$$

Finally, calculate the velocity at cross-section 2, v_2. For steady flow,

$$q_1 = q_2 = 0.031\,\text{m}^3/\text{s}$$

Since

$$v_2 S_2 = v_1 S_1$$

$$v_2 = v_1 \frac{S_1}{S_2} = v_1 \frac{D_1^2}{D_2^2} = 2\frac{14^2}{7^2} = 8\,\text{m/s}$$

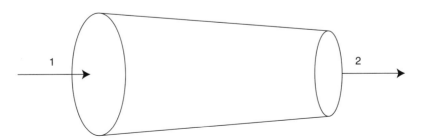

Figure 7.3 Converging pipe.

As expected, the decrease in cross-section area results in an increase in the flow velocity for steady-state flow of an incompressible fluid.

Illustrative Example 7.3 A fluid device has four openings, as shown in Fig. 7.4. The fluid has a constant density of $800 \, \text{kg/m}^3$. The steady-state conditions are listed in Fig. 7.4. Determine the magnitude and direction of the velocity, v_4. What is the mass flow rate at section 4?

Solution Calculate the volumetric flow rate through each section

$$q_1 = v_1 S_1 = 5(0.2) = 1 \, \text{m}^3/\text{s}$$

$$q_2 = v_2 S_2 = 7(0.3) = 2.1 \, \text{m}^3/\text{s}$$

$$q_3 = v_3 S_3 = 12(0.25) = 3 \, \text{m}^3/\text{s}$$

Apply the continuity equation assuming the flow is exiting section 4

$$q_1 + q_2 = q_3 + q_4$$

$$q_4 = q_1 + q_2 - q_3 = 1 + 2.1 - 3 = 0.1 \, \text{m}^3/\text{s}$$

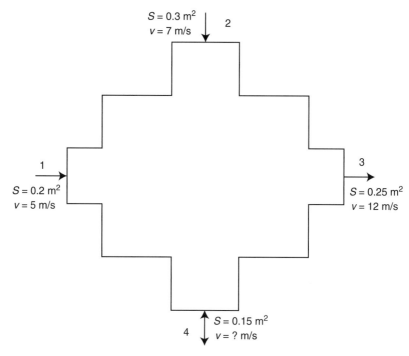

Figure 7.4 Fluid flow device.

The positive value of the calculated mass flow rate indicates the correctness of the assumption that the flow is out (and down) through section 4.

Calculate the velocity, v_4:

$$v_4 = \frac{q_4}{S_4} = \frac{0.1}{0.15} = 0.667 \, \text{m/s}$$

Also calculate the mass flow rate at section 4

$$\dot{m} = \rho q_4 = 800(0.1) = 80 \, \text{kg/s}$$

Illustrative Example 7.4   A liquid stream contaminated with a pollutant is being cleansed with a control device. If the liquid has 600 ppm (parts per million) of pollutant, and it is permissible to have 50 ppm of this pollutant in the discharge stream, what fraction of the liquid, B, can bypass the control device?

Solution   Use a basis of 1 lb of liquid fed to the control device. The flow diagram in Fig. 7.5 applies to this system. Note that:

$$B = \text{fraction of liquid bypassed}$$
$$1 - B = \text{fraction of liquid treated}$$

Performing a pollutant mass balance around point 2 in Fig. 7.5 yields

$$(1 - B)(0) + 600B = (50)(1.0)$$

Solving gives

$$B = 0.0833 = 8.33\%$$

Illustrative Example 7.5   A vertical tank 1.4 m in diameter and 1.9 m high, contains water to a height of 1.5 m. Water flows into the tank through a 9 cm pipe

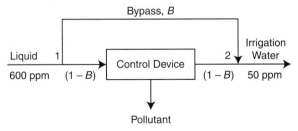

Figure 7.5   Flow diagram.

with a velocity of 4 m/s. Water leaves the tank through a 4 cm pipe at a velocity of 3 m/s. Is the level in the tank rising or falling?

Solution   Select the control volume (CV). Take the CV to be the instantaneous mass (or volume V) of water in the tank and apply the continuity principle to the CV. Since the generation rate $= 0$,

$$\left(\frac{dm}{dt}\right) = \dot{m}_{in} - \dot{m}_{out}$$

For an incompressible fluid of volume V:

$$\frac{dV}{dt} = q_{in} - q_{out}$$

since

$$V = Sz; \quad \text{where} \quad z = \text{fluid height (water)}$$

$$\frac{dV}{dt} = S\frac{dz}{dt}$$

Therefore

$$S\frac{dz}{dt} = q_{in} - q_{out}$$

Calculate the cross-section area and volumetric flow rates

$$q_{in} = \frac{\pi D^2}{4}v = \frac{\pi (0.09 \,\text{m})^2}{4}\, 4\,\text{m/s}$$

$$q_{in} = 0.0255\,\text{m}^3/\text{s}$$

$$q_{out} = \frac{\pi D^2}{4}v = \frac{\pi (0.04 \,\text{m})^2}{4}\, 3\,\text{m/s}$$

$$q_{out} = 0.0038\,\text{m}^3/\text{s}$$

$$S = \pi (1.4\,\text{m})^2/4$$

$$S = 1.539\,\text{m}^2$$

Substitute in the above material balance differential equation

$$1.539\frac{dz}{dt} = 0.0255 - 0.0038$$

$$\frac{dz}{dt} = 0.0141\,\text{m/s}$$

$$\text{at } t = 0,\ z = 1.5\,\text{m}$$

Because dz/dt is positive, the water level is rising in the tank from its initial height of 1.5 m.

7.3 MICROSCOPIC APPROACH

The equation of continuity describes the variation of density with position and time in a moving or stationary fluid, and may be viewed as an extension of the conservation law for mass. It can also be used to simplify the equations of energy transfer and momentum transfer. In addition, it serves as an excellent warm-up for the presentation of the equation of momentum transfer in the next chapter. The continuity equation can be developed by applying the conservation law for mass to a fixed volume element in a moving one-component one-phase fluid. The derivation is available in the literature.[3,4]

The continuity equation is written in rectangular, cylindrical and spherical coordinates. The results are presented in Table 7.1.

Table 7.1   Equation of continuity for incompressible fluids

Rectangular coordinates (x, y, z):

$$\frac{\partial v_x}{\partial x} + \frac{\partial v_y}{\partial y} + \frac{\partial v_z}{\partial z} = 0 \qquad (1)$$

Cylindrical coordinates (r, ϕ, z):

$$\frac{1}{r}\frac{\partial}{\partial r}(rv_r) + \frac{1}{r}\frac{\partial v_\phi}{\partial \phi} + \frac{\partial v_z}{\partial z} = 0 \qquad (2)$$

Spherical coordinates (r, θ, ϕ):

$$\frac{1}{r^2}\frac{\partial}{\partial r}(r^2 v_r) + \frac{1}{r\sin\theta}\frac{\partial}{\partial \theta}(v_\theta \sin\theta) + \frac{1}{r\sin\theta}\left(\frac{\partial v_\phi}{\partial \phi}\right) = 0 \qquad (3)$$

Illustrative Example 7.6   The velocity of incompressible fluid in a steady-state system is directed along the y rectangular coordinate. Prove that the velocity is not a function of y.

Solution   For an incompressible fluid, see Equation (1) in Table 7.1:

$$\frac{\partial v_x}{\partial x} + \frac{\partial v_y}{\partial y} + \frac{\partial v_z}{\partial z} = 0$$

One concludes from the problem statement that

$$v_x = 0$$
$$v_z = 0$$

Therefore

$$\frac{\partial v_x}{\partial x} = 0 \quad \text{since } v_x = 0$$

$$\frac{\partial v_z}{\partial z} = 0 \quad \text{since } v_z = 0$$

Finally,

$$\frac{\partial v_y}{\partial y} = 0$$

If v_y is a function of y, then $\partial v_y / \partial y$ cannot equal zero. The velocity therefore, is not a function of y.

REFERENCES

1. J. Santoleri, J. Reynolds, and L. Theodore, "Introduction to Hazardous Waste Incineration," 2nd edition, John Wiley & Sons, Hoboken, NJ, 2000.
2. J. Reynolds, J. Jeris, and L. Theodore, "Handbook of Chemical and Environmental Engineering Calculations," John Wiley & Sons, Hoboken, NJ, 2004.
3. R. Bird, W. Stewart, and E. Lightfoot, "Transport Phenomena," 2nd edition, John Wiley & Sons, Hoboken, NJ, 2002.
4. L. Theodore, "Transport Phenomena for Engineers," International Textbook Company, Scranton, PA, 1971.

NOTE: Additional problems are available for all readers at www.wiley.com. Follow links for this title.

CONSERVATION LAW
FOR ENERGY

8.1 INTRODUCTION

This chapter is concerned with the conservation law for energy. The presentation to follow once again includes a review of some key pressure terms. A general introduction to the conservation of energy is in turn followed by the development of a general total energy balance for steady-state flow. The chapter concludes by extending the total energy equation to include mechanical energy; this has come to be defined as the *mechanical energy balance equation*. It is this equation that is employed in the solution of most real-world fluid flow problems.

One of the most critical parameters in fluid flow is pressure. This was briefly defined in Chapter 3. Three additional pressure terms should be defined before proceeding to the body of this chapter. These are the *static pressure* (P_s), the *velocity pressure* (P_v), and the sum of the two—the total pressure (P_t).

Any fluid confined in a stationary enclosure has static pressure simply because the molecules of that fluid are in constant random motion and are continually colliding with the container walls. The bulk velocity of this stationary fluid is zero, and the total pressure is then equal to the static pressure. If the same fluid is flowing and the temperature has not changed, it possesses the same static pressure since its molecules still have the same degree of random motion. Its total pressure is now higher, however, because it also possesses the second pressure component, velocity pressure. If the fluid flow were to suddenly change direction because of a solid obstruction (e.g., a plate), an extra pressure on the plate (over and above the static pressure) would be exerted because of the momentum of the bulk flow against the

Fluid Flow for the Practicing Chemical Engineer. By J. Patrick Abulencia and Louis Theodore
Copyright © 2009 John Wiley & Sons, Inc.

plate. This extra pressure is the velocity pressure and the total fluid pressure is the sum of the static and velocity pressures. Static pressure is therefore the result of motion on the molecular level, while velocity pressure is due to motion at the macroscopic or bulk level.[1]

The difference in total pressure between two different points along the stream is called the pressure loss or the pressure drop. Pressure losses from fluid flow are due to any effect that can change fluid momentum at either the molecular or macroscopic levels; the two main contributing factors are skin friction and form friction. *Skin friction* losses are caused by fluid moving along (parallel to) a solid surface such as a pipe or duct wall. The layers of fluid immediately adjacent to the wall are in laminar flow and moving much slower than the bulk of the fluid. The pressure drop caused by the drag effect of the wall on the fluid is due to skin friction. *Form friction* losses are due to the acceleration or deceleration of the fluid. These include changes in bulk fluid velocity that occur because of changes in either flow direction or flow speed. An example of a change in flow direction is fluid flowing through a 90° elbow; alternatively, a change in flow speed occurs when the cross-section of a conduit changes. Besides changes in bulk fluid velocity, form friction losses also include changes in velocity that occur locally, i.e., internal to the bulk motion of the fluid. This occurs in turbulent flow (see Chapter 14 for more details), which is characterized by rapidly swirling masses of fluid called *eddies*.[1]

8.2   CONSERVATION OF ENERGY

A presentation of the conservation law for energy would be incomplete without a brief review of some introductory thermodynamic principles. *Thermodynamics* is defined as that science that deals with the relationships among the various forms of energy. A system may possess energy due to its temperature (internal energy), velocity (kinetic energy), position (potential energy), molecular structure (chemical energy), surface (surface energy), etc. Engineering thermodynamics is founded on three basic laws. As described earlier, energy, like mass and momentum, is conserved. Application of the conservation law for energy gives rise to the first law of thermodynamics. This law for batch processes, is presented below.

For batch processes

$$\Delta E = Q + W \tag{8.1}$$

where potential, kinetic, and other energy effects have been neglected and Q is energy in the form of heat transferred across the system boundaries, W is energy in the form of work transferred across system boundaries, E (often denoted as U), the internal energy of the system, and ΔE is the change in the internal energy of the system. In accordance with the recent change in convention, both Q and W are treated as *positive* terms if *added* to the system.

By definition, a flow process involves material streams entering and exiting a system. Work is done on the system at the stream entrance when the fluid is pushed into the system. Work is performed by the system to push the fluid out at the stream exit. The net work on the system is called flow work, W_f, and is given by

$$W_f = \sum P_{out} V_{out} - \sum P_{in} V_{in} = -\Delta(PV) \tag{8.2}$$

where P_{out} is the pressure of the outlet stream, P_{in} is the pressure of the inlet stream, V_{out} is the volume of fluid exiting the system during a given time interval, and V_{in} is the volume of fluid entering the system during a given time interval. If the volume term is represented as the specific volume (i.e., volume/mass), the work term carries the units of energy/mass.

For practical purposes, the total work term, W, in the first law may be regarded as the sum of shaft work, W_s, and flow work, W_f

$$W = W_s + W_f \tag{8.3}$$

where W_s is work done on the fluid by some moving solid part within the system such as the rotating vanes of a centrifugal pump. Note that in Equation (8.3), all other forms of work such as electrical, surface tension, and so on are neglected. The first law for steady-state flow processes is then:

$$\Delta H = Q + W_s \tag{8.4}$$

where H is the enthalpy of the system and ΔH is the change in the system's enthalpy.

The internal energy and enthalpy in Equations (8.1) and (8.2), as well as other equations in this section may be on a mass basis, on a mole basis, or represent the total internal energy and enthalpy of the entire system. They may also be written on a time-rate basis as long as these equations are dimensionally consistent—it makes no difference. For the sake of clarity, upper case letters (e.g., H, E) represent properties on a mole basis, while lower-case letters (e.g., h, e) represent properties on a mass basis. Properties for the entire system will rarely be used and therefore require no special symbols.

Perhaps the most important thermodynamic function the engineer works with is the above mentioned *enthalpy*. This is a term that requires additional discussion. The enthalpy is defined by the equation

$$H = E + PV \tag{8.5}$$

where P is once again the pressure of the system and V is the volume of the system. The terms E and H are state or point functions. By fixing a certain number of variables upon which the function depends, the numerical value of the function is automatically fixed; that is, it is single-valued. For example, fixing the temperature and pressure of a one-component single-phase system immediately specifies the enthalpy and internal energy. This last statement can be verified by Gibbs's Phase Rule.[2]

The change in enthalpy as it undergoes a change in state from (T_1, P_1) to (T_2, P_2) is given by

$$\Delta H = H_2 - H_1 \tag{8.6}$$

Note that H and ΔH are independent of the path. This is a characteristic of all state or point functions; that is, the state of the system is independent of the path by which the state is reached. The terms Q, W, and W_s in Equations (8.4) and (8.5) are path functions; their values depend on the path used between the two states.

The following can be written for a mathematical representation of this important thermodynamic point function

$$H = H(T, P)$$

By the rules of partial differentiation, a differential change in H is given by

$$dH = \left(\frac{\partial H}{\partial T}\right)_P dT + \left(\frac{\partial H}{\partial P}\right)_T dP \tag{8.7}$$

The term $(\partial H/\partial P)_T$ is assumed to be negligible in most engineering applications. It is exactly zero for an ideal gas and is small for solids and liquids, and gases near ambient conditions. The term $(\partial H/\partial T)_P$ is defined as the *heat capacity at constant pressure*

$$C_P = \left(\frac{\partial H}{\partial T}\right)_P \tag{8.8}$$

Equation (8.8) may also be written as

$$dH = C_P \, dT \tag{8.9}$$

If average molar heat capacity data are available, this equation may be integrated to yield

$$\Delta H = \overline{C_P} \, \Delta T \tag{8.10}$$

where $\overline{C_P}$ is the average value of C_P in the temperature range ΔT. Calculations involving enthalpy changes also finds extensive application in heat transfer.

Many industrial applications operate in a steady-state flow mode with no significant mechanical or shaft work added (or withdrawn) from the system. For this condition Equation (8.4) reduces to

$$Q = \Delta H \tag{8.11}$$

This equation is routinely used in many calculations. If a unit or system is operated adiabatically, $Q = 0$ and Equation (8.11) becomes

$$\Delta H = 0 \tag{8.12}$$

Although the topics of material and energy balances have been covered separately in this and the previous chapter, it should be emphasized that this segregation does not exist in reality. Many processes are accompanied by heat effects, and one must work with both energy and material balances simultaneously.[3]

Illustrative Example 8.1 5.5 MW of heat is transferred from a gas as it flows through a cooler. The average heat capacity of the gas is 1090 J/(kg · °C), the gas mass flow rate, $\dot{m}$, is 9 kg/s and the gas inlet temperature, T_1, is 650°C. For this example, kinetic and potential energy effects are neglected. Furthermore, there is no shaft work. Determine the gas outlet temperature.

Solution Since there are no kinetic, potential, or shaft work effects in this flow process, Equations (8.10) and (8.11) applies

$$Q = \Delta H$$

where $\Delta H = \dot{m}\overline{C}_P\,\Delta T = \dot{m}\overline{C}_P(T_2 - T_1)$.
Solving for the gas outlet temperature, T_2,

$$T_2 = \frac{Q}{\dot{m}\overline{C}_P} + T_1 = \frac{-5.5 \times 10^{-6}}{9(1090)} + 650 = 89°C$$

Note that the sign of Q is negative since the heat is transferred out from the gas.

8.3 TOTAL ENERGY BALANCE EQUATION

Equations (8.1) and (8.4) find application in many chemical process units such as heat exchangers, reactors, and distillation columns, where shaft work plus kinetic and potential energy changes are negligible compared with heat flows and either internal energy or enthalpy changes. Energy balances on such units therefore reduce to $Q = \Delta E$ (closed system) or $\dot{Q} = \Delta\dot{H}$ (open system).

Another important class of operations is one for which the opposite is true—heat flows and internal energy changes are secondary in importance to kinetic and potential energy changes and shaft work. Most of these operations involve the flow of fluids to, from, and between tanks, reservoirs, wells, and process units. Accounting for energy flows in such processes is most conveniently accomplished with mechanical energy balances.[4] Details of this approach follow.

Consider the steady-state flow of a fluid in the process pictured in Fig. 8.1. The mass entering at location 1 brings in with it a certain amount of energy, existing in various forms. Thus, because of its elevation, z_1 ft above any arbitrarily chosen horizontal reference plane, for example, $z = 0$, it possesses a potential energy $(g/g_c)z_1$ (which can be recovered by allowing the fluid to fall from the height at location 1 to that of the reference point). Because of its velocity, v_1, the mass possesses and brings into location 1 of the system an amount of kinetic energy, $v_1^2/2g_c$. It also brings its so-called internal energy, E_1, because of its temperature. Furthermore, the mass of fluid in question entering at point 1 is forced into the section by the

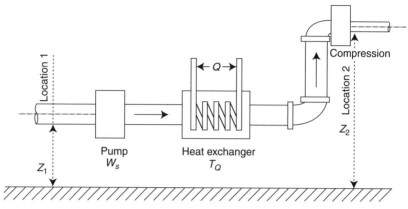

Figure 8.1 Process flow.

pressure of the fluid behind it and this form of flow energy must also be included. The amount of this energy is given by the force exerted by the flowing fluid times the distance through which it acts, and this force is clearly the pressure per unit area, P_1, times the area S_1 of the cross-section. The distance through which the force acts is the volume, V_1 of the fluid divided by the cross-sectional area S_1. Since the work is the force times the distance, that is, $(P_1 S_1)(V_1/S_1) = P_1 V_1$, the energy expended is the product of the pressure times the volume of the fluid. This was referred to earlier as *flow work* (see Eq. (8.2)).

Two additional energy terms need to be included in the analysis. These two involve energy exchange in the form of heat (Q) and work (W) between the fluid and the surroundings. In the development to follow, it will be assumed (consistent with the notation recently adopted by the scientific community) that any energy in the form of heat or work *added* to the system is treated as a *positive* term.

Applying the conservation law of energy mandates that all forms of energy entering the system equal that of those leaving. Expressing all terms in consistent units (e.g., energy per unit mass of fluid flowing), results in the total energy balance:

$$P_1 V_1 + \frac{v_1^2}{2g_c} + \frac{g}{g_c} z_1 + E_1 + Q + W_s = P_2 V_2 + \frac{v_2^2}{2g_c} + \frac{g}{g_c} z_2 + E_2 \qquad (8.13)$$

As written, each term in Equation (8.13) represents a mechanical energy effect. For this reason, it is defined as a form of the mechanical energy balance equation and is essentially a special application of the conservation law for energy. Also note that, as written, the volume term, V, (for necessity) is the specific volume. In terms of the density, the above equation becomes

$$\frac{P_1}{\rho} + \frac{v_1^2}{2g_c} + \frac{g}{g_c} z_1 + E_1 + Q + W_s = \frac{P_2}{\rho} + \frac{v_2^2}{2g_c} + \frac{g}{g_c} z_2 + E_2 \qquad (8.14)$$

and

$$V = 1/\rho \qquad (8.15)$$

Note once again that Q and W_s can be written on a time rate basis in the above equation by simply dividing by the mass flowrate though the system; the above equation then dimensionally reduces to an energy/mass balance.

Three points need to be made before leaving this subject.

1. The term Q should represent the total net heat added to the fluid, but in this analysis it includes only the heat passing into the fluid across the walls of the containing walls from an external source. This excludes heat generated by friction, by the fluid or otherwise, within the unit. However, this effect can normally be safely neglected.

2. The work, W_s, similar to Q, must pass though the retaining walls. While it could conceivably enter in other ways, it is supplied in most applications by some form of moving mechanism, such as a pump, or a fan, and is often referred to as *shaft work*.

3. The internal energy term E corresponds to the thermodynamic definition provided earlier. For convenience, the sum of E and PV may be treated as the single function defined above as the enthalpy, H,

$$H = E + PV \qquad (8.5)$$

It too is a property of the fluid, uniquely determined by point conditions. Like E, its absolute value is arbitrary; differences in value are often given above a reference. With this revision, and assuming $\alpha = 1$, Equation (8.13) becomes:

$$\frac{v_1^2}{2g_c} + \frac{g}{g_c}z_1 + H_1 + Q + W_s = \frac{v_2^2}{2g_c} + \frac{g}{g_c}z_2 + H_2 \qquad (8.16)$$

or simply

$$\frac{\Delta v^2}{2g_c} + \frac{g}{g_c}\Delta z + \Delta H = Q + W_s \qquad (8.17)$$

As noted in the presentation of Equation (8.13), each term is dimensional with units of energy/mass. If this equation is multiplied by the fluid flow rate, that is, mass/time, the units of each term become energy/time. In the absence of both kinetic and potential energy effects, the above equation reduces to Equation (8.4). Also note that Δ, the difference term, refers to a difference between the value at station 2 (the usual designation for the outlet) minus that at station 1 (the inlet).

Illustrative Example 8.2   A fluid flow device has three openings, as shown in Fig. 8.2. The flow within the control volume is steady and the fluid has a constant density of $800 \, \text{kg/m}^3$. The flow properties at each opening are provided below in Table 8.1. What is the rate of change of the system energy in the control volume? For steady-state adiabatic operation, is work being done on the system?

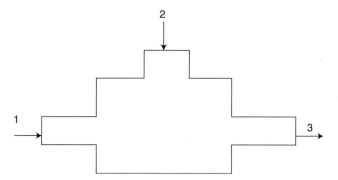

Figure 8.2   Fluid flow device for Illustrative Example 8.2.

Table 8.1   Flow/energy data for Illustrative Example 8.2

Section	Flow Rate, q, m³/s	Enthalpy, h, J/kg	Flow Direction (Relative to the Device)
1	8	250	In
2	6	150	In
3	14	200	Out

Solution   Confirm the mass balance. For incompressible flow,

$$q_1 + q_2 = q_3$$
$$8 + 6 = 14 \, \text{m}^3/\text{s}$$

Apply the total energy balance, noting that only enthalpy effects need to be considered for this flow system

$$\Delta \dot{H} = 800[(8)(250) + (6)(150) - (14)(200)] = 80{,}000 \, \text{J/s}$$

For adiabatic steady operation

$$\dot{Q} = 0$$

so that

$$\Delta \dot{H} = \dot{W}_s = 80 \, \text{kW} = 107.2 \, \text{hp}$$

Since work is positive, the surroundings must be doing work on the system through some device.

8.3.1 The Mechanical Energy Balance Equation

As noted, the solutions to many fluid flow problems are based on the mechanical energy balance equation. This equation is derived, in part, from the general (or total) energy equation developed in the previous subsection. Equation (8.13) is shown again below:

$$P_1 V_1 + \frac{v_1^2}{2g_c} + \frac{g}{g_c} z_1 + E_1 + Q + W_s = P_2 V_2 + \frac{v_2^2}{2g_c} + \frac{g}{g_c} z_2 + E_2$$

Certain "changes" to the above equation are now made:

1. Assume adiabatic flow, that is, $Q = 0$.
2. For isothermal, or near isothermal, flow (valid in most applications), the internal energy is constant, so that $E_1 = E_2$.
3. A term $\sum F$, representing the total friction arising due to fluid flow, is added to the equation. This is treated as a positive term in Equation (8.18) below.
4. An efficiency (fractional) term, η, is combined with the shaft work term, W_s. If work is imparted on the system, the term becomes ηW_s; if work is extracted (with an engine or turbine) the term appears as W_s / η. The efficiency term needs to be included since part of the work added to or extracted from the system is lost due to irreversibilities associated with the mechanical device. The notation h_s will be employed for this term in Chapter 9.

Equation (8.14) now becomes

$$\frac{\Delta P}{\rho} + \frac{\Delta v^2}{2g_c} + \frac{g}{g_c} \Delta z - \eta W_s + \sum F = 0 \tag{8.18}$$

This equation is defined as the mechanical energy balance equation; it will receive extensive attention later in the book.

8.3.2 The Bernoulli Equation

The Bernoulli equation has come to mean different things to different people. One definition of this equation is obtained by neglecting both work and friction effects. Under this condition, Equation (8.18) reduces to

$$\frac{\Delta P}{\rho} + \frac{\Delta v^2}{2g_c} + \frac{g}{g_c} \Delta z = 0 \tag{8.19}$$

This is often referred to as the Field equation in other disciplines.

This equation, which applies to flow in the absence of friction, has some interesting ramifications. If one of the three terms is increased, either of the other two terms must decrease; alternately, both of the other two terms can change but the sum of the two changes must decrease. For example, if the Bernoulli equation is applied along a horizontal streamline (path) of a fluid, an increase in the velocity results in a decrease in pressure. The phenomenon is "exploited" by birds during flight, and at the industrial level in the design of airplane wings.

The above effect can also explain why roofs are lifted off some buildings during a hurricane or tornado; the high velocity on top of the roof creates a lower pressure at the outer surface relative to the inner surface. This difference in pressure—force per unit area—across the roof's top surface produces a net upward force lifting the roof off its foundation. This can be prevented in many instants by simply opening all windows and doors; the high velocity within the structure produces a lower pressure and consequently a smaller or zero upward force.

The reader might like to test the validity of the proposed explanation by taking a sheet of $8\frac{1}{2}''$ by $11''$ paper and holding it by its sides while allowing the paper to droop. Blowing across the top of the paper does in fact result in the paper rising to a near horizontal level.

Bernoulli's equation is valid for steady-state flow. However, if the flow is not steady, but the changes in flow rate are slow enough to be ignored, then Bernoulli's equation may still be applied. In tank flow problems for example, the velocity of the fluid in the tank is taken to be the rate of change of liquid height with time, that is, $v = dz/dt$. When this velocity is combined with Bernoulli's equation, the result is a differential equation. The solution of the equation normally requires integration. To justify the assumption that the flow rate is slow, the flow acceleration, a, is calculated and compared to the gravity acceleration, g. If a/g is $\ll 1$, the above assumption may be assumed valid.

Illustrative Example 8.3   Refer to Illustrative Example 3.5 in Chapter 3. What is the kinetic energy of the water just before it reaches ground level conditions, that is, strikes the bottom?

Solution   Apply the conservation law for energy to the 1 kg of water. During free fall, $\Delta E = 0$, and the only terms that remain are the KE and PE. Therefore,

$$\Delta(PE) + \Delta(KE) = 0$$

since both

$$KE_1 = PE_2 = 0$$
$$KE_2 = PE_1 = 98 \text{ J}$$

Illustrative Example 8.4   Refer once again to Illustrative Example 3.5. If the 1 kg of water enters a river upon reaching ground conditions, what change has occurred to the water.

Solution   Physically, it is still liquid. Energy-wise, it has lost the PE it started with. That PE has been converted to the internal energy of the entire river; however, the temperature change of the river would be neglible.

Illustrative Example 8.5   A cylindrical tank (see Fig. 8.3) with a diameter of 3 m has an outlet hole of 0.3 m in diameter at its bottom. The initial water level is 9 m. How long does it take the liquid level to drop to 1 m above the tank outlet? Justify the assumption of using Bernoulli's equation. The density of water is $1000\ kg/m^3$.

Solution   Assume the control volume (CV) to be the liquid in the tank. A mass balance on the CV yields

$$\rho S_1 v_1 = \rho S_2 v_2$$

$$v_2 = v_1 \frac{S_1}{S_2} = v_1 \left(\frac{D_1}{D_2}\right)^2$$

Set

$$v_1 = \frac{dz}{dt}$$

Substituting into the above equation gives

$$v_2 = -\left(\frac{D_1}{D_2}\right)^2 \frac{dz}{dt}$$

From Bernoulli's equation

$$\frac{P_1}{\rho} + \frac{v_1^2}{2g_c} + \frac{g}{g_c}z_1 = \frac{P_2}{\rho} + \frac{v_2^2}{2g_c} + \frac{g}{g_c}z_2$$

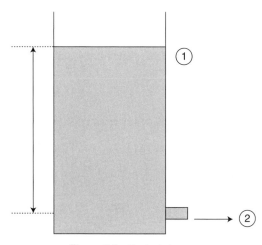

Figure 8.3 Tank drainage.

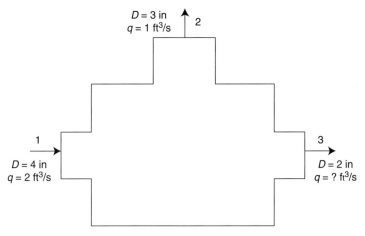

Figure 8.4 Fluid flow device.

From Fig. 8.3, it is clear that $P_1 = P_2$, since both ends are open to the atmosphere. Likewise, $v_1 = 0$, and z_2 is arbitrarily set to equal zero for convenience. Therefore,

$$\frac{v_2^2}{2g_c} = \frac{g}{g_c} z_1$$

or

$$v_2 = \sqrt{2gz_1}$$

where z_1 is the height at location 1. However, since z_1 varies, it is replaced by h, the height of water in the tank. Equating the two velocity terms for v_2 leads to

$$\frac{dh}{dt} = -\left(\frac{D_2}{D_1}\right)^2 \sqrt{2gh}$$

The initial conditions are at $t = 0$, $h = z_1$. The differential equation may be solved by separating the variables and integrating:

$$\int_{z_1}^{h_2} \frac{dh}{h^{1/2}} = -\left(\frac{D_2}{D_1}\right)^2 \sqrt{2g} \int_0^t dt$$

$$2(h_2^{1/2} - z_1^{1/2}) = -\sqrt{2g}\left(\frac{D_2}{D_1}\right)^2 t$$

$$t = \frac{2(z_1^{1/2} - h_2^{1/2})}{\sqrt{2g}\left(\frac{D_2}{D_1}\right)^2} = \frac{2(9^{1/2} - 1^{1/2})}{\sqrt{2(9.807)}(0.1)^2} = 90.3 \text{ s}$$

Justify the use of Bernoulli's equation.

$$a = \frac{dv_2}{dt} = \frac{d}{dt}(\sqrt{2gh}) = \frac{1}{2}\sqrt{\frac{2g}{h}\frac{dh}{dt}}$$

Substituting for dh/dt and noting

$$D\sqrt{h} = \frac{dh}{2\sqrt{h}}$$

$$\frac{a}{g} = -\left(\frac{D_2}{D_1}\right)^2 - \left(\frac{0.3}{3.0}\right)^2 = -0.01$$

For this example, the maximum acceleration is 1% of g. One can therefore safely use Bernoulli's equation.

Illustrative Example 8.6   Explain in layman terms, why the $\sum F$ term in Equation (8.18) is positive.

Solution   This steady-state conservation law, as written, is derived on an

$$\text{out} - \text{in} = 0$$

basis. Thus, the work term is positive only if it is removed (lost) from the system. The same applies to any energy term. Frictional effects give rise to energy that is lost from the system and must therefore be retained as a positive term.

REFERENCES

1. W. Badger and J. Banchero, "Introduction to Chemical Engineering," McGraw-Hill, New York, 1955.
2. D. Green and R. Perry (editors) "Perry's Chemical Engineers' Handbook," 8th edition, McGraw-Hill, New York, 2008.
3. J. Santoleri, J. Reynolds, and L. Theodore, "Introduction to Hazardous Waste Incineration," 2nd edition, John Wiley & Sons, Hoboken, NJ, 2000.
4. R. Felder and R. Rousseau, "Elementary Principles of Chemical Processes," 3rd edition, John Wiley & Sons, Hoboken, NJ, 2000.

NOTE: Additional problems are available for all readers at www.wiley.com. Follow links for this title.

9

CONSERVATION LAW FOR MOMENTUM

Momentum transfer is introduced by reviewing the units and dimensions of momentum, time rate of change of momentum, and force. The phenomenological law governing the transfer of momentum by molecular diffusion—Newton's second law—was briefly discussed in Chapter 5. In addition to molecular diffusion, momentum (and energy) may also be transferred by bulk motion. Since bulk motion involves transfer of mass from one point in a system to another, the equation of continuity (conservation law for mass) was also discussed earlier. These serve as an excellent warm-up for the equation of motion (equation of momentum transfer or conservation law for momentum that receives treatment in Section 9.2 of this chapter).

9.1 MOMENTUM BALANCES

A momentum balance (also termed the *impulse-momentum principle*) is important in flow problems where forces need to be determined. This analysis is inherently more complicated than those previously presented (i.e., forces possess both magnitude and direction), because the force, F, and momentum, M, are vectors. In order to describe force and momentum vectors, both direction and magnitude must be specified; for mass and energy, only the magnitude is required.

Newton's law is applied in order to derive the linear momentum balance equation. Newton's law states that the sum of all forces equals the rate of change of linear

Fluid Flow for the Practicing Chemical Engineer. By J. Patrick Abulencia and Louis Theodore
Copyright © 2009 John Wiley & Sons, Inc.

momentum

$$\sum F = \frac{d}{dt}\left(\frac{mv}{g_c}\right) = \frac{dM}{dt} = \dot{M} \tag{9.1}$$

Here $\dot{M}$ is the rate (with respect to time) of linear momentum, and m and v represent the mass and velocity, respectively. Newton's law must be applied in a specified direction (e.g., horizontal or vertical). The product $(m)(v)$ is called the *linear momentum*. When this is applied to a fluid entering or leaving a control volume, the following terms may be defined:

$\dot{M}_{out}$ = momentum rate of the fluid leaving the control volume

$\dot{M}_{in}$ = momentum rate of the fluid entering the control volume

Equation (9.1) may be rewritten in finite form

$$\sum F = \dot{M}_{out} - \dot{M}_{in} \tag{9.2}$$

This balance essentially means that for steady-state flow, the force on the fluid equals the net rate of outflow of momentum across the control surface. Equation (9.2) also may be rewritten as

$$\frac{d}{dt}\left(\frac{mv}{g_c}\right)_{in} = \frac{d}{dt}\left(\frac{mv}{g_c}\right)_{out} - \sum F \tag{9.3}$$

or

$$\dot{M}_{in} = \dot{M}_{out} - \sum F \tag{9.4}$$

This may be compared with the generalized steady-state balance equation for momentum:

{rate of momentum in} = {rate of momentum out}

+ {generation rate of momentum}   (9.5)

Thus, the generation rate of momentum may be viewed as the negative of the net force acting on the fluid mass.[1] When a momentum balance is used to calculate the forces in different (but perpendicular) directions (e.g., F_x and F_y), the net (or resultant) force is obtained

$$F_{res} = \sqrt{F_x^2 + F_y^2} \tag{9.6}$$

Application of the above principles is provided in the following two Illustrative Examples.

Illustrative Example 9.1   A horizontal water jet impinges on a vertical plate. The jet splits into several jets traveling in the vertical direction. The water flow rate, q, is $0.5 \text{ ft}^3/\text{s}$, the water's horizontal velocity, v, is $100 \text{ ft}/\text{s}$, and the water density, ρ, is $62.4 \text{ lb}/\text{ft}^3$. Determine the force required to hold the plate stationary.

Solution   The momentum balance equation in the horizontal direction is

$$F = \dot{M}_{\text{out}} - \dot{M}_{\text{in}}$$

The momentum rate of the inlet water in the horizontal direction is given by

$$\dot{M}_{\text{in}} = \frac{\rho q v}{g_c}$$

The horizontal momentum rate of the exit water is $\dot{M}_{\text{out}} = 0$. The net force in the horizontal direction, F, is therefore

$$F = 0 - \frac{\rho q v}{g_c} = -\frac{(62.4)(0.5)(100)}{32.2} = -97 \text{ lb}_{\text{f}}$$

The net horizontal force can be recalculated if the jet had an angle of $10°$ to the horizontal. For this case

$$\dot{M}_{\text{in}} = \frac{\rho q v}{g_c} \cos(10°) = 97(0.985) = 95.5 \text{ lb}_{\text{f}}$$

$$F = -95.5 \text{ lb}_{\text{f}}$$

The negative answer above indicates that to hold the plate in place, a force must be exerted in a direction opposite to that of the water flow.

Illustrative Example 9.2   A 10 cm diameter horizontal line carries saturated steam at a velocity of 420 m/s. Water is entrained (carried along) by the steam at the rate 0.15 kg/s. The line has a $90°$ bend. Calculate the force required to hold the bend in place due to the entrained water (see Fig. 9.1).

Solution   Select the control volume as the fluid in the bend and apply a mass balance.

$$\dot{m}_1 = \dot{m}_2$$

In addition,

$$v_1 = v_2$$

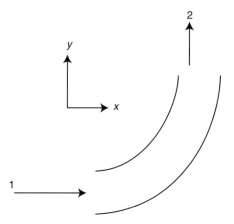

Figure 9.1 90° turn.

Apply a linear momentum balance in the horizontal (x) direction, neglecting the momentum of the steam

$$F_x = \frac{d}{dt}(mv)_{out,x} - \frac{d}{dt}(mv)_{in,x} = 0 - \dot{m}v_{in,x} = -0.15(420) = -63\,\text{N}$$

The x-direction force acting on the 90° elbow is therefore $F_x = +63\,\text{N}$.
Apply a linear momentum balance in the vertical (y) direction

$$F_y = \dot{M}_{out,y} - \dot{M}_{in,y} = \dot{m}v_{out,y} - 0 = 0.15(420) = 63\,\text{N}$$

The y-direction force acting on the 90° elbow is therefore $F_y = -63\,\text{N}$.
The resultant force may now be calculated from Equation (9.6)

$$F_{res} = \sqrt{F_x^2 + F_y^2} = \sqrt{(-63)^2 + 63^2} = 89.1\,\text{N}$$

The resultant force is the force required to hold the elbow in place.

Illustrative Example 9.3 Water (density $= 62.4\,\text{lb/ft}^3$) flows in a 2 inch diameter pipe. The pipe has a 90° bend. The bend support can withstand a maximum force in the x-direction of 5 lb$_f$. Determine the maximum water flow rate in the pipe bend.

Solution Select the control volume to be the fluid in the bend and apply a mass balance.

$$\dot{m}_1 = \dot{m}_2 = \dot{m} = \rho S v$$

For steady incompressible flow,

$$q_1 = q_2 = q = \rho v$$

Therefore,

$$v_1 = v_2 = v$$

Apply a linear momentum balance on a rate basis in the horizontal x-direction [see Equation (9.4)]

$$\dot{M}_{\text{in},x} = \dot{M}_{\text{out},x} + F_x = \dot{m}\frac{v}{g_c} = 0 - (-5)$$

$$\frac{\rho S v^2}{g_c} = 5$$

$$v = \sqrt{\frac{5g_c}{\rho S}}$$

The use of g_c is necessary to obtain the proper units on both sides of the equation. Substitute numerical values to generate the flow velocity.

$$v = \sqrt{\frac{5g_c}{\rho(\pi D^2/4)}} = \sqrt{\frac{5(32.174)}{62.4(\pi)(0.167^2/4)}} = 10.8\,\text{ft/s}$$

Finally, the volumetric and mass flow rates can be calculated

$$q = Sv = (0.0219)(10.8) = 0.238\,\text{ft}^3/\text{s}$$
$$\dot{m} = \rho q = 62.4(0.238) = 14.8\,\text{lb/s}$$

This represents the maximum water flow rate that the elbow can handle. However, the practicing engineer employs a safety factor so that the possibility of a failure or problem arising is decreased.

Illustrative Example 9.4   Water (density $= 1000\,\text{kg/m}^3$, viscosity $= 0.001\,\text{kg/}$ $(\text{m}\cdot\text{s})$) is discharged through a horizontal fire hose (see Fig. 9.2) at a rate of

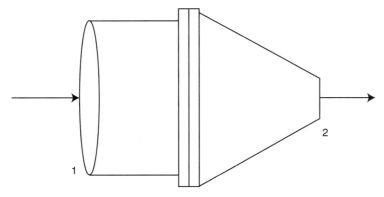

Figure 9.2   Fire hose.

1.5 m^3/min. The fire hose is 10 cm in diameter. The nozzle's diameter reduces from 10 cm to 3 cm. The nozzle discharges the water into the atmosphere. Calculate water velocities and the pressures in the fire hose and at the nozzle tip, the x-direction momentum at both ends of the nozzle, the force required to hold the hose, and the type of flow in the fire hose.

Solution   Apply a mass balance on the CV.

$$q = q_1 = v_1 S_1 = 0.025 \, \text{m}^3/\text{s}$$
$$\dot{m} = \rho q = 1000(0.025) = 25 \, \text{kg/s}$$

Calculate the velocities v_1 and v_2.

$$v_1 = \frac{q}{S_1} = \frac{0.025}{\pi(0.1)^2/4} = 3.2 \, \text{m/s}$$

$$v_2 = \frac{q}{S_2} = \frac{0.025}{\pi(0.03)^2/4} = 35.4 \, \text{m/s}$$

Determine the pressure, P_1, by applying Bernoulli's equation between points 1 and 2 (see Fig. 9.2).

$$z_1 = z_2$$
$$P_2 = 0 \, \text{Pag (Pascal gauge)}$$

$$P_1 = \frac{\rho(v_2{}^2 - v_1{}^2)}{2} \frac{1}{g_c} = \frac{1000}{2}[(35.4)^2 - (3.2)^2] = 620{,}000 \, \text{Pag}$$

Calculate the x-direction momentum rates.

$$\dot{M}_{1,x} = (\dot{m}_1 v_1)_x = (25)(3.2) = 80 \, \text{N}$$
$$\dot{M}_{2,x} = (\dot{m}_2 v_2)_x = (25)(35.4) = 885 \, \text{N}$$

Obtain the force from the momentum balance in the x-direction.

$$F_x = \dot{M}_{2,x} - \dot{M}_{1,x} - P_1 S_1 = 885 - 80 - (620{,}000)\left(\frac{\pi}{4}(0.1)^2\right) = -4067 \, \text{N}$$
$$= -915 \, \text{lb}_f$$

The magnitude of the force (915 lb$_f$) explains why it often takes several firefighters to hold a fire hose steady at full discharge.

9.2 MICROSCOPIC APPROACH: EQUATION OF MOMENTUM TRANSFER

The equation of momentum transfer—more commonly called the equation of motion—describes the velocity distribution and pressure drop in a moving fluid. It is derived from momentum considerations by applying a momentum balance on a rate basis in conjunction with Newton's law to a volume element in a moving field. Once again, this microscopic derivation is available in the literature.[2,3]

If the fluid is Newtonian, the components of the shear-stress may be replaced by the shear-stress components given by Newton's law (see Table 5.1). In addition, the density and the viscosity of the fluid are often constant, and the only significant external force concerned is that due to gravity. The resulting equation has been referred to as the Navier–Stokes equation. This equation is also expanded into rectangular, cylindrical and spherical coordinates; the results are presented in Tables 9.1, 9.2, and 9.3.

Illustrative Example 9.5   Derive Equation (5.16), as presented in Chapter 5. A fluid is flowing through a long vertical cylindrical duct of radius R under steady-state laminar flow conditions (see Fig. 9.3). Calculate the velocity profile as a function of the pressure drop per unit length in the direction of motion. Also, calculate the volumetric flow rate, the average velocity, the maximum velocity, and the ratio of the average to the maximum velocity.

Solution   This problem is solved using cylindrical coordinates. The describing equations are now "extracted" from Table 9.2. Since the flow is one-dimensional

$$v_r = 0$$

$$v_\phi = 0$$

$$v_z \neq 0$$

Table 9.1   The equation of motion: expansion in rectangular coordinates

x-component:

$$\frac{\rho}{g_c}\left(\frac{\partial v_x}{\partial t} + v_x\frac{\partial v_x}{\partial x} + v_y\frac{\partial v_x}{\partial y} + v_z\frac{\partial v_x}{\partial z}\right)$$

$$= -\frac{\partial P}{\partial x} + \frac{\mu}{g_c}\left(\frac{\partial^2 v_x}{\partial x^2} + \frac{\partial^2 v_x}{\partial y^2} + \frac{\partial^2 v_x}{\partial z^2}\right) + \rho\frac{g_x}{g_c}$$

y-component:

$$\frac{\rho}{g_c}\left(\frac{\partial v_y}{\partial t} + v_x\frac{\partial v_y}{\partial x} + v_y\frac{\partial v_y}{\partial y} + v_z\frac{\partial v_y}{\partial z}\right)$$

$$= -\frac{\partial P}{\partial y} + \frac{\mu}{g_c}\left(\frac{\partial^2 v_y}{\partial x^2} + \frac{\partial^2 v_y}{\partial y^2} + \frac{\partial^2 v_y}{\partial z^2}\right) + \rho\frac{g_y}{g_c}$$

z-component:

$$\frac{\rho}{g_c}\left(\frac{\partial v_z}{\partial t} + v_x\frac{\partial v_z}{\partial x} + v_y\frac{\partial v_z}{\partial y} + v_z\frac{\partial v_z}{\partial z}\right)$$

$$= -\frac{\partial P}{\partial z} + \frac{\mu}{g_c}\left(\frac{\partial^2 v_z}{\partial x^2} + \frac{\partial^2 v_z}{\partial y^2} + \frac{\partial^2 v_z}{\partial z^2}\right) + \rho\frac{g_z}{g_c}$$

Table 9.2   The equation of motion: expansion in cylindrical coordinates

r-component:

$$\frac{\rho}{g_c}\left(\frac{\partial v_r}{\partial t}+v_r\frac{\partial v_r}{\partial r}+\frac{v_\phi}{r}\frac{\partial v_r}{\partial \phi}-\frac{v_\phi^2}{r}+v_z\frac{\partial v_r}{\partial z}\right)=-\frac{\partial P}{\partial r}$$

$$+\frac{\mu}{g_c}\left[\frac{\partial}{\partial r}\left(\frac{1}{r}\frac{\partial}{\partial r}\{rv_r\}\right)+\frac{1}{r^2}\frac{\partial^2 v_r}{\partial \phi^2}-\frac{2}{r^2}\frac{\partial v_\phi}{\partial \phi}+\frac{\partial^2 v_r}{\partial z^2}\right]+\rho\frac{g_r}{g_c}$$

ϕ-component:

$$\frac{\rho}{g_c}\left(\frac{\partial v_\phi}{\partial t}+v_r\frac{\partial v_\phi}{\partial r}+\frac{v_\phi}{r}\frac{\partial v_\phi}{\partial \phi}+\frac{v_r v_\phi}{r}+v_z\frac{\partial v_\phi}{\partial z}\right)=-\frac{1}{r}\frac{\partial P}{\partial \phi}$$

$$+\frac{\mu}{g_c}\left[\frac{\partial}{\partial r}\left(\frac{1}{r}\frac{\partial}{\partial r}\{rv_\phi\}\right)+\frac{1}{r^2}\frac{\partial^2 v_\phi}{\partial \phi^2}+\frac{2}{r^2}\frac{\partial v_r}{\partial \phi}+\frac{\partial^2 v_\phi}{\partial z^2}\right]+\rho\frac{g_\phi}{g_c}$$

z-component:

$$\frac{\rho}{g_c}\left(\frac{\partial v_z}{\partial t}+v_r\frac{\partial v_z}{\partial r}+\frac{v_\phi}{r}\frac{\partial v_z}{\partial \phi}+v_z\frac{\partial v_z}{\partial z}\right)=-\frac{\partial P}{\partial z}$$

$$+\frac{\mu}{g_c}\left[\frac{1}{r}\frac{\partial}{\partial r}\left(r\frac{\partial v_z}{\partial r}\right)+\frac{1}{r^2}\frac{\partial^2 v_z}{\partial \phi^2}+\frac{\partial^2 v_z}{\partial z^2}\right]+\rho\frac{g_z}{g_c}$$

Table 9.3   The equation of motion: expansion in spherical coordinates

r-component:

$$\frac{\rho}{g_c}\left(\frac{\partial v_r}{\partial t}+v_r\frac{\partial v_r}{\partial r}+\frac{v_\theta}{r}\frac{\partial v_r}{\partial \theta}+\frac{v_\phi}{r\sin\theta}\frac{\partial v_r}{\partial \phi}-\frac{v_\theta^2+v_\phi^2}{r}\right)=\frac{-\partial P}{\partial r}$$

$$+\frac{\mu}{g_c}\left(\nabla^2 v_r-\frac{2}{r^2}v_r-\frac{2}{r^2}\frac{\partial v_\theta}{\partial \theta}-\frac{2}{r^2}v_\theta\cot\theta-\frac{2}{r^2\sin\theta}\frac{\partial v_\phi}{\partial \phi}\right)+\rho\frac{g_r}{g_c}$$

θ-component:

$$\frac{\rho}{g_c}\left(\frac{\partial v_\theta}{\partial t}+v_r\frac{\partial v_\theta}{\partial r}+\frac{v_\theta}{r}\frac{\partial v_\theta}{\partial \theta}+\frac{v_\phi}{r\sin\theta}\frac{\partial v_\theta}{\partial \phi}+\frac{v_r v_\theta}{r}-\frac{v_\phi^2\cot\theta}{r}\right)=-\frac{1}{r}\frac{\partial P}{\partial \theta}$$

$$+\frac{\mu}{g_c}\left(\nabla^2 v_\theta+\frac{2}{r^2}\frac{\partial v_r}{\partial \theta}-\frac{v_\theta}{r^2\sin^2\theta}-\frac{2\cos\theta}{r^2\sin^2\theta}\frac{\partial v_\phi}{\partial \phi}\right)+\rho\frac{g_\theta}{g_c}$$

ϕ-component:

$$\frac{\rho}{g_c}\left(\frac{\partial v_\phi}{\partial t}+v_r\frac{\partial v_\phi}{\partial r}+\frac{v_\theta}{r}\frac{\partial v_\phi}{\partial \theta}+\frac{v_\phi}{r\sin\theta}\frac{\partial v_\phi}{\partial \phi}+\frac{v_\phi v_r}{r}+\frac{v_\theta v_\phi}{r}\cot\theta\right)=-\frac{1}{r\sin\theta}\frac{\partial P}{\partial \phi}$$

$$+\frac{\mu}{g_c}\left(\nabla^2 v_\phi-\frac{v_\phi}{r^2\sin^2\theta}+\frac{2}{r^2\sin\theta}\frac{\partial v_r}{\partial \phi}+\frac{2\cos\theta}{r^2\sin^2\theta}\frac{\partial v_\theta}{\partial \phi}\right)+\rho\frac{g_\phi}{g_c}$$

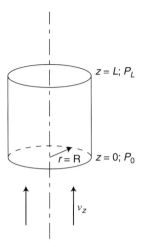

Figure 9.3 Tubular flow.

The terms v_r, v_ϕ, and all their derivatives must be zero. From Table 7.1,

$$\frac{\partial v_z}{\partial z} = 0$$

Based on physical grounds

$$\frac{\partial v_z}{\partial \phi} = 0$$

Based on the problem statement

$$\partial v_z / \partial t = 0$$

It is reasonable to conclude that v_z might vary with r, i.e.,

$$v_z = v_z(r)$$

This means

$$\frac{\partial v_z}{\partial r} \neq 0$$

or perhaps

$$\frac{\partial^2 v_z}{\partial r^2} \neq 0$$

Examining the equation of motion in cylindrical coordinates in Table 9.2, one notes that

$$\frac{\partial P}{\partial r} = 0$$

$$\frac{\partial P}{\partial \phi} = 0$$

$$\frac{\partial P}{\partial z} = \frac{\mu}{g_c} \left[\frac{1}{r} \frac{\partial}{\partial r} \left(r \frac{\partial v_z}{\partial r} \right) \right]$$

The last equation may be rewritten

$$\frac{dP}{dz} = \frac{\mu}{g_c} \left[\frac{1}{r} \frac{d}{dr} \left(r \frac{dv_z}{dr} \right) \right] \qquad (9.7)$$

The left-hand side is a constant or a function of z. The right-hand side is either a constant or a function of r. One can then conclude that both must equal a constant. Since dP/dz is a constant, it is written in the finite form

$$\frac{dP}{dz} = +\frac{\Delta P}{\Delta z}$$

$$= -\frac{\Delta P}{L}$$

The negative sign appears because P decreases as z increases. Equation (9.7) now becomes

$$\frac{1}{r} \frac{d}{dr} \left(r \frac{dv_z}{dr} \right) = -\frac{g_c \Delta P}{\mu L}$$

It would be wise to multiply both sides of the equation by $r\,dr$; otherwise, some difficulty would be encountered on integrating the equation.

$$d\left(r \frac{dv_z}{dr} \right) = -\frac{g_c \Delta P}{\mu L} r\,dr$$

Integrating once

$$r \frac{dv_z}{dr} = -\frac{g_c \Delta P}{2\mu L} r^2 + A \qquad (9.8)$$

Multiplying both sides by dr/r

$$dv_z = -\frac{g_c \Delta P}{2\mu L} r \, dr + \frac{A}{r} dr$$

and integrating

$$v_z = -\frac{g_c \Delta P}{4\mu L} r^2 + A \ln r + B \tag{9.9}$$

What about the BCs? Note that the procedure for the evaluation of integration constants A and B is also available in Chapter 5.
   BC(1)

$$v_z = 0 \quad \text{at } r = R$$

   BC(2)

or the equivalent

$$\left.\begin{array}{l} v_z = \text{finite} \quad \text{at } r = 0 \\[2mm] \dfrac{dv_z}{dr} = 0 \qquad \text{at } r = 0 \end{array}\right\} \text{based on physical grounds}$$

Substituting BC(2) into Equation (9.8) or (9.9) yields

BC(1) gives

$$A = 0$$

$$0 = -\frac{g_c \Delta P}{4\mu L} R^2 + B$$

$$B = \frac{g_c \Delta P}{4\mu L} R^2$$

Substitution of A and B leads to Equation (5.16), as given in Chapter 5 and shown again below

$$v_z = \frac{g_c \Delta P}{4\mu L}(R^2 - r^2)$$

Illustrative Example 9.6   With reference to Illustrative Example 9.5, comment on the nature of the velocity profile.

Solution   An examination of Equation (5.16) indicates that the velocity profile is parabolic. Parabolic velocity profiles are the norm for laminar flow in pipes. The reader is left the exercise of plotting v_z as a function of r in order to verify the above statement.

REFERENCES

1. I. Farag and J. Reynolds, "Fluid Flow," A Theodore Tutorial, East Williston, NY, 1995.
2. R. Bird, W. Stewart, and E. Lightfoot, "Transport Phenomena," 2nd edition, John Wiley & Sons, Hoboken, NJ, 2002.
3. L. Theodore, "Transport Phenomena for Engineers," International Textbook Company, Scranton, PA, 1971.

NOTE: Additional problems are available for all readers at www.wiley.com. Follow links for this title.

10

LAW OF HYDROSTATICS[1]

10.1 INTRODUCTION

When a fluid is at rest, there is no shear stress and the pressure at any point in the fluid is the same in all directions. The pressure is also the same across any longitudinal section parallel with the Earth's surface; it varies only in the vertical direction, that is, from height to height. This phenomenon gives rise to hydrostatics, the subject title for this chapter. Following this introduction, this chapter addresses (once again) pressure principles, buoyancy effects (including Archimedes' Law), and manometry principles.

10.2 PRESSURE PRINCIPLES

Consider a differential element of fluid of height, dz, and uniform cross-section area, S. The pressure, P, is assumed to increase with height, z. The pressure at the bottom surface of the differential fluid element is P; at the top surface, it is $P + dP$. Thus, the net pressure difference, dP, on the element is acting downward. A force balance on this element in the vertical direction yields:

$$\text{downward pressure force} - \text{upward pressure force} + \text{gravity force} = 0 \quad (10.1)$$

Fluid Flow for the Practicing Chemical Engineer. By J. Patrick Abulencia and Louis Theodore
Copyright © 2009 John Wiley & Sons, Inc.

so that

$$(P + dP)S - PS + \rho S \frac{g}{g_c} dz = 0$$

$$-(dP)S - \rho \frac{g}{g_c} S(dz) = 0 \qquad (10.2)$$

As described in Chapter 2, one has a choice as to whether to include g_c in the describing equation(s). As noted, the term g_c is a conversion constant with a given magnitude and units, e.g., 32.2 $(lb/lb_f)(ft/s^2)$ or dimensionless with a value of unity, for example, $g_c = 1.0$. In this development, g_c is retained. Rearrangement of Equation (10.2) yields

$$\frac{dP}{dz} = -\rho \frac{g}{g_c} = -\gamma \qquad (10.3)$$

The term γ is the specific weight of fluid with units of lb_f/ft^3 or N/m^3. This equation is the hydrostatic or barometric differential equation. The term dP/dz is often referred to as the pressure gradient.

Equation (10.3) is a first-order ordinary differential equation. It may be integrated by separation of variables

$$\int dP = -\int \rho \frac{g}{g_c} dz = -\frac{g}{g_c} \int \rho \, dz \qquad (10.4)$$

For most engineering applications involving liquids, and many applications involving gases, the density may be considered constant, i.e., the fluid is incompressible. Taking ρ outside the integration sign and integrating between any two limits in the fluid (Station 1 is where the pressure equals P_1 and the elevation is z_1, and Station 2 has a pressure of P_2 and elevation z_2), the pressure–height relationship is

$$P_2 - P_1 = -\rho \frac{g}{g_c} (z_2 - z_1) \qquad (10.5)$$

Equation (10.3) may also be written as:

$$\frac{P_1}{\rho(g/g_c)} + z_1 = \frac{P_2}{\rho(g/g_c)} + z_2 \qquad (10.6)$$

The term $P/\rho(g/g_c)$ is defined as the *pressure head* of the fluid, with units of m (or ft) of fluid. Equation (10.6) states that the sum of the pressure head and potential head is constant in hydrostatic "flow". Equation (10.6) is sometimes termed Bernoulli's hydrostatic equation. It is useful in calculating the pressure at any liquid depth.

Illustrative Example 10.1 Given the height of a column of liquid whose top is open to the atmosphere, determine the pressure exerted at the bottom of the column and calculate the pressure difference. The height of the liquid (mercury) column is 2.493 ft, the density of mercury is 848.7 lb/ft³, and atmospheric pressure is 2116 psf.

Solution   Refer to Equation (10.5). If P_1 is assumed to represent atmospheric pressure, $P_2 - P_1$ reduces to the gauge pressure at the bottom of the column. The equation describing the gauge pressure in terms of the column height and liquid density is:

$$P_g = \rho \frac{g}{g_c} h; \ h = \Delta z$$

Calculate the gauge pressure, a term that also represents the pressure difference in psf.

$$P_g = \rho \frac{g}{g_c} h = (848.7)(1)(2.493) = 2116 \text{ psf}$$

Determine the pressure in psfa (psf absolute)

$$P = P_g + P_a = 2116 + 2116 = 4232 \text{ psfa} = 29.4 \text{ psia}$$
$$= 14.7 \text{ psfg} = 2 \text{ atm absolute (atma)}$$

Illustrative Example 10.2   Suppose that one is interested in determining the depth in the Atlantic Ocean at which the pressure is equal to 10 atm absolute (atma). Assume that sea water is incompressible with a density of 1000 kg/m³, and the pressure at the ocean surface is 1.0 atm absolute (atma) or 0.0 atm gauge (atmg).

Solution   Take Station 1 to be at the ocean surface ($z_1 = 0$) with Station 2 at a depth equal to z_2 where the pressure, P_2, equals 10 atma $= (10)(101,325) = 1.013$ MPa abs. Apply Equation (10.6) between points 1 and 2

$$\frac{P_1}{\rho(g/g_c)} + z_1 = \frac{P_2}{\rho(g/g_c)} + z_2$$

$$z_2 = z_1 - \frac{(P_2 - P_2)}{\rho g} = 0 - \frac{(10 - 1)(101,325)}{(1000)(9.807)} = -93 \text{ m} = -305 \text{ ft}$$

The pressure is 10 atma, or 9 atmg at a depth of 93 m or 305 ft.

Calculations of fluid pressure force on submerged surfaces is important in selecting the proper material and thickness. If the pressure on the submerged surface is not uniform, then the pressure force is calculated by integration. Also note that a nonmoving, simple fluid exerts only pressure forces; moving fluids exert both pressure and shear forces.

Illustrative Example 10.3   A cylindrical tank is 20 ft (6.1 m) in diameter and 45 ft high. It contains water ($\rho = 1000$ kg/m³) to a depth of 9 ft (2.74 m) and 36 ft

(10.98 m) of an immiscible oil (SG = 0.89) above the water. The tank is open to the atmosphere (see Fig. 10.1). Calculate the density of the oil, the gauge pressure at the oil–water interface, the gauge pressure and pressure force at the bottom of the tank and the resultant pressure force on the bottom 9 ft of the side of the tank. Assume the elevation, z, to be zero at the oil–air interface.

Solution Calculate the density of oil in kg/m^3

$$\rho_{oil} = (SG)\rho_w = (0.89)(1000) = 890 \ kg/m^3$$

Apply Bernoulli's equation between points 1 and 2 to calculate the gauge pressure at the water–oil interface. Note that since this is a static fluid application, the velocity is zero and Equation (10.5) applies

$$P_1 + \rho_{oil}\frac{g}{g_c}z_1 = P_2 + \rho_{oil}\frac{g}{g_c}z_2$$

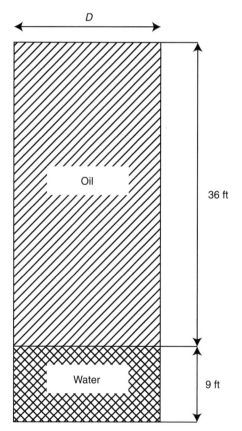

Figure 10.1 Oil–water open tank.

Note that $z_1 = 0$, $P_1 = 1$ atm, and $z_2 = -36$ ft $= -10.98$ m. Substitution yields

$$P_{2,g} = \rho_{oil} \frac{g}{g_c}(z_1 - z_2) = (890)(9.807)(10.98)$$

$$= 95{,}771 \text{ Pag} = 0.945 \text{ atmg}$$

The gauge pressure at the bottom (z_3) of the tank is

$$P_2 + \rho_w \frac{g}{g_c}z_2 = P_3 + \rho_w \frac{g}{g_c}z_3$$

$$P_3 = P_2 + \rho_w \frac{g}{g_c}(z_2 - z_3) = 95{,}771 + (1000)(9.807)(2.74)$$

$$= 122{,}673 \text{ Pag} = 1.2 \text{ atmg}$$

The pressure force at the bottom of the tank is

$$F = P_3 S = (122{,}673 + 101{,}325)\frac{\pi(6.1)^2}{4}$$

$$= 6{,}537{,}684 \text{ N} = 1{,}469{,}671 \text{ lb}_f$$

Obtain an equation describing the variation of pressure with height in the water layer.

$$P = P_2 + \rho_w \frac{g}{g_c}(z_2 - z) = (890)(9.807)(10.98)$$

$$= 95{,}771 + (1000)(9.807)(-10.98 - z)$$

$$P = 95{,}771 + 9807(-10.98 - z) = -11{,}910 - 9807z$$

Calculate the force on the side of the tank, within the water layer.

$$dF = P\,dS$$

$$F = \int dF = \int P(\pi D)\,dz = (\pi D)\int_{z_3}^{z_2} P\,dz = (\pi)(6.1)\int_{z_3}^{z_2}(-11{,}910 - 9807z)\,dz$$

$$F = 6.1\pi[-11{,}910(z_2 - z_3) - 4903.5(z_2^2 - z_3^2)]$$

Setting $z_3 = -13.72$ m and $z_2 = -10.98$ m, results in

$$F = 6.1\pi[-11{,}910(-10.98 + 13.72) - 4903.5(-10.98^2 - (-13.72)^2)]$$

$$= 5.73 \times 10^6 \text{ N}$$

10.2.1 Buoyancy Effects; Archimedes' Law

Buoyancy force is the force exerted by a fluid on an immersed or floating body. Archimedes' Law states that for a body floating in a fluid, the volume of fluid displaced equals the volume of the immersed portion of the body, and the weight (force) of fluid displaced equals the weight (force) of the body. The buoyancy force on the body, F_B, is

$$F_B = \text{(displaced volume of fluid)(fluid density)}(g/g_c)$$
$$= \text{(displaced volume of fluid)(fluid specific weight)}(1/g_c)$$

Thus,

$$F_B = V_{disp}\rho_{fl}\frac{g}{g_c} = V_{disp}\gamma_{fl}\frac{1}{g_c} \tag{10.7}$$

where γ_{fl} is the specific weight of the fluid (see Equation 10.3).

In the case where the density of the fluid and the body are equal, the body remains at its point or location in the fluid where it is placed. This is termed neutral buoyancy. In the case where the body density, ρ_{body}, is greater than the fluid density, ρ_{fluid}, the body will sink in the fluid.

Illustrative Example 10.4   A block of some material weighs 200 lb$_f$ in air. When placed in water (specific weight $\gamma_{H_2O} = 62.4$ lb$_f$/ ft^3), it weighs 120 lb$_f$. Determine the density of the material. Assume that the material density of the block is greater than the water density so that the block sinks in water, and that the volume of water displaced equals the volume of the block. Also calculate the buoyancy force and the block volume, V.

Solution   The buoyant force is

$$F_B = 200 - 120 = 80 \text{ lb}_f$$

$$F_B = V_{disp}(\rho g)_{H_2O}\frac{1}{g_c} = V_{disp}\gamma_{H_2O}\frac{1}{g_c} = V(62.4)/(1)$$

$$V = 80/62.4 = 1.282 \text{ ft}^3$$

Next, use the block weight in air and its volume to calculate the density of the block, since the density is the mass (in air) per unit volume of the material

$$\text{Weight in air} = 200 \text{ lb}_f = F_B = V\rho_{block}\frac{g}{g_c} = 1.282\rho_{block}(1)$$

$$\rho_{block} = \frac{200}{1.282} = 150 \text{ lb/ft}^3 = 2500 \text{ kg/m}^3$$

The assumption of $\rho_{block} > \rho_{water}$ (1000 kg/m^3) is justified.

Illustrative Example 10.5   A hydrometer is a liquid specific gravity indicator, with the value being indicated by the level at which the free surface of the liquid intersects the stem when floating in a liquid (see Fig. 10.2). Three hydrometer scales are commonly used. The API scale is used for oils, and the two Baumé scales are used for liquids—one for liquids heavier than water and the other for liquids lighter than water. The relationship between the hydrometer API scales and the specific gravity, SG, is

$$ SG = \frac{141.15}{131.5 + \deg \text{ API}} $$

The relationship between the hydrometer degree Baumé scale and the specific gravity, SG, of liquids lighter than water is

$$ SG = \frac{140}{130 + \deg \text{ Baumé}} $$

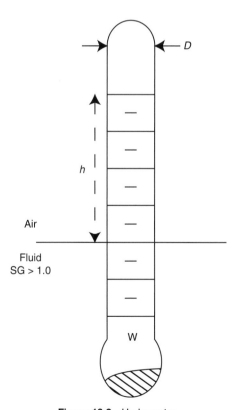

Figure 10.2   Hydrometer.

For liquids heavier than water the relation is

$$SG = \frac{145}{145 + \deg \text{ Baumé}}$$

When placed in a liquid, the hydrometer floats at a level which is a measure of the specific gravity of the liquid. The 1.0 mark is the level when the liquid is distilled water ($\rho = 1000 \text{ kg/m}^3$). The stem is of constant diameter and has a weight in the bottom of the bulb, W.

If the total hydrometer weight, W, is 0.13 N and the stem diameter is 8 mm, calculate the height, h, where it will float. The liquid is heavier than water and has a specific gravity of 33.5 deg. Baumé.

Solution   Apply Achimedes' Law

$$F = V_{\text{disp}}(SG)\rho_{\text{water}} \frac{g}{g_c}$$

For water, ($SG = 1.0$),

$$V_{\text{disp}} = \frac{\pi D^2}{4} z_1$$

where z_1 is the hydrometer reading. In the case of liquid of $SG > 1.0$,

$$V_{\text{disp}} = \frac{\pi D^2}{4} z_2$$

Therefore, two equations may be written

$$F = \frac{\pi D^2}{4} z_1 \rho_{\text{water}} \frac{g}{g_c}$$

$$F = \frac{\pi D^2}{4} z_2 (SG)\rho_{\text{water}} \frac{g}{g_c}$$

Divide the above two equations to obtain

$$\frac{z_1}{z_2} = SG$$

This equation may be rearranged to give

$$\frac{z_1 - z_2}{z_1} = \frac{h}{z_1} = 1 - \frac{1}{SG}$$

or

$$h = z_1 \left(1 - \frac{1}{SG} \right)$$

Substitute for z_1 (see above) to obtain

$$h = \frac{4F}{\pi D^2 \left(\rho_{\text{water}} \dfrac{g}{g_c} \right)} \left(1 - \frac{1}{\text{SG}} \right)$$

This is the hydrometer equation. Substitute the numerical values provided to calculate h.

$$h = \frac{4(0.13)}{\pi (0.008)^2 (9807)} \left(1 - \frac{1}{1.3} \right) = 61 \text{ mm}$$

The hydrometer is a simple device to estimate liquid densities. It is used widely in various industries. By varying the stabilizing weight or the stem diameter it is possible to design the hydrometer to be sensitive to different ranges of specific gravities.

10.3 MANOMETRY PRINCIPLES

As noted earlier, the fundamental equation of fluid statics indicates that the rate of change of the pressure P is directly proportional to the rate of change of the depth z, or

$$\frac{dP}{dz} = -\rho \frac{g}{g_c} \tag{10.3}$$

where z = vertical displacement (upward is considered positive)
ρ = fluid density
g = acceleration due to gravity
g_c = unit conversion factor

For constant density, the above equation may be integrated to give the hydrostatic equation

$$P_2 = P_1 + \frac{\rho g h}{g_c}; \quad h = z_1 - z_2 \tag{10.5}$$

Here point 2 is located at a distance h below point 1.

Manometers are often used to measure pressure differences. This is accomplished by a direct application of the above equation. Pressure differences in manometers may be computed by systematically applying the above equation to each leg of the manometer.

Illustrative Example 10.6   Consider the system pictured in Fig. 10.3.

Solution   Since the density of air is effectively zero, the contribution of the air to the 3-ft manometer reading can be neglected. The contribution to the pressure

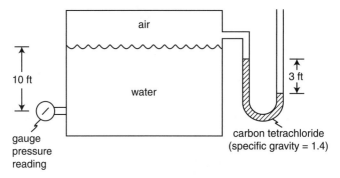

Figure 10.3   Diagram for Illustrative Example 10.6.

due to the carbon tetrachloride in the manometer is found by using the hydrostatic equation.

$$\Delta P = \rho g h / g_c$$
$$= (62.4)(1.4)(3) = 262.1 \text{ psf}$$
$$= 1.82 \text{ psi}$$

Since the right leg of the manometer is open to the atmosphere, the pressure at that point is atmospheric:

$$P = 14.7 \text{ psia}$$

Note that this should be carried as a positive term.

The contribution to the pressure due to the height of water above the pressure gauge is similarly calculated using the hydrostatic equation.

$$\Delta P = (62.4)(3) = 187.2 \text{ psf} = 1.3 \text{ psi}$$

The pressure at the gauge is obtained by summing the results of the steps above, but exercising care with respect to the sign(s):

$$P = 14.7 - 1.82 + 1.3 = 14.18 \text{ psia}$$
$$= 14.18 - 14.7 = -0.52 \text{ psig}$$

The pressure may now be converted to psfa and psfg:

$$P = (14.18)(144) = 2042 \text{ psfa}$$
$$= (-0.52)(144) = -75 \text{ psfg}$$

Care should be exercised when providing pressure values in gauge and absolute pressure. The key equation is once again

$$P(\text{gauge}) = P(\text{absolute}) - P(\text{ambient}); \quad \text{consistent units}$$

The subject of manometry will be revisited in Chapter 19.

REFERENCE

Much of the material in this chapter was adopted from:

1. J. Reynolds, J. Jeris, and L. Theodore. "Handbook of Chemical and Environmental Engineering Calculations," John Wiley & Sons, Hoboken, NJ, 2004.

NOTE: Additional problems are available for all readers at www.wiley.com. Follow links for this title.

11

IDEAL GAS LAW

11.1 INTRODUCTION

Observations based on physical experimentation can often be synthesized into simple mathematical equations called *laws*. These laws are never perfect and hence are only an approximate representation of reality. The *ideal gas law* (IGL)[1,2] was derived from experiments in which the effects of pressure and temperature on gaseous volumes were measured over moderate temperature and pressure ranges. This law works well in the pressure and temperature ranges that were used in taking the data; extrapolations outside of the ranges have been found to work well in some cases and poorly in others. As a general rule, this law works best when the molecules of the gas are far apart, i.e., when the pressure is low and the temperature is high. Under these conditions, the gas is said to behave *ideally*, that is, its behavior is a close approximation to the so-called *perfect* or *ideal gas*, a hypothetical entity that obeys the ideal gas law perfectly. For engineering calculations, and specifically for most fluid flow applications, the ideal gas law is almost always assumed to be valid, since it generally works well (usually within a few percent of the correct result) up to the highest pressures and down to the lowest temperatures used in many applications.

Fluid Flow for the Practicing Chemical Engineer. By J. Patrick Abulencia and Louis Theodore
Copyright © 2009 John Wiley & Sons, Inc.

11.2 BOYLE'S AND CHARLES' LAWS

The two precursors of the ideal gas law were *Boyle's* and *Charles'* laws. Boyle found that the volume of a given mass of gas is inversely proportional to the absolute pressure if the temperature is kept constant:

$$P_1 V_1 = P_2 V_2 \qquad (11.1)$$

where $V_1 =$ volume of gas at absolute pressure P_1 and temperature T
       $V_2 =$ volume of gas at absolute pressure P_2 and temperature T

    Charles found that the volume of a given mass of gas varies directly with the absolute temperature at constant pressure:

$$\frac{V_1}{T_1} = \frac{V_2}{T_2} \qquad (11.2)$$

where $V_1 =$ volume of gas at pressure P and absolute temperature T_1
       $V_2 =$ volume of gas at pressure P and absolute temperature T_2

11.3 THE IDEAL GAS LAW

Boyle's and Charles' laws may be combined into a single equation in which neither temperature nor pressure need be held constant:

$$\frac{P_1 V_1}{T_1} = \frac{P_2 V_2}{T_2} \qquad (11.3)$$

For Equation (11.3) to hold, the mass of gas must be constant as the conditions change from (P_1, T_1) to (P_2, T_2). This equation indicates that for a given mass of a specific gas, PV/T has a constant value. Since, at the same temperature and pressure, volume and mass must be directly proportional, this statement may be extended to

$$\frac{PV}{mT} = C \qquad (11.4)$$

where $m =$ mass of a specific gas
       $C =$ constant that depends on the gas

    Moreover, experiments with different gases showed that Equation (11.4) could be expressed in a far more generalized form. If the number of moles (n) is used in place of the mass (m), the constant is the same for all gases:

$$\frac{PV}{nT} = R \qquad (11.5)$$

where R = universal gas constant.

Equation (11.5) is called the ideal gas law. Numerically, the value of R depends on the units used for P, V, T, and n (see Table 11.1). In this text, all gases are assumed to approximate ideal gas behavior. As is generally the case in engineering practice, the ideal gas law is assumed to be valid for all illustrative and assigned problems. If a case is encountered in practice where the gas behaves in a very nonideal fashion, e.g., a high-molecular-weight gas (such as a chlorinated organic) under high pressure, one of the many *real gas* correlations found in the literature[3] should be used.

Illustrative Example 11.1   Explain why the molar volumes of all ideal gases are the same at a given temperature and pressure.

Table 11.1   Values of R in various units

R	Temperature scale	Units of V	Units of n	Units of P	Unit of PV (energy)
10.73	°R	ft^3	lbmol	psia	—
0.7302	°R	ft^3	lbmol	atm	—
21.85	°R	ft^3	lbmol	in Hg	—
555.0	°R	ft^3	lbmol	mm Hg	—
297.0	°R	ft^3	lbmol	in H_2O	—
0.7398	°R	ft^3	lbmol	bar	—
1545.0	°R	ft^3	lbmol	psfa	—
24.75	°R	ft^3	lbmol	ft H_2O	—
1.9872	°R	—	lbmol	—	Btu
0.0007805	°R	—	lbmol	—	hp · h
0.0005819	°R	—	lbmol	—	kW · h
500.7	°R	—	lbmol	—	cal
1.314	K	ft^3	lbmol	atm	—
998.9	K	ft^3	lbmol	mm Hg	—
19.32	K	ft^3	lbmol	psia	—
62.361	K	L	gmol	mm Hg	—
0.08205	K	L	gmol	atm	—
0.08314	K	L	gmol	bar	—
8314	K	L	gmol	Pa	—
8.314	K	m^3	gmol	Pa	—
82.057	K	cm^3	gmol	atm	—
1.9872	K	—	gmol	—	cal
8.3144	K	—	gmol	—	J

Solution   From the ideal gas law

$$\frac{V}{n} = \frac{RT}{P}$$

The proportionality constant, R, is the universal gas constant. The molar volumes of all ideal gases are the same at the same temperature and pressure because R is a universal constant.

Other useful forms of the ideal gas law are shown in Equations (11.6) and (11.7). Equation (11.6) applies to gas flow rather than to gas confined in a container:[4]

$$Pq = \dot{n}RT \tag{11.6}$$

where q = gas volumetric flow rate (ft^3/h)

P = absolute pressure (psia)

$\dot{n}$ = molar flow rate (lbmol/h)

T = absolute temperature (°R)

$R = 10.73$ psia-ft^3/lbmol°R

Equation (11.7) combines n and V from Equation (11.5) to express the law in terms of density:

$$P(MW) = \rho RT \tag{11.7}$$

where MW = molecular weight of gas (lb/lbmol)

ρ = density of gas (lb/ft^3)

Volumetric flow rates are often given not at the actual conditions of pressure and temperature but at arbitrarily chosen standard conditions (STP, standard temperature and pressure). To distinguish between flow rates based on the two conditions, the letters "a" and "s" are often used as part of the unit. The units acfm and scfm stand for actual cubic feet per minute and standard cubic feet per minute, respectively. The ideal gas law can be used to convert from *standard* to *actual* conditions, but, since there are many standard conditions in use, the STP being used must be known. Standard conditions most often used are shown in Table 11.2. The reader is cautioned on the incorrect use of acfm and/or scfm. The use of standard conditions is a convenience; when predicting the performance of or designing equipment, the actual conditions must be employed. Designs based on standard conditions can lead to disastrous results, with the unit usually underdesigned. For example, for a flue gas stream at 2140°F, the ratio of acfm to scfm (standard temperature = 60°F) for an incinerator application is 5.0. Equation (11.8), which is a form of Charles'

Table 11.2 Common standard conditions

System	Temperature	Pressure	Molar Volume
SI	273K	101.3 kPa	22.4 m^3/kmol
Universal scientific	0°C	760 mm Hg	22.4 L/gmol
Natural gas industry	60°F	14.7 psia	379 ft^3/lbmol
American engineering	32°F	1 atm	359 ft^3/lbmol
Hazardous waste	60°F	1 atm	379 ft^3/lbmol
Incinerator industry	70°F	1 atm	387 ft^3/lbmol

law, can be used to correct flow rates from standard to actual conditions:

$$q_a = q_s(T_a/T_s) \tag{11.8}$$

where q_a = volumetric flow rate at actual conditions (ft^3/h)
  q_s = volumetric flow rate at standard conditions (ft^3/h)
  T_a = actual absolute temperature (°R)
  T_s = standard absolute temperature (°R)

The reader is again reminded that absolute temperatures and pressures must be employed in all ideal gas law calculations.

Illustrative Example 11.2   Given the following pressure, temperature, and molecular weight data of an ideal gas, determine its density:

Pressure = 1.0 atm
Temperature = 60°F
Molecular weight of gas = 29

Solution   As noted earlier in this chapter, an ideal gas is an imaginary gas that exactly obeys certain simple laws (e.g., Boyle's law, Charles' law, and the ideal gas law). No real gas obeys the ideal gas law exactly, although the "lighter" gases (hydrogen, oxygen, air, and so on) at ambient conditions approach ideal gas law behavior. The "heavier" gases such as sulfur dioxide and hydrocarbons, particularly at high pressures and low temperatures, deviate considerably from the ideal gas law. Despite these deviations, the ideal gas law is routinely used in engineering calculations. The ideal gas law in terms of density, ρ, is (see Eq. 11.7):

$$\rho = m/V = n(MW)/V$$
$$= P(MW)/RT$$

where $MW = $ molecular weight

   $m = $ mass of gas

   $\rho = $ density of gas

The choice of R is arbitrary, provided consistent units are employed. From Table 11.1 the density of the gas using the appropriate value of R may now be calculated

$$\rho = P(MW)/RT$$
$$= (1)(29)/(0.73)(60 + 460)$$
$$= 0.0764 \, \text{lb}/\text{ft}^3$$

Since the molecular weight of the given gas is 29, this calculated density may be assumed to apply to air.

Also note that the effect of pressure, temperature, and molecular weight on density can be obtained directly from the ideal gas law equation. Increasing the pressure and molecular weight increases the density; increasing the temperature decreases the density.

Illustrative Example 11.3 Given a standard volumetric flowrate, determine the actual volumetric flowrate. Data are provided below:

Standard volumetric flowrate of a gas stream $= 2000 \, \text{scfm}$

Standard conditions $= 60°F$ and 1 atm

Actual operating conditions $= 700°F$ and 1 atm

Solution As noted earlier, the actual volumetric flowrate, usually in acfm (actual cubic feet per minute), is the volumetric flowrate based on actual operating conditions (temperature and pressure of the system). The standard volumetric flowrate, usually in scfm (standard cubic feet per minute), are $60°F$ and 1 atm or $32°F$ and 1 atm.

It should be noted again that Charles' law states that the volume of an ideal gas is directly proportional to the temperature at constant pressure. Boyle's law states that the volume of an ideal gas is inversely proportional to the pressure at constant temperature. One can combine Boyle's law and Charles' law to relate the actual volumetric flowrate to the standard volumetric flowrate:

$$q_a = q_s(T_a/T_s)(P_s/P_a) \tag{11.9}$$

where $q_a = $ actual volumetric flowrate

   $q_s = $ standard volumetric flowrate

   $T_a = $ actual operating temperature, $°R$ or K

   $T_s = $ standard temperature, $°R$ or K

P_s = standard pressure, absolute

P_a = actual operating pressure, absolute

This equation may be used to calculate the actual volumetric flowrate in acfm:

$$q_a = q_s(T_a/T_s)$$
$$= 2000(700 + 460)/(60 + 460)$$
$$= 4462 \, \text{acfm}$$

If it is desired to convert from acfm to scfm, one can reverse the procedure and use the following equation:

$$q_s = q_a(T_s/T_a)(P_a/P_s) \tag{11.10}$$

Illustrative Example 11.4   Given a mass flowrate, determine the standard volumetric flowrate. Data are provided below:

Mass flowrate of flue gas, $\dot{m} = 50 \, \text{lb/min}$

Average molecular weight of flue gas, MW = 29 lb/lbmol

Standard conditions = 60°F and 1 atm

Solution   Another application of the ideal gas law arises when one is interested in converting a mass (or molar) flowrate to a volumetric flowrate (actual or standard), or vice versa. The ideal gas equation is rearranged and solved for one variable in terms of the others. For example, the volume of 1 lbmol of ideal gas is given by

$$PV = nRT \quad R = 0.73 \, \text{atm} \cdot \text{ft}^3/(\text{lbmol} \cdot °R)$$
$$V/n = RT/P$$
$$= (0.73)(60 + 460)/1.0$$
$$= 379 \, \text{scf/lbmol}$$

Thus, if the standard conditions are 60°F and 1 atm, there are 379 standard cubic feet of gas per lbmol for any ideal gas (see Table 11.2).

The standard volumetric flowrate, q_s, in scfm is then

$$q_s = (m/\text{MW})(379 \, \text{scf/lbmol})$$
$$= (50/29)(379)$$
$$= 653 \, \text{scfm}$$

The previous result is an important number to remember in many engineering calculations: 1 lbmol of any ideal gas at 60°F and 1 atm occupies 379 ft^3; and, equally important, 1 lbmol of any ideal gas at 32°F and 1.0 atm occupies 359 ft^3. In SI units, 1 gmol of any ideal gas occupies 22.4 liters at 0°C and 1.0 atm.

Illustrative Example 11.5   A certain pure component, one-element ideal gas has a specific volume of 12.084 ft^3/lb at 70°F and 14.696 psia. Determine the molecular weight of the gas and state its name.

Solution   Rewrite the ideal gas in terms of the specific volume

$$PV = (m/\text{MW})RT$$

$$V/m = RT/(P)(\text{MW}); \ V/m = 12.084 \ \text{ft}^3/\text{lb}$$

Since

$$T = 70°F = \text{*}530°R$$
$$P = 14.7 \ \text{psia} = 1 \ \text{atm}$$
$$12.084 = (0.73)(530)/(1)(\text{MW})$$

Solving for MW,

$$\text{MW} = 32.0$$

Since this is a pure component, one-element gas, the gas is OXYGEN!

11.4 NON-IDEAL GAS BEHAVIOR

As noted earlier, an *ideal* gas is a hypothetical entity that obeys the ideal gas law perfectly. But, in industrial applications, one deals with *real* gases. Although most fluid flow applications involving gases occur at conditions approaching ideal gas behavior, there are rare occasions when the deviation from ideality is significant. Detailed calculation procedures are available[3-5] to account for these deviations.

As noted, deviations from ideality increase at higher pressures and lower temperatures, where the density increases. The aforementioned law does not describe the behavior of ideal gases in some cases because the fluid molecules themselves occupy a finite volume, and they exert forces of attraction and repulsion on each other.

Numerous attempts have been made to develop an all-purpose gas law. Although it is beyond the scope of this book to review these theories in any great detail, a brief outline and discussion is presented below.

One (and perhaps the most popular) approach to account for deviations from ideality is to include a "correction factor," Z, which is defined as the *compressibility coefficient* or *factor.*

$$PV = ZnRT \qquad\qquad (11.11)$$

Note that Z approaches 1.0 as P approaches a vacuum. For an ideal gas, Z is exactly unity. This equation may also be written as

$$Pv = ZRT \qquad (11.12)$$

where v is now the *specific* volume (not the total volume or velocity) with units of volume/mole.

The Virial equation, is another equation of state that can be used to describe gas behavior. (The definition of an equation of state is that it relates pressure, molar or specific volume, and temperature for any pure fluid.) Equations of state can take many forms. The Virial equation is one of the most important of the non-ideal gas modelling correlations because it is the one upon which many other equations of state are based. Its power series representation is:

$$Z = PV/RT = 1 + B/V + C/V^2 + D/V^3 + \cdots \qquad (11.13)$$

It can also be written as a function of pressure:

$$Z = PV/RT = 1 + B'P + C'P^2 + D'P^3 + \cdots \qquad (11.14)$$

where $B' = B/RT$
$\quad\quad\ C' = (C - B^2)/(RT)^2$

In both cases, B and C above are temperature-dependent virial coefficients. Although this power series could be infinitely long, data is generally available only up to the second virial coefficient. However, as one increases the number of terms, the accuracy of the equation improves. The two-term equation can be used up to a pressure of about 15 atmospheres total pressure while with three terms, it can be used up to 50 atmospheres.

Many equations of state for gases have been proposed, but the Redlich–Kwong equation is one of the most widely used for engineering calculations. This is due to the accuracy it provides for many gases. The equation, developed in part from the earlier and less accurate Virial equation, is:

$$P = [RT/(V - b)] - a/[T^{1/2}V(V + b)] \qquad (11.15)$$

where $a = 0.42748R^2T_c^2/P_c$
$\quad\quad\ b = 0.08664RT_c/P_c$

and the subscript c refers to a critical property. This equation has been verified from actual experimental PVT data.

Illustrative Example 11.6   The first and second virial coefficients for sulfur dioxide at 400K are specified as:

$$B = -0.159 \, \text{m}^3/\text{kgmol}$$
$$C = 0.009 \, (\text{m}^3/\text{kgmol})^2$$

Calculate the specific volume of SO_2 in L/gmol at 40 atm and 400K.

Solution   The Virial equation is

$$Z = PV/RT = 1 + B/V + C/V^2$$

Use

$$R = 0.082 \, \text{L} \cdot \text{atm/gmol} \cdot \text{K} = 82.06 \, \text{cm}^3 \cdot \text{atm/gmol} \cdot \text{K}$$

Insert the appropriate values of the terms and coefficients:

$$(40)(V)/(0.082)(400) = 1 + (-0.159)/V + (0.009)/V^2$$
$$(1.22)(V) = 1 + (-0.159)/V + (0.009)/V^2$$

Solve for V iteratively.

For ideal gas conditions, $V = 1.0/1.22 = 0.820$. Guess $V = 0.820 \, \text{L/gmol}$, substitute into the right-hand side (RHS) of the equation, and calculate the "updated" V from the left-hand side (LHS):

$$\text{For } V_{\text{guess}} = 0.820, \ V_{\text{new}} = 0.672$$
$$\text{For } V_{\text{guess}} = 0.672, \ V_{\text{new}} = 0.642$$
$$\text{For } V_{\text{guess}} = 0.642, \ V_{\text{new}} = 0.635$$

V is approximately $0.635 \, \text{L/gmol}$.

Illustrative Example 11.7   Calculate the molar volume of methane gas (ft^3/lbmol) at 373K and 10 atm. Employ the Redlich–Kwong (R–K) equation. The critical temperature and pressure of methane are 190.6K and 45.4 atm, respectively.

Solution   Convert the temperature and the critical temperature to °R:

$$T = (373)(1.8)$$
$$= 671°R$$
$$T_c = 190.6\text{K}$$
$$= 343°R$$

Calculate the numerical values of the a and b constants

$$a = (0.42748)(0.73)^2(343)^{2.5}/45.4$$
$$= 10{,}933$$
$$b = (0.08664)(0.73)(343)/45.4$$
$$= 0.478$$

Solve iteratively for the molar volume, v, in ft^3/lbmol in the R–K equation.

$$10 = [490/(v - 0.478)] - 10{,}933/(25.9)(v)(v + 0.478)$$

By trial and error

$$v = 48.8 \ \text{ft}^3/\text{lbmol}$$

Illustrative Example 11.8   Briefly discuss Van der Waals' equation.

Solution   Van der Waals' equation of state takes the form

$$P = RT/(V - b) - a/V^2$$

The two constants a and b are characteristics of the gas and are called *van der Waals constants*. This equation and many others are loosely based on the previously discussed Virial equation. It is discussed here simply as an academic exercise, i.e., as a bridge between the simplest equation of state (ideal gas law) and the more complicated equations that are available in the literature.

REFERENCES

1. J. Reynolds, J. Jeris, and L. Theodore, "Handbook of Chemical and Environmental Engineering Calculations," John Wiley & Sons, Hoboken, NJ, 2004.
2. J. Santoleri, J. Reynolds, and L. Theodore, "Introduction to Hazardous Waste Incineration," 2nd edition, John Wiley & Sons, Hoboken, NJ, 2000.
3. J. Smith, H. Van Ness, and M. Abbott, "Chemical Engineering Thermodynamics," 6th edition, McGraw-Hill, New York, 2001.
4. L. Theodore and J. Reynolds, "Thermodynamics," A Theodore Tutorial, East Williston, NY, 1995.
5. L. Theodore, "Thermodynamics for the Practicing Engineer," John Wiley & Sons, Hoboken, NJ, 2009.

NOTE: Additional problems are available for all readers at www.wiley.com. Follow links for this title.

FLUID FLOW CLASSIFICATION

This part of the book is concerned with Fluid Flow Classification. It contains five chapters and each serves a unique purpose in an attempt to treat nearly all important aspects of fluid motion. From a practical point-of-view, many systems and plants involve the movement of liquids and gases from one point to another; hence, the student and/or practicing engineer is concerned with several key topics in this area including:

1. Chapter 12: Fluid Flow Mechanisms
2. Chapter 13: Laminar Flow
3. Chapter 14: Turbulent Flow
4. Chapter 15: Compressible and Sonic Flow
5. Chapter 16: Two-Phase Flow

All receive some measure of treatment in the material contained in this part.

Fluid Flow for the Practicing Chemical Engineer. By J. Patrick Abulencia and Louis Theodore
Copyright © 2009 John Wiley & Sons, Inc.

12

FLOW MECHANISMS

12.1 INTRODUCTION

When fluids move through a closed conduit of any cross-section, one of two different types of flow may occur. These two flow types are most easily visualized by referring to a classic experiment first performed by Osborne Reynolds in 1883. In Reynolds' experiment, a glass tube was connected to a reservoir of water in such a way that the velocity of the water flowing through the tube could be varied. A nozzle was inserted in the inlet end of the tube through which a fine stream of colored dye could be introduced.

Reynolds found that when the velocity of the water was low, the "thread" of dye color maintained itself throughout the tube. By locating the nozzle at different points in the cross-section, it was shown that there was no mixing of the dye with water and that the dye flowed in parallel, straight lines.

At high velocities, it was found that the "line" or "thread" of dye disappeared and the entire mass of flowing water was uniformly colored with the dye. In other words, the liquid, instead of flowing in an orderly manner parallel to the long axis of the tube, was now flowing in an erratic manner and so there was complete mixing.

These two forms of fluid motion are known as *laminar* or *viscous flow* (low velocity), and *turbulent flow* (high velocity). The velocity at which the flow changes from laminar to turbulent is defined as the *critical velocity*.[1]

Fluid Flow for the Practicing Chemical Engineer. By J. Patrick Abulencia and Louis Theodore
Copyright © 2009 John Wiley & Sons, Inc.

12.2   THE REYNOLDS NUMBER

Reynolds, in a later study of the conditions under which the two types of flow might occur, showed that the critical velocity depended on the diameter of the tube, the velocity of the fluid, its density, and its viscosity. Further, Reynolds showed that the term representing these four quantities could be combined in a manner that later came to be defined as the *Reynolds number*.

The Reynolds number, Re, is a *dimensionless* quantity, and can be shown to be the ratio of inertia to viscous forces in the fluid:

$$\text{Re} = \frac{L\rho v}{\mu} = \frac{Lv}{\nu} \tag{12.1}$$

where L is a characteristic length, v is the average velocity, ρ is the fluid density, μ is the dynamic (or absolute) viscosity, and ν is the kinematic viscosity. In flow through round pipes and tubes, L is the diameter, D.

The Reynolds number provides information on flow behavior. It is particularly useful in scaling up bench-scale or pilot data to full-scale applications. Laminar flow is always encountered at a Reynolds number, Re, below 2100 in a circular duct, but it can persist up to higher Reynolds numbers in very smooth pipes. However, the flow is unstable and small disturbances may cause a transition to turbulent flow. Very slow flow (in circular ducts) for which Re is less than 1 is termed *creeping* or *Stokes* flow. Under ordinary conditions of flow (in circular ducts), the flow is turbulent at a Reynolds number above 4000. A transition region is observed between 2100 and 4000, where the type of flow may be either laminar or turbulent, and predictions are unreliable. The Reynolds numbers at which the fluid flow changes from laminar to transition or to turbulent are termed *critical Reynolds numbers*. In other geometries, different critical Re criteria exist.

Illustrative Example 12.1   The inlet flue to a furnace is at 200°F. It is piped through a 6.0-ft inside diameter duct at 25 ft/s. The furnace heats the gas to 1900°F. In order to maintain a velocity of 40 ft/s, what size duct would be required at the outlet of the furnace?

Solution   Applying the continuity equation, the volumetric flowrate into the furnace is

$$q_1 = A_1 v_1$$

Since

then

$$A_1 = [\pi(6.0)^2]/4 = 28.3 \text{ ft}^2$$

$$q_1 = (28.3)(25) = 707.5 \text{ ft}^3/\text{s}$$

The volumetric flowrate out of the scrubber, using Charles' law, is

$$q_2 = q_1(T_2/T_1) \quad (T \text{ in absolute units})$$

$$= (707.5)(2360/660)$$

$$= 2530 \text{ ft}^3/\text{s}$$

The cross-sectional area of outlet duct is given by

$$A_2 = q_2/v_2$$

$$= 2530/40$$

$$= 63.25 \text{ ft}^2$$

The diameter of the duct is therefore

$$D = (4A_2/\pi)^{0.5}$$

$$= (4 \times 63.25/\pi)^{0.5}$$

$$= 8.97 \text{ ft} = 108 \text{ in}$$

These calculations become important and are necessary for determining the Reynolds number. The reader is left the exercise of calculating the Reynolds number. However, since the flowing fluid is a gas, one can be virtually certain that the flow will be turbulent.

Illustrative Example 12.2   A liquid with a viscosity of 0.78 cP and a density of 1.50 g/cm^3 flows through a 1-inch diameter tube at 20 cm/s. Calculate the Reynolds number. Is the flow laminar or turbulent?

Solution   By definition, the Reynolds number (Re) is equal to:

$$Re = L\rho v/\mu \tag{12.1}$$

where ρ = fluid density
v = fluid velocity
L = characteristic length, usually the conduit diameter D
μ = fluid viscosity

Since

$$1 \text{ cP} = 10^{-2} \text{ g}/(\text{cm} \cdot \text{s})$$

$$\mu = 0.78 \times 10^{-2} \text{ g}/(\text{cm} \cdot \text{s})$$

$$1 \text{ in} = 2.54 \text{ cm}$$

$$Re = (1.50)(20)(2.54)/(0.78 \times 10^{-2})$$

$$= 9770$$

The flow is therefore turbulent. Once again, the value of the Reynolds number indicates the nature of the fluid flow in a duct or pipe and generally:

Re < 2100; flow is streamline (laminar or viscous)
Re > 4000; flow is turbulent
$2100 \leq Re \leq 4000$; transition region

Illustrative Example 12.3   Given the physical properties and velocity of a gas stream flowing through a circular duct, determine the Reynolds number of the gas stream. The velocity through the duct is 3.8 m/s, the duct diameter is 0.45 m, the gas viscosity 1.73×10^{-5} kg/m · s, and the gas density is 1.2 kg/m^3.

Solution   Substitution into Equation (12.1) gives

$$Re = \frac{Dv\rho}{\mu} = \frac{(0.45)(3.8)(1.2)}{1.73 \times 10^{-5}} = 118,600$$

The flow is in the turbulent regime and for most engineering applications, one can assume turbulent (high Reynolds number) flow for gases. The reader should also note that the Reynolds number appears in many semi-empirical and empirical equations that involve fluid flow, heat transfer, and mass transfer applications. For flow in non-circular conduits, some other appropriate length (termed the *hydraulic diameter*) replaces the diameter in Re. This is discussed later in the next chapter.

12.3   STRAIN RATE, SHEAR RATE, AND VELOCITY PROFILE

When a fluid flows past a stationary solid wall, the fluid adheres to the wall at the interface between the solid and fluid. This condition is referred to as "no slip." Therefore, the local velocity, v, of the fluid at the interface is zero. At some distance, y, normal to and displaced from the wall, the velocity of the fluid is finite. Therefore, there is a velocity variation from point to point in the flowing fluid. This causes a velocity field, in which the velocity is a function of the normal distance from the wall, that is, $v = f(y)$. If $y = 0$ at the wall, $v = 0$, and v increases with y. The rate of change of velocity with respect to distance is defined the velocity gradient,

$$\frac{dv}{dy} = \frac{\Delta v}{\Delta y} \tag{12.2}$$

This velocity derivative (or gradient) is also referred to as the shear rate, time rate of shear, or rate of deformation. See Chapter 5 for more details.

Illustrative Example 12.4   The local velocity v (ft/s or fps) near a wall varies with the normal distance, y (ft), from a stationary wall according to the equation

$$V = 20y - y^2$$

Generate an equation describing the shear rate.

Solution   Check the consistency of the given profile at the wall where $y = 0$:

$$v = 20y - y^2 = 20(0) - (0)^2 = 0$$

The profile is consistent with respect to the boundary condition.

Generate the equation of the strain rate (see Eq. (12.2)) using the above velocity profile:

$$\frac{dv}{dy} = 20 - 2y$$

The units of the strain rate are s^{-1}. Strain rate is important in the classification of real fluids. The relationships between shear stress and strain rate are presented in diagrams called *rheograms*.

12.4   VELOCITY PROFILE AND AVERAGE VELOCITY

The velocity profile for either laminar and turbulent flow is provided in Fig. 12.1. In laminar flow, the velocity profile approaches a true parabola slightly pointed in the middle and tangent to the walls of the pipe. The average velocity over the whole cross-section (volumetric flow rate divided by the cross-sectional area) is 0.5 times the maximum velocity. This fact will be derived in the next chapter. In turbulent

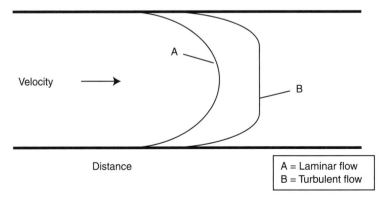

Figure 12.1   Velocity profile.

flow, the profile approaches a flattened parabola and the average velocity is usually approximately 0.8 times the maximum.

Because of its viscosity, a real fluid in contact with a nonmoving wall will have a velocity of zero at the wall. Similarly, a fluid in contact with a wall moving at a velocity, v, will move at the same velocity. This earlier described "no-slip" condition of real fluids flowing in a duct results in a fluid velocity at the wall of zero.

To calculate the volumetric flow rate, q, of the fluid passing through a perpendicular surface, S, one must integrate the product of the component of the velocity that is normal to the area and the area, over the whole cross-section area of the duct, i.e.,

$$q = \int_S v \, dS \tag{12.3}$$

In accordance with the definition of average values, the average velocity of the fluid passing through the surface, S, is then given by

$$\bar{v} = v_{av} = \frac{\int_S v \, dS}{\int_S dS} = \frac{q}{S} \tag{12.4}$$

Illustrative Example 12.5   A liquid has a specific gravity (SG) of 0.96 and an absolute viscosity of 9 cP. The liquid flows through a long circular tube of radius, $R = 3$ cm. The liquid has the following linear distribution of the axial velocity, v, (the velocity in the direction of the flow):

$$v \, (\text{m/s}) = 6 - 200r$$

where r is the radial position (in meters) measured from the tube centerline. A total of $20 \, \text{m}^3$ of liquid passes through the tube.

Calculate the average velocity of the fluid and the volumetric flow rate. Also, calculate the time for a specified volume (or mass) of fluid to pass through a section of the duct.

Solution   Write the equation for q in a differential equation form in terms of r, the radial coordinate. By definition,

$$dq = v \, dS$$

and in cylindrical coordinates,

$$dS = 2\pi r \, dr$$

Therefore,

$$dq = 2\pi r v \, dr = 2\pi r(6 - 200r) \, dr = 2\pi (6r - 200r^2) \, dr$$

Integrate the above equation between the limits of $r = 0$ and $r = R$

$$q = 2\pi \int_0^R (6r - 200r^2) \, dr = 2\pi \left(3R^2 - \frac{200}{3} R^3 \right)$$

Calculate the volumetric flow rate. Set $R = 3 \, \text{cm} = 0.03 \, \text{m}$.

$$q = 2\pi \left[3(0.03)^2 - \frac{200}{3} (0.03)^3 \right] = 0.00565 \, \text{m}^3/\text{s}$$

Calculate the mass flow rate, $\dot{m}$

$$\dot{m} = \rho q = (SG)(1000 \, \text{kg/m}^3)q$$
$$\dot{m} = 0.96(1000)(0.00565) = 5.42 \, \text{kg/s}$$

Calculate the average velocity. Since

$$S = \pi R^2$$

and (see Eq. (12.4))

$$v_{av} = \frac{q}{S} = \frac{2\pi \left[3(R)^2 - \frac{200}{3} (R)^3 \right]}{\pi R^2}$$
$$= 2\left[3 - \left(\frac{200}{3} \right) R \right]$$

By setting $R = 3 \, \text{cm} = 0.03 \, \text{m}$
$$v_{av} = 2.0 \, \text{m/s}$$

Illustrative Example 12.6   Refer to Illustrative Example 12.5. Calculate the time to pass $20 \, \text{m}^3$ of the liquid through the cross-section of the pipe.

Solution   The time, t, to pass the liquid is given by

$$t = \frac{V}{q} = \frac{20}{0.00565} = 3540 \, \text{s} = 59 \, \text{min}$$

As noted in Chapter 8, applying the conservation law of energy mandates that all forms of energy entering the system equal that of those leaving. See Equation (8.14) in Chapter 8. Expressing all terms in consistent units, e.g., energy per unit mass of fluid flowing, resulted in the total energy balance equation (rewritten below)

$$\frac{P_1}{\rho} + \frac{v_1^2}{2g_c} + \frac{g}{g_c}z_1 + E_1 + Q + W_s = \frac{P_2}{\rho} + \frac{v_2^2}{2g_c} + \frac{g}{g_c}z_2 + E_2 \qquad (12.5)$$

An important point needs to be made before leaving this subject. By definition, the kinetic energy of a small parcel of fluid with local velocity, v, is $v^2/2g_c$ (ft $\cdot$ lb$_f$/lb). If the local velocities at all points in the cross-section were uniform, v_{av} would be equal to v, and the kinetic energy term can be retained as written. Ordinarily there is a velocity gradient across the passage; this introduces an error, the magnitude of which depends on the nature of the velocity profile and the shape of the cross-section. For the usual case where the velocity is approximately uniform (i.e., turbulent flow), the error is not serious, and since the error tends to cancel because of the appearance of kinetic terms on each side of any energy balance equation, it is customary to ignore the effect of velocity gradients. When the error cannot be ignored, the introduction of a correction factor that is used to multiply the v^2/g_c term is needed. The term α, called the kinetic energy correction factor, is employed, where

$$\alpha = \frac{\displaystyle\int_S v^3 \, dS}{v_{av}^3 S} \qquad (12.6)$$

For most engineering applications, the flow is turbulent and α may be assumed to be unity. Where the velocity distribution is parabolic, as in laminar flow, it can be shown that the exact value of α is 2. For transition state flow, $1 \le \alpha \le 2$.[1]

Illustrative Example 12.7   Given 1000 scfm (28.3 scmm) of gas flowing in a circular duct with a 1.2 ft (0.366 m) diameter at 300°F and 1 atm, calculate the average velocity, and the Reynolds number. Standard conditions are 60°F and 1.0 atm. The viscosity of the gas at 300°F is 2.2×10^{-5} kg/m $\cdot$ s and its molecular weight is 33.

Solution   Calculate the actual volumetric flow rate, q, using the ideal gas law (see Chapter 11)

$$q = q_s \left(\frac{T}{T_s}\right)\left(\frac{P_s}{P}\right) = 1000 \frac{(300 + 460)}{(60 + 460)}\left(\frac{1}{1}\right)$$

$$= 1461.5 \, \text{acfm} = 41.36 \, \text{acmm}$$

The cross-sectional area of the duct, S, is

$$S = \frac{\pi D^2}{4} = \frac{\pi (1.2)^2}{4} = 1.131 \, \text{ft}^2 = 0.105 \, \text{m}^2$$

The velocity, v, is

$$v = \frac{q}{S} = \frac{1461.5}{1.131}$$
$$= 1292.2 \text{ ft/min}$$
$$= 21.5 \text{ ft/s}$$
$$= 6.55 \text{ m/s}$$

Note that the average velocity was simply calculated by dividing the actual volumetric gas flow rate by the cross-sectional area through which the gas flows. The velocity calculated here is therefore the bulk, or average, velocity. Note that the same calculational procedure may be employed to calculate the average velocity in any process unit, including chemical reactors, pipes, stacks, etc.

Calculate the gas density from the ideal gas law:

$$\rho = \frac{P(MW)}{RT} = \frac{(1)(33)}{(0.7302)(760)} = 0.0595 \text{ lb/ft}^3 = 0.952 \text{ kg/m}^3$$

The mass flow rate is calculated as follows:

$$\dot{m} = \rho v S = (0.0595)(21.5)(0.105) = 0.134 \text{ lb/s} = 0.656 \text{ kg/s}$$

Calculate the Reynolds number

$$\text{Re} = \frac{Dv\rho}{\mu} = \frac{(0.366)(6.55)(0.952)}{2.2 \times 10^{-5}} = 103,670$$

As noted above, the velocity calculated is an average value. Plug flow, characterized by a uniform velocity distribution, is often assumed. In actual operation, the following velocity profiles might develop (see Fig. 12.1):

1. Parabolic—laminar flow.
2. "Flattened" parabola—turbulent flow, wherein velocities are low (often near zero at the perimeter/walls) and high (often near 20% above the average velocity) at the center.
3. Random distribution—following a bend, valve, or disturbance.

These profiles are discussed further in the next two chapters.

REFERENCE

1. W. Badger and J. Banchero, "Introduction to Chemical Engineering," McGraw-Hill, New York, 1955.

NOTE: Additional problems are available for all readers at www.wiley.com. Follow links for this title.

13

LAMINAR FLOW IN PIPES

13.1 INTRODUCTION

The following equation was derived in Chapter 8 (see Eq. (8.18))

$$\frac{\Delta P}{\rho} + \frac{\Delta v^2}{2g_c} + \frac{g}{g_c}\Delta z - \eta W_s + \sum F = 0 \tag{13.1}$$

This was defined as the mechanical energy equation. The above equation was later rewritten without the work and friction terms

$$\frac{\Delta P}{\rho} + \frac{\Delta v^2}{2g_c} + \frac{g}{g_c}\Delta z = 0 \tag{13.2}$$

This equation was defined as the basic form of the Bernoulli equation.

In applying the Bernoulli equation to a prime mover (e.g., a centrifugal pump application), Equation (8.18) was written as

$$\frac{P_1 g_c}{\rho\,g} + \frac{v_1{}^2}{2g} + z_1 = \frac{P_2 g_c}{\rho\,g} + \frac{v_2{}^2}{2g} + z_2 - h_s\frac{g_c}{g} + h_f\frac{g_c}{g} \tag{13.3}$$

where h_s and h_f have effectively replaced ηW_s and $\sum F$, respectively, in Equation (13.1). Note that h_s is a positive term as is h_f. The units of both h_s and h_f are therefore ft-lb$_f$/lb. However, each term in this equation, as written, has units of ft of flowing

Fluid Flow for the Practicing Chemical Engineer. By J. Patrick Abulencia and Louis Theodore
Copyright © 2009 John Wiley & Sons, Inc.

133

fluid (a pressure term). If the equation is multiplied by g/g_c, each term returns to units of energy/mass (e.g., ft $\cdot$ lb$_f$/lb). The result is provided in Equation (13.4)

$$\frac{\Delta P}{\rho} + \frac{\Delta v^2}{2g_c} + \Delta z \frac{g}{g_c} - h_s + h_f = 0 \tag{13.4}$$

Note that the prime notation is retained if the units of h'_s and h'_f are in height of flowing fluid, e.g., ft H_2O. The choice of units for h_s and h_f is, of course, optional and/or arbitrary.

The reader should note that in the process of converting Equation (13.3) to (13.4), the Δ term (representing a difference) applies to outlet minus inlet conditions (i.e., $P_2 - P_1$, etc.). One can now examine Equation (13.4) at the following extreme or limiting conditions.

1. If only the first ($\Delta P/\rho$) and last term (h_f) are present, an increase in the latter's frictional losses would result in a corresponding decrease in the former pressure term P_2. Thus, a large pressure drop results, which is to be expected.
2. If only the first ($\Delta P/\rho$) and fourth term (h_s) are present, an increase in the latter's input mechanical (shaft) work term would result in a corresponding increase in the frame pressure term P_2. Thus, a smaller pressure drop results, and again, this is in agreement with what one would expect.
3. If only the latter two terms are present, $h_s = h_f$, and this too agrees with one's expectation since both terms are positive and any frictional effect is compensated by the mechanical (shaft) work introduced to the system.

Care should be exercised in the interpretation of the term ΔP. Although the notation Δ represents difference, ΔP can be used to describe the difference between the inlet minus the outlet pressure (i.e., $P_1 - P_2$), or it can describe the difference between the outlet minus the inlet pressure (i.e., $P_2 - P_1$). When a fluid is flowing in the $1 \rightarrow 2$ direction, the term $P_1 - P_2$ is a positive and represents a decrease in pressure that is defined as the pressure drop. The term $P_2 - P_1$, however, also represents a pressure change whose difference is negative and is also defined as a pressure drop. One's wording and interpretation of this pressure change is obviously a choice that is left to the user.

13.2 FRICTION LOSSES

As indicated above, the h_f term was included to represent the loss of energy due to friction in the system. These frictional losses can take several forms. An important engineering problem is the calculation of these losses. It has been shown (earlier) that the fluid can flow in either of two modes—laminar or turbulent. For laminar flow, an equation is available from basic theory to calculate friction loss in a pipe. In practice, however, fluids (particularly gases) are rarely moving in laminar flow.

Since two methods of flow are so widely different, a different equation describing frictional resistance is to be expected in the case of turbulent flow from that which applies in the case of laminar flow. On the other hand, it will be shown in the next chapter that both cases may be handled by one relationship in such a way that it is not necessary to make a preliminary calculation to determine whether the flow is taking place above the critical Reynolds number or below it.[1]

One can theoretically derive the h_f term for laminar flow.[2,3] The equation takes the form

$$h_f = \frac{32\mu v L}{\rho g_c D^2} \tag{13.5}$$

for a fluid flowing through a straight cylinder of diameter D and length L. A friction factor, f, that is dimensionless may now be defined as (for laminar flow)

$$f = \frac{16}{\text{Re}} \tag{13.6}$$

so that Equation (13.5) takes the form

$$h_f = \frac{4fLv^2}{2g_c D} \tag{13.7}$$

Although the above equation describes friction loss or the pressure drop across a conduit of length L, it can also be used to provide the pressure drop due to friction per unit length of conduit, for example, $\Delta P/L$ by simply dividing the above equation by L.

It should also be noted that another friction factor term exists that differs from that of Equation (13.6). In this other case, f_D is defined as

$$f_D = \frac{64}{\text{Re}} \tag{13.8}$$

The f_D is used to distinguish the value of Equation (13.6) from that of Equation (13.8). In essence, then

$$f_D = 4f \tag{13.9}$$

The term f is defined as the *Fanning friction* factor while f_D is defined as the *Darcy* or *Moody friction factor*. Care should be taken as to which of the friction factors are being used and this will become more apparent in the next chapter. In general, chemical engineers employ the Fanning friction factor; other engineers prefer the Darcy (or Moody) factor. This book employs the Fanning friction factor.

With reference to Equation (13.6), one should note that this is an equation of a straight line with a slope of -1 if f is plotted versus Re on log-log coordinates.

Note (once again) that the equation for f applies only to laminar flow, i.e., when Re is <2100 for pipe flow.

Employing Equation (13.7), Equation (13.4) may be extended and rewritten as

$$\frac{\Delta P}{\rho} + \frac{\Delta v^2}{2g_c} + \Delta z \frac{g}{g_c} - h_s + \sum \frac{4fLv^2}{2g_cD} = 0 \qquad (13.10)$$

Note that a summation sign has been inserted before the new term for the loss in a straight pipe because this loss may result from flow through several sections in a series of pipes of various lengths and diameters. The symbols $\sum h_c$ and $\sum h_e$, representing the sum of the contraction and expansion losses, respectively, may also be added to the equation, as provided in Equation (13.11). (These effects will be discussed in the next part of this book.)

$$\frac{\Delta P}{\rho} + \frac{\Delta v^2}{2g_c} + \Delta z \frac{g}{g_c} - h_s + \sum \frac{4fLv^2}{2g_cD} + \sum h_c + \sum h_e = 0 \qquad (13.11)$$

13.3 TUBE SIZE

Fluids are usually transported in pipes or tubes. Generally speaking, pipes are heavy-walled and have a relatively large diameter. Tubes are thin-walled and often come in coils.

Pipes are specified in terms of their diameter and wall thickness. The nominal diameters range from 1/8 to 30 inches for steel pipe. Standard dimensions of steel pipe are provided in Table A.5 in the Appendix and are known as IPS (iron pipe size) or NPS (nominal pipe size). The wall thickness of the pipe is indicated by the schedule number, which can be approximated from $1000(P/S)$ where P is the maximum internal service pressure (psi) and S is the allowable bursting stress in the pipe material (psi). (The S value varies by material, grade of material and temperature; allowable S values may be found in Piping Handbooks).

Tube sizes are indicated by the outside diameter. The wall thickness is usually given a BWG (Birmingham Wire Gauge) number. The smaller the BWG, the heavier the tube. Table A.6 in the Appendix lists the sizes and wall thicknesses of condenser and heat exchanger tubes. For example, a 3/4-inch 16 BWG tube has an outside diameter (OD) of 0.75 in, an inside diameter (ID) of 0.62 in, a wall thickness of 0.065 in, and a weight of 0.476 lb/ft.

Illustrative Example 13.1   Consider the following three cases. Calculate the average velocities below for which the flow will be viscous (laminar).

1. Water at 60°F in a 2-inch standard pipe.
2. Air at 60°F and 5 psig in a 2-inch standard pipe.
3. Oil of a viscosity of 300 cP and SG of 0.92 in a 4-inch standard pipe.

Solution   For laminar flow, Re < 2100, so the equation

$$Re = \frac{Dv\rho}{\mu} = 2100$$

can be solved for the velocity term

$$v = \frac{2100\mu}{D\rho}$$

1. For water, $\mu = 6.72 \times 10^{-4}$ lb/(ft · s), $\rho = 62.4$ lb/ft³. In addition, from Table A.5 in the Appendix, $D = 2.067$ in. Therefore,

$$v = \frac{2100(6.72 \times 10^{-4})}{(2.067/12)(62.4)} = 0.13 \text{ ft/s}$$

2. For air, $\mu = 12.1 \times 10^{-6}$ lb/ft · s, $\rho = 0.1024$ lb/ft³ (from ideal gas law), and $D = 2.067$ in. Therefore,

$$v = \frac{2100(12.1 \times 10^{-6})}{(2.067/12)(0.1024)} = 1.44 \text{ ft/s}$$

3. For oil, $\mu = 300(6.72 \times 10^{-4})$ lb/ft · s, $\rho = 0.92(62.4)$ lb/ft³, and, from Table A.5 in the Appendix, $D = 4.026$ in. Therefore,

$$v = \frac{2100(300)(6.72 \times 10^{-4})}{(4.026/12)(0.92)(62.4)} = 22.0 \text{ ft/s}$$

Illustrative Example 13.2   Refer to Illustrative Example 13.1. Determine the pressure drop per unit length of pipe for part (1).

Solution   To determine the pressure drop per unit length of pipe, $\Delta P/L$, in psf/ft, apply a modified form of the Hagen–Poiseuille equation. For these units, Equation (13.5) reduces to:

$$\frac{\Delta P}{L} = \frac{32\mu g v}{g_c D^2}$$

Substituting

$$\frac{\Delta P}{L} = \frac{32(6.72 \times 10^{-4})(0.13)}{(2.067/12)^2}$$

$$= 0.0293 \text{ psf/ft}$$

Illustrative Example 13.3   Given the nominal size and schedule number of a 1-inch schedule 80 steel pipe, determine its inside diameter (ID), outside diameter (OD), wall thickness and pipe weight (lb/ft).

Solution   Using Table A.5 in the Appendix, obtain the pipe inside diameter, outside diameter, wall thickness and weight

$$ID = 0.957 \, in.$$
$$OD = 1.315 \, in.$$
$$Wall \; thickness = 0.179 \, in.$$
$$Pipe \; weight = 2.17 \, lb/ft$$

13.4 OTHER CONSIDERATIONS

As discussed earlier, flow in conduits that are not cylindrical (e.g., a rectangular parallel pipe), are treated as if the flow occurs in a pipe. For this situation, a hydraulic radius, r_h, is defined as:

$$r_h = \frac{\text{Cross-sectional area perpendicular to flow}}{\text{Wetted perimeter}} \tag{13.12}$$

For flow in a circular tube

$$r_h = \frac{(\pi D^2/4)}{\pi D} = \frac{D}{4}$$

and

$$D = 4r_h \tag{13.13}$$

One may extend this concept to any cross-section such that

$$D_{eq} = 4r_h \tag{13.14}$$

It is then possible to use this equivalent diameter in the circular pipe expression presented in Equation (13.7) for pressure drop

$$h_f = \frac{4fLv^2}{2g_cD_{eq}} \tag{13.15}$$

The hydraulic diameter approach is usually valid for laminar flow and always valid for turbulent flow.

Illustrative Example 13.4   An air-conditioning duct has a rectangular cross-section of 1 m by 0.25 m. If the kinematic viscosity of the air is approximately $1 \times 10^{-5} \, m^2/s$, determine the maximum air velocity before the flow becomes turbulent. Assume the critical Reynolds number is 2300.

Solution   Compute the equivalent or hydraulic diameter

$$D_h = \frac{2wh}{w+h} = \frac{2(1)(0.25)}{1+0.25}$$

$$= 0.4\,\text{m}$$

The equation for the "critical" Reynolds number is

$$\text{Re}_{\text{crit}} \approx 2300 = \frac{vD_h}{\nu}$$

Solve for v and substitute

$$v = 2300\,\frac{1 \times 10^{-5}}{0.4} = 0.0575\,\text{m/s}$$

$$= 5.8\,\text{cm/s}$$

Another important concept is that referred to as a "calming," "entrance," or "transition" length. This is the length of conduit required for a velocity profile to become fully developed following some form of disturbance in the conduit. This disturbance can arise because of a valve, a bend in the line, an expansion in the line, etc. This is an important concern when measurements are conducted in the cross-section of the pipe or conduit. An estimate of this "calming" length, L_c, for laminar flow is

$$\frac{L_c}{D} = 0.05\,\text{Re} \qquad (13.16)$$

For turbulent flow (see the next chapter), one may employ

$$L_c = 50\,D \qquad (13.17)$$

Illustrative Example 13.5   A circular 2-inch diameter horizontal tube contains liquid asphalt. For the purposes of this problem, assume asphalt to be a Newtonian fluid of density 70 lb/ft^3. The tube radius is 1 in. When a steady pressure gradient of 1.0 psi/ft is applied, the steady-state flow rate of the asphalt is 0.486 ft^3/s. Calculate the asphalt viscosity in cP and kg/m-s. Determine if the flow is laminar. Calculate the Darcy and Fanning friction factors. How long must the pipe be to ensure a fully developed flow?

Solution   Apply the continuity equation to obtain the flow velocity

$$q = vS = v(\pi D^2/4)$$

$$v = \frac{4q}{\pi D^2} = \frac{4(0.486)}{\pi(0.1667)^2}$$

$$= 22.3\,\text{ft/s}$$

To determine the dynamic viscosity, assume laminar flow and use Equation (13.7) with $v = q/(\pi D^2/4)$ and $f = 16/\text{Re}$.

$$\mu = \frac{\pi \Delta P D^4 g_c}{128 q L} = \frac{\pi(144)(0.1667)^4(32.174)}{128(0.486)(1)}$$

$$= 0.1806 \, \text{lb/ft} \cdot \text{s}$$

Check on the assumption of laminar flow:

$$\text{Re} = \frac{D v \rho}{\mu} = \frac{(0.1667)(22.3)(70)}{0.1806} = 1440 < 2300$$

As expected, the flow is laminar. The Fanning friction factor is given by

$$f = \frac{16}{\text{Re}} = \frac{16}{1440}$$

$$= 0.0111$$

As expected, the friction factor is large.

The pipe must be longer than the entrance length to have fully developed flow. Calculate the entrance length from Equation (13.16):

$$L_c = 0.05D \, \text{Re} = 0.05(1440)(0.1667)$$

$$= 12.0 \, \text{ft}$$

If $L > 12$ ft the flow may be assumed to be fully developed.

Illustrative Example 13.6   Liquid glycerin at 20°C flows in a tube of diameter 4 cm. The velocity profile of a slow moving fluid through a long circular tube is parabolic. Two possible expressions of the parabolic velocity distribution are available:

$$v = 16(1 - 2500r^2)$$

or

$$v = 16(1 - 8000r^2)$$

where v is the glycerin velocity at any radial distance and r is the radial distance measured from the center line. What is the correct velocity distribution and why? Using the correct velocity distribution, calculate the average velocity (m/s), volumetric flow rate (m³/s and gpm), mass flow rate (kg/s), mass flux (kg/m² · s), and linear momentum flux ($\dot{M}$) of glycerin.

Solution   Obtain the properties of liquid glycerine. Use Table A.2 in the Appendix

$$\rho = 1260 \, \text{kg/m}^3$$

$$\mu = 1.49 \, \text{kg/m-s}$$

$$v = \mu/\rho = 0.00118 \, \text{m}^2/\text{s}$$

Because of no-slip conditions at the wall, the correct velocity distribution must produce $v = 0$ when $R = 0.02$ m. This is satisfied only for:

$$v = 16(1 - 2500r^2)$$

where, for $R = 0.02$ m

$$v = 16(1 - 2500(0.02)^2) = 0$$

This must be the correct velocity distribution.

To calculate q, v, $\dot{m}$, G, and $\dot{M}$, first write a differential equation for the volumetric flow rate, dq

$$dq = v \, dS = 2\pi r v \, dr = 2\pi r[16(1 - 2500r^2)] \, dr$$

$$= 32\pi(r - 2500r^3) \, dr$$

Integrating

$$q = \int dq = 32\pi \int_{0}^{R=0.02} (r - 2500r^3) \, dr = 32\pi \left(\frac{R^2}{2} - \frac{2500}{4} R^4 \right)$$

$$= 32\pi \left[\frac{(0.02)^2}{2} - \frac{2500}{4}(0.02)^4 \right]$$

$$= 0.010 \, \text{m}^3/\text{s} = 158.5 \, \text{gpm}$$

The average velocity for laminar flow (see next section) is given by

$$\bar{v} = \frac{v_{max}}{2} = \frac{16(1 - 2500r^2)}{2}$$

Substituting

$$\bar{v} = 8(1 - 2500(0)^2) = 8 \, \text{m/s}$$

Finally,

$$\dot{m} = q\rho = 0.01(1260) = 12.6 \, \text{kg/s}$$

$$G = \frac{\dot{m}}{S} = \rho v = (1260)(8) = 10{,}080 \, \text{kg/m}^2\text{-s}$$

$$\dot{M} = \dot{m}v = 12.6(8) = 100.8 \, \text{N}$$

Illustrative Example 13.7   Refer to Illustrative Example 13.6. Calculate the Reynolds number of the flow and determine how many hours it will take for 14,000 gallons to pass through the cross-section of the tube.

Solution   Calculate the flow Reynolds number:

$$\text{Re} = \frac{\rho v D}{\mu} = \frac{1260(8)(0.02)}{1.49} = 135.3$$

Since $135.3 < 2100$, the flow is laminar.

Calculate the time, t, to pass 14,000 gallons of glycerine through a cross-section of the tube

$$t = \frac{V}{q} = \frac{14,000}{159.6} = 87.7 \text{ min}$$

13.5   MICROSCOPIC APPROACH

It is important to note that the equations provided in Tables 9.1–9.3 in Chapter 9 are valid only for laminar flow. As such, they can be used to develop equations describing laminar flow for various systems under different conditions. For example, refer to Equation (5.16) provided in Chapter 5

$$v_z = \frac{g_c \Delta P}{4\mu L} \left(R^2 - r^2 \right) \tag{5.16}$$

This equation can be derived from the aforementioned tables using the approach provided below. This system requires using cylindrical coordinates. The describing equations are now "extracted" from Table 9.2. Since the flow is one-dimensional

$$v_r = 0$$
$$v_\phi = 0$$
$$v_z \neq 0$$

The terms v_r, v_ϕ, and all their derivatives must be zero. From equation (2) in Table 9.1, one concludes

$$\frac{\partial v_z}{\partial z} = 0$$

Based on physical grounds

$$\frac{\partial v_z}{\partial \phi} = 0$$

Based on the problem statement

$$\partial v_z / \partial t = 0$$

it is reasonable to conclude that v_z might vary with r, that is

$$v_z = v_z(r)$$

This means

$$\frac{\partial v_z}{\partial r} \neq 0$$

or perhaps

$$\frac{\partial^2 v_z}{\partial r^2} \neq 0$$

Examining the equation of motion in cylindrical coordinates in Table 9.2, one notes that

$$\frac{\partial P}{\partial r} = 0$$

$$\frac{\partial P}{\partial \phi} = 0$$

$$\frac{\partial P}{\partial z} = \frac{\mu}{g_c} \left[\frac{1}{r} \frac{\partial}{\partial r} \left(r \frac{\partial v_z}{\partial r} \right) \right]$$

The last equation may be rewritten

$$\frac{dP}{dz} = \frac{\mu}{g_c} \left[\frac{1}{r} \frac{d}{dr} \left(r \frac{dv_z}{dr} \right) \right]$$

The left-hand side (LHS) is a constant or a function of z. The right-hand side (RHS) is either a constant or a function of r. It is therefore concluded that both must equal a constant. Since $\Delta P / dz$ is a constant, it is written in finite form:

$$\frac{dP}{dz} = +\frac{\Delta P}{\Delta z}$$

$$= -\frac{\Delta P}{L}$$

The negative sign appears because P decreases as z increases. The above equation is now written as

$$\frac{1}{r} \frac{d}{dr} \left(r \frac{dv_z}{dr} \right) = -\frac{g_c \Delta P}{\mu L}$$

It would be wise to multiply both sides of the equation by $r\,dr$; otherwise, some difficulty would be encountered upon integration.

$$d\left(r\frac{dv_z}{dr}\right) = -\frac{g_c\Delta P}{\mu L}r\,dr$$

Integrating once

$$r\frac{dv_z}{dr} = -\frac{g_c\Delta P}{2\mu L}r^2 + A$$

Multiplying both sides by dr/r

$$dv_z = -\frac{g_c\Delta P}{2\mu L}r\,dr + \frac{A}{r}\,dr$$

and integrating

$$v_z = -\frac{g_c\Delta P}{4\mu L}r^2 + A\ln r + B$$

The integration constants were evaluated in Chapter 5. This ultimately resulted in

$$v_z = \frac{g_c\Delta P}{4\mu L}(R^2 - r^2) \tag{5.16}$$

Illustrative Example 13.8   Refer to Equation (5.16). Express the local velocity in terms of the volumetric flow rate q.

Solution   The volumetric flow rate is given by

$$q = \iint_f v_z\,df: \quad f = \text{area available for flow}$$

where $df = r\,dr\,d\phi = $ differential area in cylindrical coordinates. Substituting Equation (5.16) into the above equation leads to

$$q = \int_{\phi=0}^{2\pi}\int_{r=0}^{R}\frac{g_c\Delta P}{4\mu L}(R^2 - r^2)r\,dr\,d\phi$$

$$= \frac{\pi g_c\Delta P}{2\mu L}\int_0^R(R^2 r - r^3)\,dr$$

$$= \frac{\pi g_c\Delta P}{2\mu L}\left[\frac{R^2 r^2}{2} - \frac{R^4}{4}\right]_0^a$$

$$= \frac{\pi g_c\Delta P R^4}{8\mu L}$$

and

$$\frac{\Delta P}{L} = \frac{8\mu q}{g_c \pi R^4}$$

Substituting the above equation into Equation (5.10) leads to

$$v_z = \frac{2q}{\pi R^2}\left[1 - \left(\frac{r}{R}\right)^2\right]$$

The flow profile is therefore parabolic.

Illustrative Example 13.9   Refer to Illustrative Example (13.8). Calculate the maximum velocity, v_{max}, the average velocity, v_{av}, and the ratio of the average to the maximum velocity

Solution   The maximum velocity is located at $r = 0$. Therefore

$$q_{max} = \frac{2q}{\pi R^2}$$

The average velocity is defined as

$$\bar{v} = v_{av} = \frac{\displaystyle\int_0^{2\pi}\int_0^R \frac{2q}{\pi R^2}\left[1 - \left(\frac{r}{R}\right)^2\right] r\,dr\,d\phi}{\displaystyle\int_0^{2\pi}\int_0^R r\,dr\,d\phi}$$

$$= \frac{\displaystyle\frac{2q}{\pi R^2}(2\pi)\int_0^R \left[1 - \left(\frac{r}{R}\right)^2\right] r\,dr}{\pi a^2}$$

$$= \frac{4q}{\pi R^4}\left[\frac{r^2}{2} - \frac{r^4}{4R^2}\right]_0^R = \frac{q}{\pi R^2}$$

$$= \frac{q}{\pi R^2} = \frac{q}{S}; \text{ as expected}$$

Therefore

$$v_z = 2v_{av}\left[1 - \left(\frac{r}{R}\right)^2\right] = v_{max}\left[1 - \left(\frac{r}{R}\right)^2\right]$$

Finally

$$\frac{v_{av}}{v_{max}} = \frac{1}{2}$$

REFERENCES

1. W. Badger and J. Banchero, "Introduction to Chemical Engineering," McGraw-Hill, New York, 1955.
2. C. Bennett and J. Meyers, "Momentum, Heat, and Mass Transfer," McGraw-Hill, New York, 1962.
3. L. Theodore, "Transport Phenomena for Engineers," International Textbook Company, Scranton, PA, 1971.

NOTE: Additional problems are available for all readers at www.wiley.com. Follow links for this title.

14

TURBULENT FLOW IN PIPES

14.1 INTRODUCTION

The development now proceeds to turbulent flow. In the previous chapter, the Fanning friction factor was defined and presented in Equation (13.6)

$$f = \frac{16}{\text{Re}} \qquad (14.1)$$

However, this only applies to laminar flow. Unlike laminar flow, the friction factor for turbulent flow cannot be derived from basic principles. Fortunately, extensive experimental data is available and this permits numerical evaluation of the friction factor for turbulent flow. Comments on turbulent flow now follow.

As described earlier, as the Reynolds number is increased above 2100 for flow in pipes, eddies and turbulence start to develop in the flowing fluid. From Re equal to 2100 to about 4000, the flow becomes more unstable. As the Reynolds number is increased to values above 4000, the turbulent state of the fluid core becomes well developed and the velocity distribution across a diameter of the pipe becomes similar to that of a flattened parabola (see profile D in Fig. 14.1 as well as Fig. 12.1). The equation of this flattened parabola can be approximately described by the following equation

$$v = v_{\max}\left(\frac{2n^2}{(n + 1)(2n + 1)}\right) \qquad (14.2)$$

Fluid Flow for the Practicing Chemical Engineer. By J. Patrick Abulencia and Louis Theodore
Copyright © 2009 John Wiley & Sons, Inc.

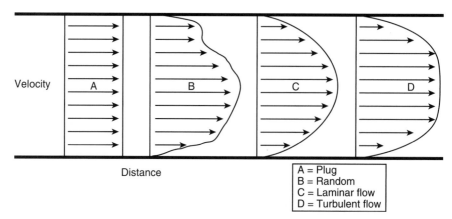

Figure 14.1   Velocity profiles.

where v_{max} is the centerline (maximum) velocity and n is 7 (the one-seventh power law applies).

Illustrative Example 14.1   A liquid with a viscosity of 0.78 cP and a density of $1.50\ g/cm^3$ flows through a 1-in. diameter tube at 20 cm/s. Calculate the Reynolds number. Is the flow laminar or turbulent?

Solution   By definition, the Reynolds number (Re) is equal to:

$$Re = Dv\rho/\mu$$

Since

$$1\,cP = 10^{-2}\,g/(cm \cdot s)$$
$$\mu = 0.78 \times 10^{-2}\,g/(cm \cdot s)$$
$$1\,in = 2.54\,cm$$
$$Re = (2.54)(20)(1.50)/(0.78 \times 10^{-2})$$
$$= 9769.23 \approx 9800$$

The flow is turbulent since Re > 2100.

Illustrative Example 14.2   A fluid is moving in laminar flow through a cylinder whose inside radius is 0.5 in. The viscosity and density of the fluid are 1.03 cP and $62.4\ lb/ft^3$, respectively. The velocity is then increased to higher values until turbulence appears. Determine the minimum velocity at which turbulence will appear (i.e., Re = 2100).

Solution   Since

$$Re = \frac{Dv\rho}{\mu}$$

$$2100 = \frac{(1.0\,\text{in})(62.4\,\text{lb/ft}^3)(V)}{1.03\,\text{cP}} \times \frac{1\,\text{ft}}{12\,\text{in}} \times \frac{1\,\text{cP}}{6.72 \times 10^{-4}\,\text{lb/ft} \cdot \text{s}}$$

$$v = \frac{(2100)(1.03)(6.72 \times 10^{-4})(12)}{62.4}$$

$$= 0.280\,\text{ft/s}$$

14.2 DESCRIBING EQUATIONS

It is important to note that almost all the key fluid flow equations presented in Chapter 13 for laminar flow apply as well to turbulent flow, provided the appropriate friction factor is employed. These key equations (Eqs. (13.7), (13.10) and (13.11)) are again provided below. Note once again that v (the average velocity) is given by $q/(\pi d^2/4)$.

$$h_f = \frac{4fLv^2}{2g_cD}$$

This equation may also be written as

$$h_f = \frac{32fLq^2}{\pi^2 g_c D^5} \tag{14.3}$$

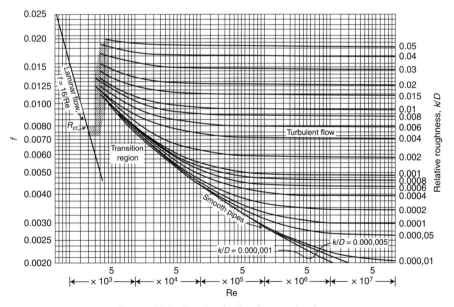

Figure 14.2 Fanning friction factor; pipe flow.

$$\frac{\Delta P}{\rho} + \frac{\Delta v^2}{2g_c} + \Delta z \frac{g}{g_c} - h_s + \sum \frac{4fLv^2}{2g_cD} = 0 \tag{14.4}$$

$$\frac{\Delta P}{\rho} + \frac{\Delta v^2}{2g_c} + \Delta z \frac{g}{g_c} - h_s + \sum \frac{4fLv^2}{2g_cD} + \sum h_c + \sum h_e = 0 \tag{14.5}$$

The effect of the Reynolds number on the Fanning friction factor is provided in Fig. 14.2. Note that Equation (14.1) appears on the far left-hand side of Fig. 14.2.

14.3 RELATIVE ROUGHNESS IN PIPES

In the turbulent regime, the "roughness" of the pipe becomes a consideration. In his original work on the friction factor, Moody[1] defined the term k as the roughness and the ratio, k/D, as the relative roughness. Thus, for rough pipes/tubes in turbulent flow

$$f = f(\text{Re}, k/D) \tag{14.6}$$

This equation reads that the friction factor is a function of *both* the Re and k/D. However, as noted above, the dependency on the Reynolds number is a weak one. Moody[1] provided one of the original friction factor charts. His data and results, as applied to the Fanning friction factor, are presented in Fig. 14.2 and covers the laminar, transition, and turbulent flow regimes. It should be noted that the laminar flow friction factor is independent of the relative roughness. Figure 14.2 also contains friction factor data for various relative roughness values.

The reader should note the following:

1. Moody's original work included a plot of the Darcy (or Moody) friction factor, not the Fanning friction factor. His chart has been adjusted to provide the Fanning friction factor, i.e., the plot in Fig. 14.2 is for the Fanning friction factor. Those choosing to work with the Darcy friction factor need only multiply the Fanning friction factor by 4, since

$$f_D = 4f \tag{14.7}$$

2. The intermediate regime of Re between 2100 and 4000 is indicated by the shaded area in Fig. 14.2.
3. The average "roughness" of commercial pipes is given in Table 14.1.[2]
4. Notice in Fig. 14.2 the relative roughness lines are nearly horizontal in the fully turbulent regime to the right of the dashed lines.
5. Roughness is a function of a variety of effects—some of which are difficult, if not impossible, to quantify. In effect, the roughness of a pipe resembling a smooth sine wave exhibits different frictional effects than a sharp sawtooth or step function.

Table 14.1 **Average roughness of commerical pipes**

Material (new)	Roughness, k	
	ft	mm
Riveted steel	0.003–0.03	0.9–9.0
Concrete	0.001–0.01	0.3–3.0
Wood stove	0.0006–0.003	0.18–0.9
Cast iron	0.00085	0.26
Galvanized iron	0.0005	0.15
Asphalted cast iron	0.0004	0.12
Commercial steel (Wrought iron)	0.00015	0.046
Drawn tubing	0.000005	0.0015
Glass	"smooth"	"smooth"

In summary, for Reynolds numbers below 2100, the flow will always be laminar and the value of f should be taken from the line at the left in Fig. 14.2. For Reynolds numbers above 4000, the flow will practically always be turbulent and the values of f should be read from the lines at the right. Between Re = 2100 and Re = 4000, no accurate calculations can be made because it is generally impossible to predict flow type in this range. If an estimate of friction loss must be made in this range, it is recommended that the figures for turbulent flow should be used, as that provides an estimate on the high side.[3]

14.4 FRICTION FACTOR EQUATIONS

Approximate equations for smooth pipe, turbulent flow, Fanning friction factors are:[2]

$$f = \frac{0.0786}{\text{Re}^{0.25}} \quad \text{for } 5000 < \text{Re} < 50{,}000, \text{ and} \tag{14.8}$$

$$f = \frac{0.046}{\text{Re}^{0.20}} \quad \text{for } 30{,}000 < \text{Re} < 1{,}000{,}000 \tag{14.9}$$

Farag[2] has also provided two formulas for the Fanning friction factor, valid for $10^{-5} < k/D < 0.02$ and $4000 < \text{Re} < 10^8$.

$$f = \frac{1}{4\left(1.8 \log_{10}\left[0.135\left(\dfrac{k}{D}\right) + \dfrac{0.8775}{\text{Re}}\right]\right)^2} \tag{14.10}$$

The other approximate equation of the Fanning friction factor in the completely turbulent region is:

$$f = \frac{1}{4\left[1.14 - 2.0 \log_{10}\left(\dfrac{k}{D}\right)\right]^2} \tag{14.11}$$

In a classic review of these equations, Churchill[4] provided a host of models for estimating the "Churchill" friction factor, f_c. The key equation is given below:

$$f_c = \left[\left(\frac{8}{Re} \right)^{12} + \frac{1}{(A+B)^{3/2}} \right]^{1/12}$$ (14.12)

where

$$A = \left[2.457 \ln \left(\frac{1}{\left(\frac{7}{Re} \right)^{0.9} + \frac{0.27k}{D}} \right) \right]^{16}$$

$$B = \left(\frac{37,530}{Re} \right)^{16}$$

Equation (14.12) is valid for all Re and k/D. A trial-and-error solution is necessary if the pressure drop rather than the flow rate is specified. Churchill also notes that the equation is a convenient and accurate replacement for all of the friction-factor plots in the literature. The equation not only reproduces the friction factor plot but also avoids interpolation and provides unique values in the transition region. These values are, of course, subject to some uncertainty because of the physical instability inherent in this region. One of the drawbacks to Churchill's work is that all the presented equations for f_c need to be converted to the Fanning or Darcy friction factors by multiplying f by 2 or 8, respectively.

Perhaps the most accurate of all the equations appearing in the literature is that attributed to Jain.[5] This equation is as follows:

$$\frac{1}{f^{0.5}} = 2.28 - 4 \log_{10} \left[\frac{k}{D} + \frac{21.25}{(Re)^{0.9}} \right]$$ (14.13)

Illustrative Example 14.3   PAT (Patrick, Abulencia, and Theodore) Consultants have proposed the following equation to predict the Fanning friction factor:

$$f = 0.0015 + [(8)(Re)^{0.30}]^{-1}$$

For a Re of 14,080 and k/D of 0.004, calculate/obtain the friction factor using

1. The above equation.
2. Equations (14.8) and (14.9).
3. Equation (14.11).
4. Equation (14.13).
5. Figure 14.2.

Comment on the results.

Solution

1. $f = 0.0015 + [(8)(14{,}080)^{0.30}]^{-1}$

    $= 0.0015 + 0.0071$

    $= 0.0086$

2. $f = 0.0786/(\mathrm{Re})^{0.25}$

    $= 0.0786/(14{,}080)^{0.25}$

    $= 0.0072$

and

$f = 0.046/(\mathrm{Re})^{0.2}$

    $= 0.046/(14{,}080)^{0.2}$

    $= 0.0068$

3. $f = 1/4[1.14 - 2.0\log{(k/D)}]^2$

    $= (1)/4[1.14 - 2.0\log{(0.004)}]^2$

    $= 0.0071$

4. $\sqrt{f} = 1 \Big/ \left[2.28 - 4\log_{10}\left(\dfrac{k}{D} + \dfrac{21.25}{(\mathrm{Re})^{0.9}}\right)\right]$

    $\sqrt{f} = 1 \Big/ \left[2.28 - 4\log_{10}\left(0.004 + \dfrac{21.25}{4.28}\right)\right]$

    $= 0.00875$

5. $f = 0.0085$

As can be seen from the above five results, there is some modest but acceptable scatter; the average value is 0.00782.

14.5 OTHER CONSIDERATIONS

Two "other" considerations are discussed in this subsection: flow in non-circular conduits and flow in parallel pipe/conduit systems.

Flow in non-circular conduits was discussed earlier but is reviewed again because of its importance in fluid flow studies. As noted in Chapter 13, the approach employed is to represent any conduit by a pipe or cylinder with an "equivalent" diameter, D_{eq}. Key equations include:

$$D_{eq} = \frac{4S}{P_p} = 4r_h \tag{14.14}$$

where S is the cross-sectional area of the conduit, P_p is the wetted perimeter and r_h is the hydraulic radius. For flow in the annular space between two concentric pipes of diameter D_1 and D_2,

$$D_{eq} = \frac{4\pi(D_1{}^2 - D_2{}^2)}{4\pi(D_1 + D_2)} = D_1 - D_2 \tag{14.15}$$

Although this approach strictly applies to turbulent flow, it may be employed for laminar flow situations if no other approaches are available.

Illustrative Example 14.4   Calculate the equivalent (or hydraulic) diameter for turbulent fluid flow in a cross-section which has:

1. a 2 in × 10 in rectangle flowing full;
2. an annulus with outer diameter $D_o = 10$ cm and inner diameter $D_i = 8$ cm; and,
3. a 10 cm diameter circle (tube) flowing half full.

Solution   First, write Equation (14.14) describing the calculation of the equivalent diameter:

$$D_{eq} = \frac{4S}{P_p} = 4r_h$$

1. Consider the 2 in × 10 in rectangle flowing full:

$$D_{eq} = \frac{4S}{P_p} = \frac{4(2)(10)}{(2 + 10 + 1 + 10)} = 3.33 \text{ in}$$

2. Consider the annulus:

$$D_{eq} = \frac{4S}{P_p} = \frac{4(\pi/4)(D_o{}^2 - D_i{}^2)}{\pi(D_o - D_i)} = D_o - D_i = 2 \text{ cm}$$

3. Consider the half-full circle:

$$D_{eq} = \frac{4S}{P_p} = 4\frac{\pi D^2/8}{\pi D/2} = D = 10 \text{ cm}$$

Note the importance of using the wetted perimeter and the cross-sectional area of the fluid flow for situations where the conduit is not full of fluid.

14.6 FLOW THROUGH SEVERAL PIPES

Flow through a number of pipes or conduits often arise in engineering practice. If flow originates from the same source and exits at the same location, the pressure drop cross each conduit must be the same. Thus, for flow through conduits 1, 2, and 3, one may write:

$$\Delta P_1 = \Delta P_2 = \Delta P_3 \tag{14.16}$$

Solutions to this type of problem usually require a trial-and-error solution since several (in this case three) simultaneous, nonlinear equations may be involved.

14.7 GENERAL PREDICTIVE AND DESIGN APPROACHES

Almost any problem involving friction losses in long pipe flows can be solved using the Fanning friction factor charts or an equivalent equation. Unless otherwise indicated, the charts are employed in the solution to illustrative examples, applications and problems to follow. Some problems can be solved directly; however, others are trial-and-error, since a knowledge of the Reynolds number is required.

There are three important fundamental pipe flow problems. These are detailed below (see Fig. 14.3).

1. Head loss problem: Given D, L, and v or q, ρ, μ, and g. Compute the head loss (h_f) or pressure drop (ΔP).
2. Flow rate problem: Given D, L, h_f (or ΔP), ρ, μ, and g. Compute the velocity, v, or flow rate q.
3. Sizing problem: Given q, L, h_f (or ΔP), ρ, μ, and g. Compute the diameter of the pipe, D.

Only a Type 1 problem involves a direct application of the chart and does not require a trial-and-error calculation. The engineer has to iterate to compute the velocity (Type 2) or diameter (Type 3) because both D and v are contained in the ordinate and abscissa of the charts or equations. The iteration proceeds as follows:

1. Make an initial guess for v or D.
2. Calculate the corresponding Reynolds number.
3. If necessary, calculate the relative roughness.
4. Use the Fanning chart to find the corresponding friction factor.
5. From the data given, generate an improved v or D.
6. Use the improved v or D from step (5) and repeat steps (2–5).

The iteration converges when v or D stops changing significantly.

An approximate explicit formula to obtain the unknown volumetric flow rate is:[2]

$$q = -2.22D^{2.5}\sqrt{\frac{gh'_f}{L}}\log_{10}\left(\frac{k/D}{3.7} + \frac{1.78\mu}{D^{1.5}\rho\sqrt{gh'_f/L}}\right) \tag{14.17}$$

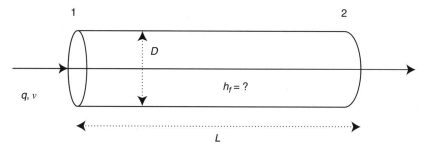

Type 1: Head Loss

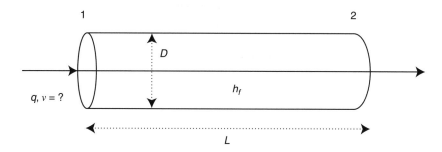

Type 2: Flow Rate

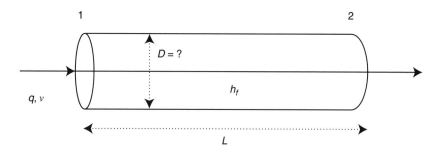

Type 3: Sizing

Figure 14.3 Three flow problems.

For a horizontal pipe, the flow rate equation simplifies to:[2]

$$q = -2.22D\sqrt{\frac{\Delta PD^3}{\rho L}}\log_{10}\left(\frac{k/D}{3.7} + \frac{1.78\mu}{D^{1.5}\rho\sqrt{\Delta PD^3/\rho L}}\right) \qquad (14.18)$$

The units of h'_f are height of flowing fluid (e.g., ft H$_2$O), while ΔP is in force per unit area (e.g., lb$_f$/ft^2). The relationship (as noted earlier) is given by $gh'_f/g_c = \Delta P/\rho$.

Consistent units must be used in the above two equations. If SI units are used, then the volumetric flow rate, q, is in m^3/s. If engineering units are used, the pressure drop, ΔP, must be in psf, the density, ρ, must be in $slug/ft^3$, the kinematic viscosity in ft^2/s, and the volumetric flow rate in ft^3/s. Note 1.0 slug = 32.174 lb.

An approximate explicit correlation has been developed to calculate the pipe diameter, D:[2]

$$D = 0.66 \left[k^{1.25} \left(\frac{Lq^2}{gh'_f} \right)^{4.75} + \frac{\mu}{q\rho} \left(\frac{Lq^2}{gh'_f} \right)^{5.2} \right]^{0.4} \tag{14.19}$$

For a horizontal pipe, $gh_f = \Delta P/\rho$, and the equation simplifies to

$$D = 0.66 \left[k^{1.25} \left(\frac{\rho Lq^2}{\Delta P} \right)^{4.75} + \frac{v}{q} \left(\frac{\rho Lq^2}{\Delta P} \right)^{5.2} \right]^{0.4} \tag{14.20}$$

Consistent units with those provided above need to be used. If engineering units are employed, the pressure drop must be expressed in psf.

Illustrative Example 14.5 Air at a temperature of 40°F and pressure of $P_1 = 0.1$ psig is to be transported horizontally through a circular conduit of length $L = 800$ ft. At the delivery point, the air pressure, P_2, is 0.01 psig, and the air rate is 500 cfm. The circular duct is made of sheet metal and has a roughness of 0.00006 in. Find the pipe diameter, D, and the average air velocity, v.

Solution Calculate the air density using the ideal gas law and employing an average pressure of 14.75 psia.

$$\rho = \frac{P(MW)}{RT} = \frac{(14.75)(28.9)}{(10.73)(500)}$$
$$= 0.08 \, lb/ft^3$$

Obtain the air viscosity from Fig. B.2 in the Appendix (estimated at 40°F).

$$\mu = 0.0173 \, cP = 1.14 \times 10^{-5} \, lb/ft \cdot s$$

Assume first the flow is laminar and calculate D using Equation (14.3) with $f = 16/Re$ and $v = q/(\pi D^2/4)$.

$$q = 500 \, ft^3/min = 8.33 \, ft^3/s$$

$$D = \sqrt[4]{\frac{128\mu Lq}{\pi(P_1 - P_2)}} = \sqrt[4]{\frac{128(3.54 \times 10^{-7})(800)(8.33)}{\pi(0.09)(144)}} = 0.293 \, ft$$

Check the flow type.

$$\text{Re} = \frac{Dv\rho}{\mu} = \frac{4q\rho}{\pi D\mu} = \frac{4(8.33)(0.08)}{\pi(0.293)(1.14 \times 10^{-5})} = 3.17 \times 10^6$$

Therefore, the assumption of laminar flow is not valid.

Assume a pipe diameter of 1 ft (0.3048 m). The relative roughness is then

$$k/D = (0.00006/12)/1 = 0.000005$$

Calculate the Reynolds number.

$$\text{Re} = \frac{Dv\rho}{\mu} = \frac{4q\rho}{\pi D\mu} = \frac{4(8.33)(0.08)}{\pi(1)(1.14 \times 10^{-5})} = 74{,}168$$

Obtain an estimate of the Fanning friction factor from Fig. 14.2.

$$f = 0.005$$

Write the pressure drop equation; see Equation (14.3).

$$\Delta P = P_1 - P_2 = \frac{32\rho f L q^2}{g_c \pi^2 D^5}$$

Solving for the diameter:

$$D = \left(\frac{32\rho f L q^2}{g_c \pi^2 \Delta P}\right)^{0.2} = \left(\frac{32(0.08)(0.005)(800)(8.33)^2}{(32.174)\pi^2(12.96)}\right)^{0.2} = 0.70\,\text{ft}$$

Start the second iteration with the newly calculated D.

$$k/D = (0.00006/12)/0.70 = 0.0000071$$

Calculate the new Reynolds number.

$$\text{Re} = \frac{Dv\rho}{\mu} = \frac{4q\rho}{\pi D\mu} = \frac{4(8.33)(0.08)}{\pi(0.7)(1.14 \times 10^{-5})}$$

$$= 106{,}000$$

Obtain the new Fanning friction factor.

$$f = 0.0045$$

Solving for the diameter:

$$D = \left(\frac{8\rho f L q^2}{g_c \pi^2 \Delta P}\right)^{0.2} = \left(\frac{8(0.08)(0.0045)(800)(8.33)^2}{g_c \pi^2(12.96)}\right)^{0.2} = 0.69\,\text{ft}$$

The iteration may now be terminated.

Calculate the flow velocity using the last calculated diameter.

$$v = \frac{q}{S} = \frac{8.33}{\pi(0.69)^2/4} = 6.8 \, \text{m/s}$$

Illustrative Example 14.6   If ethyl alcohol at 20°C is to be pumped through 60 m of horizontal drawn tubing at 10 m³/h with a friction loss of 30 m, what tube diameter in cm must be employed? What is the alcohol velocity? Is the flow turbulent?

Solution   Obtain the properties of ethyl alcohol at 20°C from Table A.2 in the Appendix.

$$\rho = 789 \, \text{kg/m}^3$$

$$\mu = 1.1 \times 10^{-3} \, \text{kg/m-s}$$

Obtain the roughness of drawn tubing using Table 14.1.

$$k = 0.0015 \, \text{mm}$$

Use the approximate explicit Equation (14.20) to calculate the diameter.

$$D = 0.66 \left[k^{1.25} A^{4.75} + \frac{\mu}{q\rho} A^{5.2} \right]^{0.04}$$

where

$$A = \frac{Lq^2}{gh'_f} = \frac{60(2.778 \times 10^{-3})^2}{(9.807)(30)}$$

$$= 1.574 \times 10^{-6} \, \text{m}$$

Substituting gives

$$D = 0.66 \left[(1.5 \times 10^{-6})^{1.25}(1.57 \times 10^{-6})^{4.75} + \frac{1.1 \times 10^{-3}}{2.778 \times 10^{-3}(789)}(1.57 \times 10^{-6})^{5.2} \right]^{0.04}$$

$$= 0.0303 \, \text{m}$$

Take $D = 3$ cm. Next, calculate the velocity of alcohol in the tube.

$$v = \frac{q}{S} = \frac{4(2.778 \times 10^{-3})}{\pi(0.03)^2} = 3.93 \, \text{m/s}$$

Characterize the flow.

$$\text{Re} = \frac{Dv}{\nu} = \frac{(3.93)(0.0377)}{1.395 \times 10^{-6}} = 106405 > 4000$$

The flow is turbulent since Re is greater than 4000.

Illustrative Example 14.7   Kerosene at 20°C (SG = 0.82, $\mu = 0.0016\,\mathrm{kg/m \cdot s}$) flows in a 9 meter long, smooth, horizontal, 2-inch schedule 80 pipe. The flow Reynolds number is 60,000. Using SI units, calculate the kerosene density, the pipe inside diameter, the average velocity of kerosene, the volumetric and mass flow rate and the maximum kerosene velocity in the pipe assuming the one-seventh power-law applies. Where will the maximum velocity occur? How good is the assumption of fully developed flow? Assume

$$L_C/D = 4.4\,\mathrm{Re}^{1/6}$$

Solution   Calculate the kerosene density.

$$\rho = \mathrm{SG}(1000) = 820\,\mathrm{kg/m^3}$$

Obtain the pipe inside diameter from Table A.5 in the Appendix.

$$D = 1.939\,\mathrm{in} = 0.0493\,\mathrm{m}$$

Calculate the average velocity from the Reynolds number equation.

$$\mathrm{Re} = \frac{Dv\rho}{\mu}$$

Solving for the average velocity, v,

$$v = \frac{(\mathrm{Re})\mu}{D\rho} = \frac{(60,000)(0.0016)}{(0.0493)(820)} = 2.38\,\mathrm{m/s}$$

Calculate the volumetric and mass flow rates.

$$q = \frac{v}{S} = \frac{v}{\pi D^2/4} = \frac{2.38}{\pi(0.0493)^2/4} = 0.00454\,\mathrm{m^3/s}$$

$$\dot{m} = \rho q = (820)(0.00454) = 3.72\,\mathrm{kg/s}$$

Calculate the maximum velocity assuming one-seventh power law applies (see Eq. 14.2).

$$\frac{v}{v_{\max}} = \frac{2n^2}{(n+1)(2n+1)}$$

For $n = 7$

$$\frac{v}{v_{\max}} = 0.817$$

$$v_{\max} = 2.92\,\mathrm{m/s}$$

As noted earlier, the maximum velocity will occur at the pipe center line.

Check on the assumption of fully developed flow.

$$\frac{L_C}{D} = 4.4\,\mathrm{Re}^{1/6} = 4.4(60{,}000^{1/6}) = 27.5$$

$$L_C = 27.5(0.0493) = 1.36\,\mathrm{m}$$

Since L_C is less than L ($= 9\,\mathrm{m}$), the assumption is valid.

Illustrative Example 14.8   Refer to Illustrative Example 14.7. Determine the Fanning friction factor, the friction loss, and the pressure drop (in Pa and atm) due to friction.

Solution   Calculate the Fanning friction factor using Equation (14.9) since Re $=$ 60,000.

$$f = \frac{0.046}{\mathrm{Re}^{0.2}} = \frac{0.046}{(60{,}000)^{0.2}}$$

$$= 0.0051$$

Calculate the friction loss due to friction.

$$h'_f = 4f\frac{L}{D}\frac{v^2}{2g} = 4(0.0051)\frac{9}{0.0493}\frac{(2.38)^2}{2(9.807)}$$

$$= 1.08\,\mathrm{m\ of\ kerosene}$$

Calculate the pressure drop using Bernoulli's equation or the hydrostatic equation, noting that $v_1 = v_2 = v$, $z_1 = z_2$, and $h_s = 0$.

$$\Delta P = \rho\frac{g}{g_c}h'_f = 820(9.807)(1.08)$$

$$= 8685\,\mathrm{Pa} = 0.086\,\mathrm{atm}$$

Illustrative Example 14.9   Refer to Illustrative Example 14.7. Calculate the force required to hold the pipe in place.

Solution   The force required to hold the pipe is:

$$F = PS = \pi\frac{(0.0493)^2}{4}(8685) = 16.6\,\mathrm{N}$$

The force direction is opposite the flow.

14.8 MICROSCOPIC APPROACH

As noted in this and the previous two chapters, fluid-flow systems are classified as either laminar or turbulent flow. This last section is concerned with a microscopic treatment of turbulent flow. However, the reader should note that a complete and fundamental understanding of turbulent flow has yet to be developed.

A fixed point in space during a given finite time interval θ can be resolved into the average velocity over the same time interval and a fluctuation or disturbance velocity term that accounts for the turbulent motion

$$v = \bar{v} + v' \tag{14.21}$$

where v = instantaneous velocity,
$\bar{v}$ = average velocity over the time interval θ,
v' = instantaneous fluctuation velocity.

This system is represented pictorially in Fig. 14.4. One can conclude from the discussion above and the definition of average values that

$$\bar{v}' = \frac{\displaystyle\int_0^\theta v' \, dt}{\displaystyle\int_0^\theta dt} = 0 \tag{14.22}$$

where $\bar{v}'$ = time average value of the fluctuation velocity during θ.

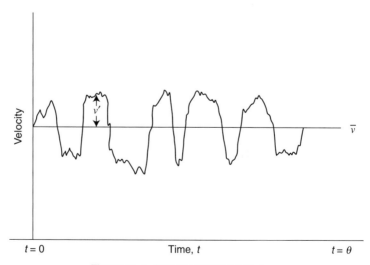

Figure 14.4 Velocity variation with time.

The intensity of turbulence I for the velocity component v_y at a given point in space is defined by

$$I = \frac{(\bar{v}_y')^2}{\bar{v}_y} \tag{14.23}$$

As its name implies, I is a measure of the intensity of turbulence at a point and is given by the ratio of the magnitude of the fluctuation and average velocities.

Illustrative Example 14.10 A fluid is moving in turbulent flow through a pipe. A hot-wire anemometer is inserted to measure the local velocity at a given point P in the system. The following readings were recorded at equal time intervals during a very short period of time:

Time Increment	1	2	3	4	5	6	7	8	9	10
Velocity v_z at P	43.4	42.1	42.0	40.8	38.5	37.0	37.5	38.0	39.0	41.7

Determine the intensity of turbulence at point P.

Solution The terms $\bar{v}_z'$ and $(\bar{v}_z')^2$ are first calculated

$$\bar{v}_z = \frac{\sum\limits_{i=1}^{i=n} v_z}{n} = \frac{400.0}{10} = 40$$

where n = number of time increments.

$$
\begin{aligned}
(\bar{v}_z')^2 &= \frac{\sum\limits_{i=1}^{10} (v_z - \bar{v}_z)^2}{10} \\
&= \frac{45.9}{10} \\
&= 4.59
\end{aligned}
$$

Substituting into Equation (14.23) gives

$$
\begin{aligned}
I &= \frac{\sqrt{4.59}}{40} \\
&= \frac{2.15}{40} \\
&= 0.0538
\end{aligned}
$$

The disturbance or fluctuation velocity components of a fluid in turbulent flow not only alters the transport equations for momentum, energy, and mass, but also gives rise to the aforementioned velocity profiles that are quite different from the corresponding profiles for laminar flow. A semiempirical equation describing the velocity profile in pipes for turbulent flow is:

$$v_z = v_{z_{max}}\left[1 - \left(\frac{r}{R}\right)\right]^{1/7}$$

(14.24)

where v_z = local velocity,
         $v_{z_{max}}$ = maximum velocity,
         R = radius of the pipe,
         r = radial cylindrical coordinate.

Illustrative Example 14.11   A fluid is flowing through a pipe whose inside diameter is 2.0 in. The maximum velocity measured is 30 ft/min. Calculate the volumetric flow rate Q if the flow characteristics are (a) laminar, (b) plug, (c) turbulent.

Solution

a. Laminar flow:

$$\bar{v}_z = \frac{1}{2}v_{z_{max}}$$

$$= \frac{1}{2}(30\,\text{ft/min})$$

$$= 15\,\text{ft/min}$$

By definition,

$$q = \bar{v}A$$

where

$$A = \pi D^2/4$$

$$= \frac{\pi(2/12)^2}{4} = 0.0218\,\text{ft}^2$$

Therefore

$$q = 15\frac{\text{ft}}{\text{min}} \times .0218\,\text{ft}^2$$

$$= 0.327\,\text{ft}^3/\text{min}$$

b. Plug flow:

$$\bar{v}_z = v_{z_{max}}$$

$$q = (30\,\text{ft/min})(.0218\,\text{ft}^2)$$

$$= 0.654\,\text{ft}^3/\text{min}$$

c. Turbulent flow: For turbulent flow,

$$v_z = v_{z_{max}} \left[1 - \left(\frac{r}{R} \right) \right]^{1/7} \tag{14.24}$$

By definition

$$\bar{v}_z = \frac{\displaystyle\int_0^{2\pi} \int_0^R v_z r \, dr \, d\phi}{\displaystyle\int_0^{2\pi} \int_0^R r \, dr \, d\phi}$$

$$= \frac{2\pi v_{z_{max}} \displaystyle\int_0^R \left(1 - \frac{r}{R} \right)^{1/7} r \, dr}{\pi R^2}$$

$$= \frac{2 v_{z_{max}}}{R^2} \int_0^R \left(1 - \frac{r}{R} \right)^{1/7} r \, dr$$

(The above integral can be evaluated from a standard table of integrals.)

$$\int_0^R \left(1 - \frac{r}{R} \right)^{1/7} r \, dr = + \frac{R^2}{\left[\frac{1}{7} + 2 \right]} \left(1 - \frac{r}{R} \right)^{(1/7)+2} \Bigg|_0^R$$

$$- \frac{R^2}{\left[\frac{1}{7} + 1 \right]} \left(1 - \frac{r}{R} \right)^{(1/7)+1} \Bigg|_0^R$$

$$= -\frac{7}{15} R^2 + \frac{7}{8} R^2$$

$$= \frac{49}{120} R^2$$

Therefore,

$$\bar{v}_z = \frac{2 v_{z_{max}}}{R^2} \left(\frac{49}{120} R^2 \right)$$

$$= \frac{49}{60} v_{z_{max}}$$

$$= (0.816) v_{z_{max}}$$

(The 0.816 term is in agreement with the value presented earlier.)

Substitution

$$\bar{v}_z = 0(.816)(30)$$
$$= 24.5 \, \text{ft/min}$$

and

$$q = (24.5)(0.0218)$$
$$= 0.535 \, \text{ft}^3/\text{min}$$

REFERENCES

1. L. Moody, "Friction Factors for Pipe Flow," *Trans. Am. Soc. Mech. Engrs.*, **66**, 671–684, 1944.

2. I. Farag, "Fluid Flow," A Theodore Tutorial, Theodore Tutorials, East Williston, NY, 1996.

3. W. Badger and J. Banchero, "Introduction to Chemical Engineering," McGraw-Hill, New York, 1955.

4. S. Churchill, "Friction Factor Equations Spans All Fluid Flow Regimes," *Chem. Eng.*, Nov. 7, 1977.

5. A. Jain, "Accurate Explicit Equation for Friction Factor," *Amer. Soc. of Civil Engrs, Journal of the Hydraulics Division*, May 1976.

NOTE: Additional problems are available for all readers at www.wiley.com. Follow links for this title.

15

COMPRESSIBLE AND SONIC FLOW

15.1 INTRODUCTION

Compressibility refers to a condition where the volume or density of a fluid varies with the pressure. In fluid flow applications, it is a consideration only when vapors/gases are involved; liquids can safely be considered incompressible in these calculations. When the pressure drop in a flowing gas system is less than (on the order of) 10–20% of the absolute pressure in the system, satisfactory engineering accuracy is obtained in pressure drop calculations by assuming the fluid incompressible at conditions corresponding to the average pressure in the system. For larger pressure drops, compressibility effects can become important. The compressible flow of a fluid is further complicated by the fact that the fluid density is dependent on temperature as well as on pressure. In such systems, temperature may vary in accordance with thermodynamic principles[1] (see Chapters 8 and 11 for more details).

Although this chapter is primarily concerned with sonic flow, it also addresses the general topic of compressible flow. The presentation that follows first examines compressible flow, which in turn is followed by sonic flow, which in turn is followed by key pressure drop equations that may be employed in engineering flow calculations for this topic.

15.2 COMPRESSIBLE FLOW

As noted above, flowing fluids are typically considered compressible when the density varies by more than 10–20% during a particular application. In practice,

Fluid Flow for the Practicing Chemical Engineer. By J. Patrick Abulencia and Louis Theodore
Copyright © 2009 John Wiley & Sons, Inc.

compressible flows are normally limited to gases, supercritical fluids, and multiphase flows containing gases; flowing liquids are normally considered incompressible. In industrial applications, one-dimensional gas flow through nozzles or orifices and in pipelines are the most important applications of compressible flow. Multidimensional external flows are of interest mainly in aerodynamic applications, a topic beyond the scope of this text.[2]

In addition to the factors discussed above, compressible flow calculations are further complicated by other system parameter variations. For example, for a given pipe diameter and mass flow rate, the friction factor depends upon the viscosity, which, in turn, depends upon temperature. This problem does not exist for isothermal flow but can be important during adiabatic operation. However, in adiabatic compressible flow, Reynolds numbers are usually high indicating turbulent flow and any variation of the friction factor due to temperature variations along the pipe length is small. Thus, the friction factor may be assumed constant.

The first step in a compressible–incompressible flow analysis is to classify the flow. One has to specify either steady or unsteady flow as well as whether the flow is compressible or incompressible. Steady and unsteady flow refers to variations with time, while incompressible and compressible flow refers to density variations. If the density is constant, or its variation is very small (most liquids, and gases with a Mach number less than 0.3), the flow is deemed *incompressible*. The Mach number is discussed in the next section.

Illustrative Example 15.1   The inlet air density to an expanding nozzle is $0.071 \, \text{lb/ft}^3$. The outlet density is $0.0049 \, \text{lb/ft}^3$. Is the flow compressible?

Solution   Since the density of the fluid changes during its residence in the nozzle, the flow can be considered compressible.

15.3 SONIC FLOW

The Mach number, Ma, is a dimensionless number defined as the ratio of fluid velocity to the speed of sound in the fluid, i.e.,

$$Ma = \frac{v}{c} \qquad (15.1)$$

where v is the average velocity of the fluid and c is the speed of sound. If the Mach number is less than or equal to 0.3, compressibility effects may usually be neglected, and one may safely assume incompressible flow.

The speed of sound in (selected) common liquids is given in Table 15.1. The speed of sound, c, in an ideal gas may be calculated from

$$c = \sqrt{\frac{kRT}{\text{MW}}} \qquad (15.2)$$

Table 15.1 Speed of Sound in Various Liquids

Liquid	Sound Velocity (m/s)
Acetone	1174
Benzene	1298
Ethanol	1144
Ethylene glycol	1644
Methanol	1103
Water	1498

where k (see Table 15.2) is the ratio of C_p/C_v, R the universal gas constant, T the absolute temperature and MW the molecular weight of the fluid. The derivation of this from basic principles is available in the literature. Note that the k values in Table 15.2 are approximate for 1 atm and 25°C; a decrease in temperature or an increase in pressure will generally result in higher values.[3]

For air, Equation (15.2) simplifies to

Table 15.2 Values of k

Gas	k
C_2H_6	1.2
CO_2, SO_2, H_2O, H_2S, NH_3, Cl_2, CH_4, C_2H_2, C_2H_4	1.3
Air, H_2, O_2, N_2, CO, NO, HCl	1.4
Monatomic gases	1.67

$$c = 20\sqrt{T(\mathrm{K})}; \ \mathrm{m/s} \tag{15.3}$$

and

$$c = 20\sqrt{T(^{\circ}\mathrm{R})}; \ \mathrm{ft/s} \tag{15.4}$$

Illustrative Example 15.2 Nitrogen gas at 20°C and 1 atm flows in a duct at a velocity of 82 m/s. Is it reasonable to neglect compressibility effects? Assume $k = 1.4$ for nitrogen.

Solution Calculate the speed of sound in N_2 at 20°C. See Equation (15.2).

$$c = \sqrt{\frac{kRT}{M}} = \sqrt{\frac{(1.4)(8314.4)(293)}{28}}$$

$$= 349 \ \mathrm{m/s}$$

The Mach number can then be calculated from Equation (15.1).

$$Ma = \frac{v}{c} = \frac{82}{349} = 0.235$$

Since $0.235 < 0.3$, compressibility effects may be neglected.

Equation (15.2) indicates that the square of the velocity of sound is proportional to the absolute temperature of the ideal gas. Thus, the velocity of sound may be viewed as being proportional to the internal kinetic energy of the gas. Since the kinetic energy of a flowing gas is proportional to v^2, the ratio v^2/c^2 provides a measure of the ratio of the kinetic energy to the internal energy. The velocity of sound in air at room temperature is approximately 1100 ft/s. Thus, for a velocity of 220 ft/s, and noting that v/c is defined as the Mach number, Ma is 0.2 and $(Ma)^2$ is only 0.04. This indicates that kinetic energy effects do not become important until somewhat higher Mach numbers are achieved.[3]

Most often, the Mach number is calculated using the speed of sound evaluated at the local pressure and temperature. When $Ma = 1$, the flow is *critical* or *sonic*, and the velocity equals the local speed of sound. For subsonic flow, $Ma < 1$, while supersonic flow has $Ma > 1$. A potential error is to assume that compressibility effects are always negligible when the Mach number is small. Proper assessment of whether compressibility is important should be based on relative density changes, not on Mach number alone.[2] However, the Mach number is usually employed in engineering calculations.

Equations developed earlier for incompressible fluids are applicable to compressible fluids—in a general sense. However, these same equations may often be applied to compressible fluids if the *fractional change* in pressure is not large. For example, compressibility effects may not be important if there is a change in pressure from 14.7 to 15.7 psia, but could be very important if the change is from 0.1 to 1.0 psia.[3]

A detailed treatment of sonic flow though a variety of process units is provided in Perry's Handbook.[1] Included in the treatment are defining equations for:

1. Flow through a frictionless nozzle
2. Adiabatic flow with friction in a duct of constant cross-section
3. Compressible flow with friction loss
4. Convergent/divergent nozzles

Illustrative Example 15.3   Propane ($k = C_p/C_v = 1.3$) at 17°C and 0.35 MPa is flowing in a tube (inside diameter of 1 in) at an average velocity of 43 m/s. Determine the speed of sound in the propane. Is the propane flow compressible? Why or why not? Is the propane flow laminar or turbulent?

Solution   The speed of sound in propane is first calculated from Equation (15.2).

$$c = \sqrt{\frac{kRT}{M}} = \sqrt{\frac{1.3(83.14)(290)}{44}}$$
$$= 267 \, \text{m/s}$$

The Mach number is therefore [see Eq. (15.1)]

$$Ma = \frac{v}{c} = \frac{43}{267} = 0.161$$

Since $0.161 < 0.3$, the flow is incompressible.

Determine the Reynolds number. First calculate the density from the ideal gas law

$$\rho = 6.39 \, \text{kg/m}^3$$

The viscosity is estimated from Fig. B.2 in the Appendix,

$$\mu = 8 \times 10^{-3} \text{cP} = 8 \times 10^{-6} \, \text{m}^2/\text{s}$$

Therefore,

$$Re = \frac{43(0.0254)(6.39)}{8 \times 10^{-6}} = 872{,}000$$

Since Re is 872,000 and >4000, the flow is turbulent.

15.4 PRESSURE DROP EQUATIONS

Two equations for pressure drop are presented in this section—one for laminar flow and one for turbulent flow. Both can be employed for most real-world applications involving compressible flow.

15.4.1 Isothermal Flow

For laminar flow of gases in pipes and other conduits, the pressure drop from P_1 to P_2 may be estimated from Equation (15.5) for laminar flow conditions.

$$P_1^2 - P_2^2 = \frac{8\mu RTG}{g_c MD}\left[\frac{8L}{D} + \frac{Re}{3}\ln\left(\frac{P_1}{P_2}\right)\right] \qquad (15.5)$$

where   Re = Reynolds number,
         μ = gas viscosity,
         T = absolute temperature,
         G = mass velocity flux,
         M = gas molecular weight,
         D = pipe/conduit diameter,
         L = pipe/conduit length.

Equation (15.5) may be used for engineering purposes provided that the Mach number is below 0.5 (i.e., $Ma < 0.5$).

Finally, one should note that if the flow rate is unknown (this is often the desired quantity), a trial-and-error solution is involved since the friction factor depends on the velocity. However, the value of f does not vary significantly over a very wide range of Reynolds numbers and the solution is therefore not sensitive to the value of f. Consequently, an (initial) assumption of an average value of 0.004 for f is satisfactory. This will yield a value of the velocity for which the Reynolds number can be calculated and the corresponding value of f determined. An iterative calculation using these new and updated values of f will provide an acceptable answer.

$$P_1^2 - P_2^2 = \frac{4fLG^2RT}{g_cD(\text{MW})}\left(1 + \frac{2D}{fL}\ln\frac{P_1}{P_2}\right) \tag{15.6}$$

In ducts of appreciable length, the second term in the parentheses can be assumed negligible unless the pressure drop is very large. When this term is omitted, Equation (15.6) becomes

$$P_1 - P_2 = \frac{2fLG^2}{2g_c\rho_{av}D} \tag{15.7}$$

where ρ_{avg} is the density at the average pressure of $(P_1 + P_2)/2$ for the mass rate of flow of the system. This equation may also be written as

$$\dot{m} = \frac{\pi}{8}\sqrt{\frac{(P_1^2 - P_2^2)g_cD^5(\text{MW})}{fLRT}} \tag{15.8}$$

It should be noted that in most applications, the flow can more appropriately be described as adiabatic rather than truly isothermal.

The equation for adiabatic flow is based on the condition that the flow arises from the adiabatic expansion of the gas through a frictionless nozzle leading from an inlet source where the velocity is negligible. Such a system is frequently encountered in practice. For this system, the describing equation is given by

$$\frac{T_1}{T_0} = \left(\frac{P_1}{P_0}\right)^{(k-1)/k}$$

$$\frac{P_0}{P_1} = \left[1 + \frac{G^2}{2g_c}\left(\frac{k-1}{k}\right)\frac{RT_1}{(\text{MW})P_1^2}\right]^{k/(k-1)} \tag{15.9}$$

Illustrative Example 15.5 Verify Equation (15.8); assume Equation (15.6), with the second term neglected, to be correct.

Solution Start with Equation (15.6). Neglecting the second term,

$$P_1^2 - P_2^2 = \frac{4fLG^2RT}{g_cD(\text{MW})}$$

Note that

$$G = \frac{\dot{m}}{(\pi/4)D^2}$$

Substituting and rearranging

$$P_1^2 - P_2^2 = \frac{4fL(\dot{m})^2 RT(16)}{g_c D(D^2)^2 (\pi)^2 (MW)}$$

Solving for $\dot{m}$

$$(\dot{m})^2 = \frac{\left(P_1^2 - P_2^2\right) g_c D^5 \pi^2 (MW)}{64fLRT}$$

$$\dot{m} = \frac{\pi}{8} \sqrt{\frac{\left(P_1^2 - P_2^2\right) g_c D^5 (MW)}{fLRT}}$$

Illustrative Example 15.6   Calculate the pressure drop accompanying the flow of natural gas (which may be assumed to be methane) at 70°F through a horizontal steel pipe 12 inches in diameter and 3 miles long. The gas enters the pipe at 75 psig and at a rate of 236 scfs (14.7 psia, 60°F). The viscosity of methane at 70°F is 0.011 centipoise and a friction factor of 0.008 may be assumed.

Solution   The mass flow rate is

$$\dot{m} = (236/379)16$$
$$= 10 \, \text{lb/s}$$

The mass velocity flux is

$$G = 10/[(\pi/4)(1)^2]$$
$$= 12.7 \, \text{lb/ft}^2 \cdot \text{s}$$

Apply Equation (15.7) since the flow is turbulent (gas). Assume an average density based on inlet conditions

$$\rho = \frac{P(MW)}{RT}$$
$$= \frac{(89.7)(16)}{(10.73)(530)}$$
$$= 0.252 \, \text{lb/ft}^3$$

$$P_1 - P_2 = \frac{2fLG^2}{g_c \rho_{av} D}$$
$$= \frac{2(0.008)(3)(5280)(12.7)^2}{(32.2)(0.252)(1)}$$
$$= 5036 \, \text{psf}$$

Rearranging

$$P_2 = P_1 - 5036$$
$$= (89.7 \times 144) - 5036$$
$$= 12{,}917 - 5036$$
$$= 7881 \text{ psf}$$
$$= 54.7 \text{ psia}$$
$$= 40.0 \text{ psig}$$

The pressure drop is approximately given by

$$\Delta P = 89.7 - 54.7 = 35.0 \text{ psia}$$

Strictly speaking, the calculation should be repeated with an updated value for the density at the average of the inlet and outlet pressure.

Illustrative Example 15.7   Refer of Illustrative Example 15.6. Calculate the Reynolds number for the system. Comment on the results.

Solution   The calculation is based on inlet conditions. The Reynolds number is

$$\text{Re} = \frac{DG}{\mu}$$
$$= \frac{(1)(12.7)}{\mu}$$

Convert cP to lb/ft · s.

$$\mu = (0.011 \text{ cP})(6.72 \times 10^{-4})$$
$$= 7.39 \times 10^{-6} \text{ lb/ft} \cdot \text{s}$$

The Reynolds number is therefore

$$\text{Re} = \frac{(1)(12.7)}{7.39 \times 10^{-6}}$$
$$= 17.2 \times 10^{5}$$

The value provided for the friction factor is reasonable for this Reynolds number.

Illustrative Example 15.8   Air at 2.7 atm and 15°C enters a horizontal 8.5 cm steel pipe that is 6.5 m long. The velocity at the entrance of the pipe is 30 m/s. What is the pressure drop across the line?

Solution   Assume isothermal flow and apply Equation (15.7). From the ideal gas law

$$\rho = \left(\frac{29}{22.4}\right)\left(\frac{2.7}{1}\right)\left(\frac{273}{288}\right)$$

$$= 3.31 \, \text{kg/m}^3$$

The mass velocity is

$$G = v\rho = (30)(3.31)$$

$$= 99.3 \, \text{kg/m}^2 \cdot \text{s}$$

Assume (initially)

$$f = 0.004$$

Rearranging Equation (15.7)

$$P_1 - P_2 = \frac{2fLG^2}{g_c\rho D}$$

$$P_2 = P_1 - \frac{2fLG^2}{g_c\rho D}$$

Substituting, while noting 1 atm $= 101,325 \, \text{kg/m} \cdot \text{s}^2$,

$$P_2 = 2.7 - \left[\frac{(2)(0.004)(65)(99.3)^2}{(3.31)(0.085)(101,325)}\right]$$

$$= 2.7 - 0.18$$

$$= 2.52 \, \text{atm}$$

The pressure drop is therefore

$$\Delta P = P_1 - P_2$$

$$= 0.18 \, \text{atm} = 2.65 \, \text{psi}$$

Illustrative Example 15.9   Refer to Illustrative Example 15.8. Is the assumption for the friction factor reasonable?

Solution   From Table A.3 in the Appendix,

$$\mu = 0.0174 \, \text{cP} = 1.74 \times 10^{-5} \, \text{kg/(m} \cdot \text{s)}$$

Calculate the Reynolds number,

$$\mathrm{Re} = \frac{DG}{\mu}$$
$$= \frac{(0.085)(99.3)}{1.74 \times 10^{-5}} = 485{,}000$$

Refer to Fig. 14.2. The assumption of $f = 0.004$ is reasonable.

REFERENCES

1. D. Green and R. Perry (editors), "Perry's Chemical Engineers' Handbook," 8th edition, McGraw-Hill, New York, 2008.
2. I. Farag, "Fluid Flow," A Theodore Tutorial, East Williston, NY, 1995.
3. C. Bennett and J. Myers, "Momentum, Heat, and Mass Transfer," McGraw-Hill, New York, 1962.

NOTE: Additional problems are available for all readers at www.wiley.com. Follow links for this title.

16

TWO-PHASE FLOW

16.1 INTRODUCTION

The simultaneous flow of two phases in pipes (as well as other conduits) is complicated by the fact that the action of gravity tends to cause settling and "slip" of the heavier phase with the result that the lighter phase flows at a different velocity in the pipe than does the heavier phase. The results of this phenomena are different depending on the classification of the two phases, the flow regime, and the inclination of the pipe (conduit).

As one might suppose, the major industrial application in this area is gas (G)–liquid (L) flow in pipes. Therefore, the subjects addressed in this chapter key on a G–L flow in pipes. The extension of much of the material to follow to flow in various conduits can be accomplished by employing the equivalent diameter of the conduit in question.

The general subject of liquid–solid flow in pipes is not considered in this chapter. Suspensions of solids in liquids fall into two general classes, Newtonian and non-Newtonian. Newtonian suspensions are characterized by a constant viscosity, independent of the rate of shear. In the case of non-Newtonian suspensions, the viscosity is a variable that is a function of the rate of shear and (in some cases) a function of the duration or period of shear for viscous flow. If the suspension is found to be Newtonian in character, the pressure drop can be calculated by standard equations available for both viscous flow and turbulent flow by employing the average density and viscosity of the mixture (see Chapters 11 and 12). The procedures for computing the pressure drop for non-Newtonian suspensions are more involved but received treatment in Chapter 6. Details on liquid–solid flow is also available from Perry and Green.[1]

Fluid Flow for the Practicing Chemical Engineer. By J. Patrick Abulencia and Louis Theodore
Copyright © 2009 John Wiley & Sons, Inc.

The general subject of flashing and boiling liquids is also not considered in this chapter. However, when a saturated liquid flows in a pipeline from a given point at a given pressure to another point at a lower pressure, several processes can take place. As the pressure decreases, the saturation or boiling temperature decreases, leading to the evaporation of a portion of the liquid. The net results that a one-phase flowing mixture is transformed into a two-phase mixture with a corresponding increase in frictional resistance in the pipe. Boiling liquids arise when liquids are vaporized in pipelines at approximately constant pressure. Alternatively, the flow of condensing vapors in pipes is complicated due to the properties of the mixture constantly changing with changes in pressure, temperature, and fraction condensed. Further, the condensate, which forms on the walls, requires energy in order to be transformed into spray, and this energy must be obtained from the main vapor stream, resulting in an additional pressure drop. An analytical treatment of these topics is beyond the scope of this book. However, information is available in the literature.[1]

The remainder of the chapter examines the following topics:

- Gas (G)–Liquid (L) Flow Principles: Generalized Approach
- Gas (Turbulent) Flow–Liquid (Turbulent) Flow
- Gas (Turbulent) Flow–Liquid (Viscous) Flow
- Gas (Viscous) Flow–Liquid (Viscous) Flow
- Gas–Solid Flow

16.2   GAS (G)–LIQUID (L) FLOW PRINCIPLES: GENERALIZED APPROACH

The suggested method of calculating the pressure drop of gas–liquid mixtures flowing in pipes is essentially that originally proposed by Lockhart and Martinelli[2] nearly 60 years ago. The basis of their correlation is that the two-phase pressure drop is equal to the single-phase pressure drop for either phase (G or L) multiplied by a factor that is a function of the single-phase pressure drops of the two phases. The equations for the total pressure drop per unit length Z $(\Delta P/Z)_T$ are written as:

$$(\Delta P/Z)_T = Y_G(\Delta P/Z)_G \tag{16.1}$$

$$(\Delta P/Z)_T = Y_L(\Delta P/Z)_L \tag{16.2}$$

The terms Y_L and Y_G are functions of the variable X:

$$Y_G = F_G(X) \tag{16.3}$$

$$Y_L = F_L(X) \tag{16.4}$$

where

$$X = \left[\frac{(\Delta P/Z)_L}{(\Delta P/Z)_G} \right]^{0.5} \tag{16.5}$$

The relationship between Y_L and Y_G is therefore given by

$$Y_G = X^2 Y_L \tag{16.6}$$

The single-phase pressure-drop gradients $(\Delta P/Z)_L$ and $(\Delta P/Z)_G$ can be calculated by assuming that each phase is flowing alone in the pipeline, and the phase in question is traveling at its superficial velocity. The superficial velocities are therefore based on the full cross-sectional area, S, of the pipe so that

$$v_L = q_L/S \tag{16.7}$$

and

$$v_G = q_G/S \tag{16.8}$$

where v_L = liquid-phase superficial velocity,
       v_G = gas-phase superficial velocity,
       q_L = liquid-phase volume flow rate,
       q_G = gas-phase volume flow rate,
       S = pipe cross-sectional area.

Note that either Equation (16.1) or (16.2) can be employed to calculate the pressure drop.

The functional relationships for Y_L and Y_G in Equations (16.3) and (16.4) in terms of X were also provided by Lockhart and Martinelli[2] for the phase classification under different flow conditions. (These relationships are provided later in this chapter.) For gas–liquid flows, semi-empirical data were provided for the following three flow categories:

gas (turbulent flow, t)–liquid (turbulent flow)
gas (turbulent flow)–liquid (viscous flow, v)
gas (viscous flow)–liquid (viscous flow)

The next three subsections address each of the above topics. Note that applications involving gas (viscous flow)–liquid (viscous flow) do not receive treatment since this type of flow rarely occurs in practice; the low viscosity of a gas (or vapor) virtually eliminates the possibility of gas moving in a laminar flow.

A variety of the above flow phenomena is possible with the two-phase flow of gases and liquids in horizontal pipes ranging from parallel (two-layer) flow at low velocities to dispersed flow at high velocities (gas carried as bubbles in a continuous liquid phase or liquid carried as spray in the gas). The pressure drop is greater in liquid–gas flow than that for the single-phase flow of either gas or liquid for several reasons. These include the irreversible work done on the liquid by the gas and that the effective cross-sectional area of flow for either phase is reduced by the flow of the other phase in the area.

The basis for the Martinelli correlations[2,3] assumes that the pressure drop for the liquid phase must equal the pressure drop for the gas phase for all types of flow, provided that no appreciable pressure differences exist across any pipe diameter and that the volume occupied by the liquid and by the gas at any instant of time must equal the total volume of the pipe. Using these assumptions, the pressure drop due to the liquid flow and that due to the gas flow was expressed in each case by standard pressure drop equations using unknown "hydraulic diameters." The hydraulic diameters were then expressed in terms of the actual cross-sectional area of flow and the ratio of the actual cross-sectional area of flow to the area of a circle of diameter equal to the unknown hydraulic diameter. The unknown hydraulic diameter for the liquid flow was eliminated in the analysis and an expression was obtained for the pressure drop as a function of the single-phase pressure drop for gas alone. The function, expressed as ϕ^2 in their study, was introduced in order to reduce the range of the variables when providing ϕ^2 vs $\sqrt{X}$. Isothermal flow in smooth pipes was assumed.

It is important to know what type of flow is occurring, although this can obviously be a difficult task. In order to establish which flow mechanisms applied, Martinelli et al.[2,3] used a set of flow conditions (as noted above) that were functions of the Reynolds number

$$Re = 4w/\pi D\mu \tag{16.9}$$

where w is the mass flow rate. Martinelli et al.[2,3] computed the Reynolds number for each using the *actual* pipe diameter; i.e., a superficial velocity was employed. For $Re < 2000$, the flow for that phase was assumed to be viscous (laminar); for $Re > 2000$, the flow is assumed to be turbulent.

Illustrative Example 16.1   Air and oil are in concurrent flow through a horizontal pipe. The following pressure drop calculations were obtained from Theodore Consultants (a group of engineers with limited technical capabilities):

$$(\Delta P/Z)_G = 2.71 \, \text{psft}/100 \, \text{ft}$$
$$(\Delta P/Z)_L = 7.50 \, \text{psft}/100 \, \text{ft}$$

Calculate the dimensional parameter X.

Solution   Refer to Equation (16.5):

$$X = [(\Delta P/Z)_L/(\Delta P/Z)_G]^{0.5}$$

Substitute

$$X = (7.50/2.71)^{0.5}$$
$$= 1.66$$

The volume fraction or holdup of a phase for two-phase flow in a horizontal pipe is also available[1]:

$$\varepsilon_L = F_3(X) \tag{16.10}$$

and

$$\varepsilon_G = F_4(X) \tag{16.11}$$

where $\varepsilon_G + \varepsilon_L = 1$ and ε_G and ε_L are, the fraction (dimensionless) of pipe volume occupied by the liquid phase and gas phase, respectively; X is the aforementioned variable defined by Equation (16.5). The relationship between ε_L and X is approximately provided by[4]:

$$\varepsilon_L = 0.298 + 0.117\ln(X) \tag{16.12}$$

Gas–liquid flow usually occurs in horizontal pipes. However, when gas–liquid mixtures flow in vertical pipes, there is an increase in liquid concentration or build-up of liquid due to the density difference in the case of upward flow, and a decrease in liquid concentration in the case of downward flow. Since information is available on the upward flow of gas–liquid mixtures, a variety of flow phenomena are possible including gas as the dispersed phase in a continuous liquid phase to gas as the continuous phase with liquid carried as spray. One of the intermediate types of flow is where the liquid flows as an annulus and the gas as a central core. The major applications are gas lifts. A gas lift is a vertical pipe (known as an eduction pipe) open at both ends, part of which is submerged below the surface of the liquid to be pumped. Compressed gas is admitted through a foot-piece inside the lower end; a mixture of liquid and gas is thus formed within the pipe. The gas reduces the average density of the mixture in the eduction pipe to a point where the weight of the mixture is less than equivalent to the pressure at the foot-piece. With the gas and liquid being supplied at a sufficient rate, the mixture rises upward through the pipe and is discharged at the upper end. Industrial use occurs with the operation of flowing oil wells. Considerable operating and experimental data have been reported but little attempt has been made to correlate them.

16.3 GAS (TURBULENT) FLOW–LIQUID (TURBULENT) FLOW

This section provides additional details of the original work of Lockhart and Martinelli.[2,3] This is followed by a simpler approach for predicting pressure drop. The simpler approach is recommended for industrial applications.

In the original work (with most of the notation retained), the ratio of the actual cross-sectional area of flow to the area of a circle of diameter equal to the unknown equivalent (or hydraulic) diameter for the gas phase was assumed to be unity and the ratio for the liquid phase was determined from experimental data. The following correlations were obtained from the ratio for the liquid phase and the properties of the liquid and gas

$$\left(\frac{\Delta P}{Z}\right)_{tt} = \phi_{tt}\left(\frac{\Delta P}{Z}\right)_{G} \tag{16.13}$$

where ϕ_{tt} is a function of a dimensionless group, X_{tt}; and,

$$X_{tt} = \left(\frac{\mu_L}{\mu_G}\right)^{0.111}\left(\frac{\rho_G}{\rho_L}\right)^{0.555}\left(\frac{\dot{m}_L}{\dot{m}_G}\right) \tag{16.14}$$

The magnitude of ϕ_{tt} for values of X_{tt} is given in Table 16.1.

Table 16.1 ϕ_{tt} vs $\sqrt{X_{tt}}$

$\sqrt{X_{tt}}$	ϕ_{tt}
0	1.00
0.10	1.50
0.20	1.68
0.40	2.13
0.70	3.03
1.00	4.08
2.00	8.30
4.00	19.6
7.00	42.3
10.0	71.0
20.0	222
40.0	770
46.2	1000
for $\sqrt{X_{tt}} > 46$	$\phi_{tt} = (\sqrt{X_{tt}})^{1.8}$

Results were later expressed in terms of Y_L and Y_G, both of which are functions of X_{tt}; see Equations (16.3) and (16.4) for more details. Van Vliet[5] subsequently regressed the data to a model of the form

$$Y_G = a + bX + cX^2 + dX^3$$

and

$$Y_L = aX^b \tag{16.15}$$

The final results for Y_L and Y_G for tt flow are presented in Equations (16.16) and (16.17):

$$Y_G(tt) = 1.7172 + 15.431X + 3.9314X^2 - 2.2952X^3; \quad X < 1 \tag{16.16a}$$

$$= 5.80 + 6.7143X + 6.9643X^2 - 0.75X^3; \quad 1 < X < 10 \tag{16.16b}$$

$$= 131 + 1.4105X + 1.9362X^2 - 0.0087X^3; \quad X > 10 \tag{16.16c}$$

$$Y_L(tt) = 11.745X^{-1.4901}; \quad X < 1 \tag{16.17a}$$

$$= 18.219X^{-0.8192}; \quad 1 < X < 10 \tag{16.17b}$$

$$= 6.3479X^{-0.3518}; \quad X > 10 \tag{16.17c}$$

As noted earlier, the pressure drop for two-phase flow can be calculated using either Equation (16.1) or (16.2). Longhand calculations have shown[5] that the two equations can, in some cases, produce different results. The authors recommend using either the gas phase value or the average of the two for design purposes

However, if the volume fraction of one phase predominates (see Eq. (16.12)), the authors suggest employing the pressure drop calculation for that phase.

Illustrative Example 16.2   Refer to Illustrative Example (16.1). Calculate the pressure drop (total) if the flow for both phases is turbulent. Base the calculation on:

a. Y_G
b. Y_L

Solution

a. Since the flow is tt and $1 < X < 10$, apply Equation (16.16b) to obtain Y_G while noting $X = 1.66$:

$$Y_G(\text{tt}) = 5.80 + 6.7143X + 6.9643X^2 - 0.75X^3$$

Substituting

$$Y_G(\text{tt}) = 5.80 + 6.7143(1.66) + 6.9643(1.66)^2 - 0.75(1.66)^3$$
$$Y_G = 5.80 + 11.145 + 19.19 - 3.431$$
$$= 32.7$$

This value is an excellent agreement with the values provided by Lockhart and Martinelli.[3] The pressure drop is therefore (from Eq. 16.1):

$$\Delta P/Z = (Y_G)(\Delta P/Z)_G$$
$$= (32.7)(2.71)$$
$$= 88.6 \text{ psf}/100 \text{ ft}$$

b. Apply Equation (16.17b) to generate Y_L

$$Y_L(\text{tt}) = 18.219X^{-0.8192}$$
$$= 18.219(1.66)^{-0.8192}$$
$$= 12.0$$

The literature value is approximately 12.[3] The pressure drop is therefore (from Eq. 16.2):

$$\Delta P/Z = (Y_L)(\Delta P/Z)_L$$
$$= (12.0)(7.50)$$
$$= 90.2 \text{ psf}/100 \text{ ft}$$

As expected, both results are in reasonable agreement.

16.4 GAS (TURBULENT) FLOW–LIQUID (VISCOUS) FLOW

The original work of Lockhart and Martinelli[2] is once again reviewed for the turbulent–viscous (tv) case. Using the same procedures as that for the turbulent–turbulent (tt) case, the final correlation took the form:

$$\left(\frac{\Delta P}{Z}\right)_{tv} = \phi_{tv}^2 \left(\frac{\Delta P}{Z}\right)_G \qquad (16.18)$$

where ϕ_{tv} is a function of a dimensionless group, X_{tv}. The magnitude of ϕ_{tv} for values of X_{tv} is provided in Table 16.2.

Table 16.2 ϕ_{tv} vs $\sqrt{X_{tv}}$

$\sqrt{X_{tv}}$	ϕ_{tv}
0	1.00
0.07	2.00
0.10	2.14
0.20	2.46
0.40	2.96
0.70	3.42
1.00	3.85
2.00	5.30
4.00	7.87
7.00	11.3
10.0	14.8
20.0	25.4
40.0	46.0
70.0	75.8
100	105
200	203
400	400
1000	1000
for $\sqrt{X_{tv}} > 1000$	$\phi_{tv} = \sqrt{X_{tv}}$

The results of Table 16.2, which are functions of X_{tv} were expressed in terms of Y_G and Y_L. Van Vliet[5] subsequently regressed the data to a model of the form as in Equations (16.16) and (16.17). The final results for Y_L and Y_G are presented in Equations (16.19) and (16.20)

$$Y_G(tv) = 1.6204 + 1.1825X + 34.778X^2 - 30.522X^3; \quad X < 1 \qquad (16.19a)$$

$$= 20 - 21.81X + 16.357X^2 - 1.8333X^3; \quad 1 < X < 10 \qquad (16.19b)$$

$$= 50.333 + 2.9782X + 1.9395X^2 - 0.0088X^3; \quad X > 10 \qquad (16.19c)$$

$$Y_L(tv) = 6.7147X^{-1.5757}; \quad X < 1 \tag{16.20a}$$

$$= 11.702X^{-0.7334}; \quad 1 < X < 10 \tag{16.20b}$$

$$= 5.5873X^{-0.3215}; \quad X > 10 \tag{16.20c}$$

Illustrative Example 16.3 Refer to Illustrative Example (16.1). Calculate the pressure drop (total) if the flow for the gas phase is turbulent and the liquid phase is viscous. Base the calculation on:

a. Y_G
b. Y_L

Solution

a. Since the flow is tv and $1 < X < 10$, apply Equation (16.19b) to obtain Y_G while noting once again $X = 1.66$:

$$Y_G(tv) = 20 - 21.81X + 16.357X^2 - 1.8333X^3$$

Substituting

$$Y_G(tv) = 20 - 21.81(1.66) + 16.357(1.66)^2 - 1.8333(1.66)^3$$
$$Y_G = 20 + 3.62 + 45.1 - 8.39$$
$$= 20.5$$

The literature value is approximately 22.[3] The pressure drop is therefore (from Eq. 16.1):

$$\Delta P/Z = (Y_G)(\Delta P/Z)_G$$
$$= (32.7)(2.71)$$
$$= 55.6 \, psf/100 \, ft$$

b. Apply Equation (16.20b) to generate Y_L.

$$Y_L(tv) = 11.702X^{-0.7334}$$
$$= 11.702(1.66)^{-0.7334}$$
$$= 8.07$$

The literature value is approximately 7.5.[3] The pressure drop is therefore (from Eq. 16.2):

$$\Delta P / Z = (Y_L)(\Delta P / Z)_L$$
$$= (8.07)(7.50)$$
$$= 60.5 \, \text{psf} / 100 \, \text{ft}$$

As expected, both results are in reasonable agreement.

16.5   GAS (VISCOUS) FLOW–LIQUID (VISCOUS) FLOW

The same procedure was employed by Lockhart and Martinelli[2,3] as in the preceding two cases except that both liquid and gas ratios of the actual cross-sectional area of flow to the area of a circle of diameter equal to the unknown hydraulic diameter for the gas phase were determined experimentally in capillary tubes. Their correlation was expressed as

$$\left(\frac{\Delta P}{Z}\right)_{vv} = \phi_{vv}^2 \left(\frac{\Delta P}{Z}\right)_G \tag{16.21}$$

where ϕ_{vv} is a function of a dimensionless group, X_{vv}. The magnitude of ϕ_{vv} for values of X_{vv} is provided in Table 16.3.

Table 16.3 ϕ_{vv} vs $\sqrt{X_{vv}}$

$\sqrt{X_{vv}}$	ϕ_{vv}
0.2	1.40
0.4	1.69
0.6	1.93
0.8	2.16
1	2.44
2	3.81
3	5.15
4	6.4
6	8.7 (limit of experimental data)
.	.
.	.
.	.
∞	$\phi_{vv} = \sqrt{X_{vv}}$

The results in Table 16.3 were later expressed in terms of Y_L and Y_G, both of which are functions of X_{vv} (see Equations (16.3) and (16.4) for more details). Van Vliet[5] subsequently regressed the data to a model of the form as in Equations

(16.16)–(16.17) and (16.19)–(16.20). The final results for Y_L and Y_G are presented in Equations (16.22) and (16.23):

$$Y_G(vv) = 1.1241 + 3.7085X + 6.7318X^2 - 11.541X^3; \quad X < 1 \qquad (16.22a)$$

$$= 10 - 10.405X + 8.6786X^2 - 0.9167X^3; \quad 1 < X < 10 \qquad (16.22b)$$

$$= -78.333 + 7.3223X + 1.8957X^2 - 0.0087X^3; \quad X > 10 \qquad (16.22c)$$

$$Y_L(vv) = 3.9794X^{-1.6583}; \quad X < 1 \qquad (16.23a)$$

$$= 6.4699X^{-0.556}; \quad 1 < X < 10 \qquad (16.23b)$$

$$= 3.7013X^{-0.2226}; \quad X > 10 \qquad (16.23c)$$

As noted earlier, this finds flow regime limited application in practice.

Illustrative Example 16.4   Refer to Illustrative Example (16.1). Calculate the pressure drop (total) if the flow for both phases is laminar. Base the calculation on:

a. Y_G
b. Y_L

Solution

a. Since the flow is tv and $1 < X < 10$, apply Equation (16.22b) to obtain Y_G, noting $X = 1.66$.

$$Y_G(vv) = 10 - 10.405X + 8.6786X^2 - 0.9167X^3$$

Substituting

$$Y_G(vv) = 10 - 10.405(1.66) + 8.6786(1.66)^2 - 0.9167(1.66)^3$$
$$Y_G = 10 - 17.3 + 23.9 - 4.19$$
$$= 12.41$$

The literature value is approximately 12.5.[3] The pressure drop is therefore (from Eq. 16.1):

$$\Delta P/Z = (Y_G)(\Delta P/Z)_G$$
$$= (12.41)(2.71)$$
$$= 33.6 \, psf/100 \, ft$$

b. Apply Equation (16.23b) to generate Y_L:

$$Y_L(vv) = 6.4699X^{-0.556}$$
$$= 6.4699(1.66)^{-0.556}$$
$$= 4.88$$

The literature value is approximately 5.0.[3] The pressure drop is therefore (from Eq. 16.2):

$$\Delta P/Z = (Y_L)(\Delta P/Z)_L$$
$$= (4.88)(7.50)$$
$$= 36.1 \, psf/100 \, ft$$

As expected, both results are in fair agreement.

16.6 GAS–SOLID FLOW

There are many gas–solid flow systems in pipes but this section solely addresses pneumatic conveying. The material to follow in this section will essentially be divided into five subsections:

1. Introduction
2. Solids Motion
3. Pressure Drop
4. Design Procedure
5. Pressure Drop Reduction in Gas Flow

16.6.1 Introduction

Conveying material pneumatically has been used for many years. The system can be either a pressure system or a suction system. The materials that have been handled include grain, wood shavings, pulverized coal, cement, staple, plastic chips, small metal parts, and money containers in department stores. Pneumatic conveyors are simple, quiet, convenient, and clean; however, pneumatic conveyors have a much lower efficiency than the belt or bucket type conveyor.

In a pressure system, the material can be fed by a screw conveyor or similar feeder and then forced through the system by compressed air or the material can be fed into a tank and then forced through the system by compressed air. In a suction system, a fan or blower is installed after the separating system thereby putting the entire system under vacuum. The material, with sufficient air to keep the material in suspension, is then drawn or "sucked" through the system.

Occasionally, it is more convenient to use a combination of pressure and suction systems. In a combination system, the material is drawn in, passes through the fan and then under pressure is forced through the remainder of the system. For cases where the material may damage the fan, an ejector may be used in place of the fan. Also, using an ejector, the material can be fed in the mixing throat of the ejector.

It should be noted that in the design of pneumatic conveying systems, neither a pneumatic or theoretical method has been developed; the design is based on practical operating experience and empirical correlations of test data. Earlier general treatments of the particle motion and pressure drop are available.[6–8]

16.6.2 Solids Motion

The path of the solids in a horizontal pipe is somewhat sinusoidal, the solids striking the bottom of the pipe at intervals and then rising again. The height and length of the rise appears to decrease as the air velocity decreases. The vertical distribution of the solids across the pipe diameter is fairly uniform at low concentrations but becomes more dense at the bottom of the pipe as the loading (ratio of weight rate of solids to weight rate of air) increases. Finally, at high loadings, a considerable portion of the solids have been reported along the bottom of the pipe. The difference between the final average velocity of the solids and that of the air stream is almost constant for both horizontal and vertical conveyors. This difference is the "slip" between the solids and the air and increases with increasing velocity of the air stream. This "slip" velocity is of the order of magnitude of the "choking" velocity and is essentially the minimum transport or conveying velocity. For estimating purposes, the "slip" velocity may be taken as equal to the "choking" velocity. It has been reported that the "choking" velocity is independent of loading for relatively large particles.

The minimum transport velocities of a material can be estimated by testing the solids in horizontal and vertical glass tubes by determining the minimum air velocity to convey the solids in a horizontal tube and the minimum air velocity to just suspend the solids in a vertical tube. The minimum transport velocity of the solids may be several times the free fall velocity.

The minimum velocity v_m to prevent the settling of some particles of diameter d_p (in inches) and specific gravity s can be estimated from the following correlation:

$$(16.24)$$

For horizontal pipes
$$k = 100$$
and
$$b = 0.40.$$

For vertical pipes
$$k = 205$$
and
$$b = 0.60.$$

See Chapter 22 for additional details.

16.6.3 Pressure Drop

The total pressure drop in the system can be considered to consist of the sum of the following pressure drops:

1. to accelerate the air to the carrying velocity
2. to overcome the friction of the air on the pipe walls
3. to supply the loss of momentum of the air in:
   a. accelerating the solids
   b. keeping the solids in suspension
4. to support the air (vertical pipes)
5. to support the solids (vertical pipes)

The total pressure drop for horizontal pipes, ΔP_{TH}, is given by

$$\Delta P_{TH} = \Delta P_{AG} + \Delta P_{AS} + \Delta P_F \qquad (16.25)$$

where ΔP_{AG} = pressure drop to accelerate the air
ΔP_{AS} = pressure drop to accelerate the solids
ΔP_F = pressure drop due to the friction of moving air

For vertical pipes, the pressure drop ΔP_{TV} is

$$\begin{aligned} \Delta P_{TV} &= \Delta P_{AG} + \Delta P_{AS} + \Delta P_F + \Delta P_V \\ &= \Delta P_{TH} + \Delta P_V \end{aligned} \qquad (16.26)$$

where ΔP_V = pressure drop to support the air and solid. Details are available in the literature.[9]

16.6.4 Design Procedure

In an actual design, the quantity of material to be conveyed and the distance are generally known. One can then assume a loading and conveying velocity, and the diameter of the pipe can be computed. Finally, the pressure drop through the system is computed. If the pressure drop is excessive, a smaller loading can be taken and the above procedure is repeated until a reasonable pressure drop is obtained. There is no reliable method to accurately calculate the conveying velocity; however, a conveying velocity of 70 ft/s can be assumed in lieu of any information.

Some order of magnitude values of loading, conveying velocities, and pressure drops for various systems are outlined below[10]:

1. Fan system
   pressure drop = 10 to 30 in H_2O (50 in H_2O is about the maximum)
   loading = 0.1 to 2.0 (possibly 5.0) lbsolids/lbair

conveying velocity $= 30$ to 100 ft/s; usually 50 to 70 ft/s
The fan system is generally used for distances less than 200 ft.

2. Vacuum system

pressure drop $= 5$ to 10 in Hg

loading $= 5$ to 20 lbsolids/lbair

conveying velocity $=$ (same as above)

3. Pressure system

pressure drop $= 10$ to 50 psia (possibly as high as 100 psia)

loading $= 5.0$ to 40 lbsolids/lbair

conveying velocity $=$ (same as above)

Large radius bends are recommended as the pressure drop will be less than with tight bends and it will also be less likely for the solids to collect and choke the bend.

16.6.5 Pressure Drop Reduction in Gas Flow

Scattered statements in the literature seem to suggest that the pressure to convey a gas can be reduced by the addition of fine particles to the moving stream. This is an area that requires more research since the pressure drop reduction effect is a function of both the particle size (and/or particle size distribution) and concentration.

Illustrative Example 16.5   Illustrative Examples 16.2, 16.3, and 16.4 were solved using the Y_G and Y_L equations presented in Equations (16.16)–(16.17), Equations (16.19)–(16.20), and Equations (16.22)–(16.23), respectively. Comment on the similarity of the equations.

Solution   A quick check of the Y_G values generated from the three equations shows little variation. The same applies to the three Y_L values. Because of the token variation of the values for each of the three equations, one might be justified to combine the three equations into one. This suggestion is addressed in one of the problems for this chapter.

Illustrative Example 16.6   A mixture of air (a) and kerosene (k) are flowing in a horizontal 2.3-inch ID pipe. Data for each component is provided below

$$\rho_a = 0.075\,\text{lb/ft}^3 \qquad\qquad \rho_k = 52.1\,\text{lb/ft}^3$$

$$\mu_a = 1.24 \times 10^{-5}\,\text{lb/ft}\cdot\text{s} \qquad \mu_k = 0.00168\,\text{lb/ft}\cdot\text{s}$$

$$q_a = 5.3125\,\text{ft}^3/\text{s} \qquad\qquad q_k = 1.790\,\text{ft}^3/\text{s}$$

Calculate the flow regime for both phases employing superficial velocities.

Solution   Calculate the cross-sectional area of the pipe.

$$S = (\pi/4)(2.3/12)^2$$
$$= 0.0288 \, \text{ft}^2$$

The superficial velocity of each phase can be obtained by applying either Equation (16.7) or (16.8).

$$v_a = 5.3125/(0.0288)(60)$$
$$= 3.07 \, \text{ft/s}$$
$$v_k = 1.79/(0.0288)(60)$$
$$= 1.036 \, \text{ft/s}$$

The Reynolds number can now be calculated by employing Equation (16.9) or the equivalent.

$$Re_a = (2.3/12)(3.07)(0.075)/(1.24 \times 10^{-5})$$
$$= 3570$$
$$Re_k = (2.3/12)(1.036)(52.1)/(0.00168)$$
$$= 6158$$

Turbulent flow exists for both phases based on the superficial velocities.

REFERENCES

1. R. Perry and D. Green (editors), "Perry's Chemical Engineers' Handbook," 6th edition, McGraw-Hill Book Co., New York, 1986.
2. R. Lockhart and R. Martinelli, "Generalized Correlation of Two-Phase, Two-Component Flow Data," *CEP*, **45**, 39–48, 1949.
3. R. Martinelli et al., "Two-Phase Two-Component Flow in the Viscous Region," *Trans. AIChE*, **42**, 681–705, 1946.
4. F. Ricci: Personal correspondence to L. Theodore, 2008.
5. T. Van Vliet: Personal correspondence to L. Theodore, 2008.
6. J. Gasterstadt, "Die Experimentelle Untersuchung des Pneumatischen Fordervorganges," *Z. V. D.I.*, **68**, 617–624 (1924).
7. J. Gasterstadt, "Die Experimentelle Untersuchung des Pneumatischen Fordervorganges," *Forschungsarb. Gebiete Ingenieurw.*, **265**, 1924.
8. S. Wood and A. Bailey, "The Horizontal Carriage of Granular Material by an Injector-Driven Air Stream," *Proc. Inst. Mech. Engrs.* (London), **142**, 149–164, 1939.

9. J. Dalla Valle, "Determining Minimum Air Velocities in Exhaust Systems," *Heating, Piping & Air Conditioning*, **4**, 639–641, 1932 Sept.

10. C. Lapple (editor), "Fluid and Particle Mechanics," John Wiley & Sons, Hoboken, NJ, 1948.

NOTE: Additional problems are available for all readers at www.wiley.com. Follow links for this title.

IV

FLUID FLOW TRANSPORT AND APPLICATIONS

This part of the book is concerned with Fluid Flow Transport and Applications. It contains six chapters and each serves a unique purpose in an attempt to treat nearly all the important aspects of fluid transport applications. From a practical point of view, all systems and plants move liquids and gases from one point to another; hence, the student and/or practicing engineer is concerned with several key topics: determining power requirements, designing and sizing pumps and blowers, reviewing the various valves and fittings, gauging and measuring the flowrate of fluid streams, and estimating costs. The part concludes with Chapters 21 and 22 on academic applications and real-world applications, respectively both chapters of which are essentially extensions of material presented earlier. All receive some measure of treatment in the six chapters contained in this part.

It should be noted that the handling and flow of either gases or liquids is much simpler, cheaper, and less troublesome than solids. Consequently, the engineer attempts to transport most quantities in the form of a gas or liquid whenever possible. It is also important to note that throughout this book, the word "fluid" will always be used to include both liquids and gases. In many operations, a solid is handled in a finely subdivided state so that it stays in suspension in a fluid. Such two-phase mixtures behave in many respects like fluids (see Chapter 16) and are often called "fluidized" solids. This latter topic is treated in Part V, Chapter 26.

Fluid Flow for the Practicing Chemical Engineer. By J. Patrick Abulencia and Louis Theodore
Copyright © 2009 John Wiley & Sons, Inc.

17

PRIME MOVERS

17.1 INTRODUCTION

If a pressure difference is required between two points in a system, a prime mover such as a fan, pump, or compressor is usually used to provide the necessary pressure and/or flow impetus. Engineers are often called on to specify prime movers more frequently than any other piece of processing equipment, particularly in the chemical industry. In a general sense, these prime movers are to a process plant what the engine is to one's automobile. Whether one is processing petrochemicals, caustic soda, acids, and so on, the fluid must usually be transferred from one point to another somewhere in the process. At a chemical plant, chemicals must be loaded or unloaded, sent to heat exchangers or cooling coils, transferred from one processing unit to another, or packaged for shipment.

To move material (either a fluid or slurry) through the various pieces of equipment at a facility (including piping and duct work) requires mechanical energy, not only to impart an initial velocity to the material, but more importantly, to overcome pressure losses that occur throughout the flow path. This energy may be imparted to the moving stream in one or more of three modes: an increase in the stream's velocity, an increase in the stream pressure, or an increase in stream height. In the first case, the additional energy takes the form of increased external kinetic energy as the bulk stream velocity increases. In the second, the internal energy (mainly potential energy, but usually some kinetic as well) of the stream increases. This pressure increase may also cause a stream temperature rise, which represents an internal

Fluid Flow for the Practicing Chemical Engineer. By J. Patrick Abulencia and Louis Theodore
Copyright © 2009 John Wiley & Sons, Inc.

energy increase. In the third case, which may be relatively small for some operations, the bulk fluid experiences an increase in external potential energy in the Earth's gravitational field.

Three devices which convert electrical energy into the mechanical energy that is to be applied to various streams are discussed in this chapter. These devices are fans, which move low pressure gases; pumps, which move liquids and liquid–solid mixtures such as slurries, suspensions, and sludges; and, compressors, which move high pressure gases.

There are three general process classifications of prime movers—centrifugal, rotary, and reciprocating—that can be selected. Except for special applications, centrifugal units are normally employed. Basically, a centrifugal unit consists of an impeller, which is a series of radial-vanes of various shapes and curvatures, spinning in a circular casing. Fluid enters the "eye" or axis of rotation and discharges radially into a peripheral chamber at a higher pressure that corresponds to the sum of the centrifugal force of rotation and the kinetic energy given to the fluid by the turning vanes. The only moving part in the unit is the impeller. The vanes of the impeller extend from the center of rotation to the periphery and the shrouds are the disks on each side of the vanes enclosing them. The vanes may be radial, may curve slightly "forward" (in the same direction as that of rotation), or may curve "backward," which is the usual case.

As indicated above, there are three classes of prime movers—fans, pumps, and compressors. The three sections that follow will treat each individually. These units are normally rated in terms of four characteristics:

1. Capacity: the quantity of fluid discharged per unit time.
2. Increase in pressure, often reported for pumps as head: head can be expressed as the energy supplied to the fluid per unit weight and is obtained by dividing the increase in pressure by the fluid density.
3. Power: the energy consumed by the mover per unit time.
4. Efficiency: the energy supplied to the fluid divided by the energy supplied to the unit.

The net effect of most prime movers is to increase the pressure of the fluid. However, as described earlier, some provide the fluid with an increase in kinetic energy (velocity) or an increase in potential energy (elevation) or both.

Noise generation is a problem common to all these devices. The noise level is strongly related to the rotative speed of the unit. Noise control can include:

1. Isolation.
2. Proper maintenance.
3. Encapsulation.
4. Piping insulation (to and from the unit).

Additional information is provided in the following section "Fans".

17.2 FANS

The term *fans* and *blowers* are often used interchangeably, and no distinction will be made between the two in the following discussion. Whatever is stated about fans applies equally to blowers. Strictly speaking, however, fans are used for low pressure (drop) operation, generally below 2 psi. Blowers are generally employed when generating pressure heads in the 2.0–14.7-psi range. Higher pressure operations require compressors.

Fans are usually classified as centrifugal or axial-flow type. In centrifugal fans (as noted earlier), the gas is introduced into the center of the revolving wheel (the eye) and discharges at right angles to the rotating blades. In axial-flow fans, the gas moves directly (forward) through the axis of rotation of the fan blades. Both types are used in industry, but it is the centrifugal fan that is employed at most facilities.

The gas in a centrifugal fan is subjected to the same centrifugal forces described earlier. These forces compress the gas giving it additional static pressure. Centrifugal fans are enclosed in a scroll-shaped housing that helps convert kinetic energy to static pressure. Gas rotating between the fan blades is compressed in the fan scroll, which increases the static pressure. Centrifugal fans are classified by blade configuration as not only radial, forward curved, backward curved, but also as air foil and radial tip-forward curved heel.

Radial or straight blade fans physically resemble a paddle wheel with long radial blades attached to the rotor and are the simplest design of all centrifugal fans. This enables most radial blade fans to be built with great mechanical strength and to be easily repaired. These fans can be used in a variety of situations, especially heavy duty applications. This type of fan can handle erosive and corrosive gases as well as very viscous gases. It is particularly well-suited for high static pressure operations and can generate pressures in excess of 50 in. H_2O. When operated properly, the horse-power efficiency range is 55–69%, with 65% as a typical value. *Forward curved* fans are the most popular for general ventilation purposes (high flow rates and low static pressures). These fans have both the heel and the tip of the blade curved forward in the direction of rotation. Blades are smaller and spaced much closer together than in other blade designs. They are generally not used with dirty gases when dust or sticky materials are present because contaminants easily accumulate on the blades and cause imbalance. Efficiencies range from 52–71%, with 65% being typical. *Backward curved* or backward inclined fans have blades inclined in a direction opposite to that of the direction of rotation. This feature causes the gas to leave the tip of the blade at a lower velocity than the wheel-tip speed, a factor that improves the mechanical efficiency. These types of fans are not suitable for a heavily particulate-laden gas, sticky material, or abrasive dust. Centrifugal forces tend to build up particulate matter on the backside of the fan blades. The airfoil is similar to the backward curved fan, except that the blade has been contoured to increase stability and operating efficiency. These fans are more expensive to construct than backward curved fans, but have lower power requirements. They are rarely used in air pollution control where the gas must be clean and noncorrosive. Since the blades are hollow, abrasion and wear could allow dust, water vapor, and so on, to enter the blade and cause imbalance. This is the most efficient of the various fan types, with typical efficiencies at the 85% level. A modification of the radial fan is the *radial tip-forward curved heel*. The blades are curved forward with this unit. This fan is reportedly more dependable than the radial or high tip speed applications (i.e., high static pressures) due to better vibrational characteristics and its ability to resist fatigue. It finds application in the processing of large flow rates (>200,000 acfm) with light to medium particulate loadings. However, sticky material and particulates, in general, can accumulate in the slight curvature of the blades and cause imbalance. Efficiencies here typically range from 52–74%, with 70% being common.

Generally, centrifugal fans are easier to control, more robust in construction, and less noisy than axial units. They have a broader operating range at their highest efficiencies. Centrifugal fans are better suited for operations in which there are flow variations and they can handle dust and fumes better than axial fans.

Fan laws are equations that enable the results of a fan test (or operation) at one set of conditions to be used to calculate the performance at another set of conditions, including differently sized but geometrically similar models of the same fan design. The fan laws can be written in many different ways. The three key laws are provided in the following equations:

$$q_a = k_1(\text{rpm})D^3 \tag{17.1}$$

$$P_s = k_2(\text{rpm})^2 D^2 \rho \tag{17.2}$$

$$\text{hp} = k_3(\text{rpm})^3 D^5 \rho \tag{17.3}$$

where q_a = volumetric flow rate
P_s = static pressure
hp = horsepower
rpm = revolutions per minute
D = wheel diameter
ρ = gas density
k_1, k_2, k_3 = proportionality constants

Thus, these three laws may be used to determine the effect of fan speed, fan size, and gas density on flow rate, developed static pressure head, and horsepower. For two conditions, where the constants k remain unchanged, Equations (17.1) and (17.3) become:

$$\left(\frac{q_a'}{q_a}\right) = \left(\frac{\text{rpm}'}{\text{rpm}}\right)\left(\frac{D'}{D}\right)^3 \tag{17.4}$$

$$\left(\frac{P_s'}{P_s}\right) = \left(\frac{\text{rpm}'}{\text{rpm}}\right)^2\left(\frac{D'}{D}\right)^2\left(\frac{\rho'}{\rho}\right) \tag{17.5}$$

$$\left(\frac{\text{hp}'}{\text{hp}}\right) = \left(\frac{\text{rpm}'}{\text{rpm}}\right)^3\left(\frac{D'}{D}\right)^5\left(\frac{\rho'}{\rho}\right) \tag{17.6}$$

Note: The *prime* refers to the new condition. It is also important to note that the fan laws are approximations and should not be used over wide ranges or changes of flow rate, size, etc.

Illustrative Example 17.1   Fan A has a blade diameter of 46 inches. It is operating at about 1575 rpm while transporting 16,240 acfm of flue gas and requires 47.5 brake horse power (bhp). Fan B is to replace Fan A. It is to operate at 1625 rpm with a blade diameter of 42 inches and is of the same homologous series as Fan A. What is the power requirement of Fan B?

Solution   The power requirement for Fan B is calculated, using the fan law, Equation (17.6):

$$\frac{hp_B}{hp_A} = \left(\frac{rpm_B}{rpm_A}\right)^3 \left(\frac{D_B}{D_A}\right)^5 \left(\frac{\rho_B}{\rho_A}\right) = \left(\frac{1625}{1575}\right)^3 \left(\frac{42}{46}\right)^5 (1) = 0.697$$

$$hp_B = (0.697)(47.5)$$
$$= 33.1 \, bhp$$

Illustrative Example 17.2   A fan operating at a speed of 1694 rpm delivers 12,200 acfm of flue gas at 5.0 in. H_2O static pressure and requires 9.25 bhp. What will be the new operating conditions if the fan speed is increased to 2100 rpm?

Solution   The new fan flow rate (superscript prime) is calculated using Equation (17.4):

$$q_a' = q_a \frac{rpm'}{rpm} \left(\frac{D'}{D}\right)^3 = (12,200)\frac{2100}{1694}(1)$$

$$= 15,124 \, acfm$$

Using Equation (17.5), the new static pressure is calculated:

$$P_s' = P_s \left(\frac{rpm'}{rpm}\right)^2 \left(\frac{D'}{D}\right)^2 \left(\frac{\rho'}{\rho}\right) = 5.0\left(\frac{2100}{1694}\right)^2 (1)(1)$$

$$= 7.68 \, in. \, H_2O$$

The required horsepower is calculated using Equation (17.6):

$$hp' = hp\left(\frac{rpm'}{rpm}\right)^3 \left(\frac{D'}{D}\right)^5 \left(\frac{\rho'}{\rho}\right) = 9.25\left(\frac{2100}{1694}\right)^3 (1)(1)$$

$$= 17.62 \, bhp$$

A rigorous, extensive treatment on fan selection is beyond the scope of this text. It is common practice among fan vendors to publish voluminous data in tabular form providing flow rate, static pressure, speed, and horsepower at a standard temperature and gas density. These are often referred to as *multirating tables*.

Note: These tables should not be used for fan selection except by those who have experience in this area. For those who do not, the proper course of action to follow is to provide the fan manufacturer with a complete description of the system and allow the manufacturer to select and guarantee the optimum fan choice.

To help in the actual selection of fan size, a typical fan rating table is given in Table 17.1. The fan size and dimensions are usually listed at the top of the table. Values of static pressure are arranged as columns that contain the fan speed and brake horsepower required to produce various volume flows. The point of maximum efficiency at each static pressure is usually underlined or printed in special type. In order to select a fan for the exact condition desired, it is sometimes necessary to interpolate between values presented in the multirating tables. Straight-line interpolation can be used with negligible error for multirating tables based on a single fan size. Some multirating tables attempt to show ratings for a whole series of geometrically similar (homologous) fans in one table; in this case, interpolation is not advised.[1]

The selection procedure is, in part, an examination of the fan curve and the system curve. A fan curve, relating static pressure with flow rate, is provided in Fig. 17.1. Note that each type of fan has its own characteristic curve. Also note that fans are usually tested in the factory or laboratory with open inlets and long smooth straight discharge ducts. Since these conditions are seldom duplicated in the field, actual operation often results in lower efficiency and reduced performance. A system curve is also shown in Fig. 17.1. This curve is calculated prior to the purchase of the fan and provides a best estimate of the pressure drop across the system through which the fan must deliver the gas. (This curve should approach a straight line with an approximate slope of 1.8 on log–log coordinates.) The system pressure (drop) is defined as the resistance through ducts, fittings, equipment, contractions, expansions, etc.

A number of methods are available to estimate the total system pressure change. These vary from very crude approximations to detailed, rigorous calculations. The simplest procedure is to obtain estimates of the pressure change associated with the movement of the gas through all of the resistances described previously. The sum of these pressure changes represents the total pressure drop across the system, and represents the total pressure (change) that must be developed by the fan. This calculation becomes more complex if branches (i.e., combining flows) are involved. However, the same stepwise procedure should be employed.

When the two curves are superimposed, the intersection is defined as the *point of operation*. The fan should be selected so that it operates just to the right of the peak on the fan curve. The fan operates most efficiently and with maximum stability at this condition. If a fan is selected for operation too close to its peak, it will surge and oscillate. Thus, the point of intersection of the two curves determines the actual volumetric flow rate. If the system resistance has been accurately specified

Table 17.1 Typical fan rating table[a–e]

	Static Pressure (in. H₂O)											
	$\frac{1}{2}$		1		$1\frac{1}{2}$		2		$2\frac{1}{2}$		3	
q_a (acfm)	rpm	bhp	rpm	bhp	rpm	bhp	rpm	bhp	rpm	bhp	rpm	bhp
5727	216	0.74	278	1.34	330	2.01	378	2.75	421	3.50	460	4.28
6873	236	0.97	291	1.67	339	2.42	384	3.24	424	4.06	462	4.91
8018	250	1.27	305	2.05	352	2.87	393	3.76	430	4.69	467	5.66
9164	271	1.63	320	2.49	366	3.42	405	4.35	441	5.36	475	6.40
10,309	293	2.12	338	3.05	381	4.06	419	5.06	453	6.14	486	7.26
11,455	315	2.72	356	3.65	396	4.76	432	5.88	468	7.03	499	8.19
12,600	337	3.46	377	4.39	413	5.58	448	6.81	482	8.04	514	9.31
13,746	360	4.39	399	5.25	430	6.48	465	7.82	496	9.16	527	10.53
14,891	382	5.43	421	6.25	451	7.52	481	8.93	512	10.35	542	11.80
16,037	405	6.66	442	7.48	473	8.67	501	10.16	529	11.69	557	13.25
17,182	429	8.08	463	8.97	496	10.05	521	11.54	547	13.18	574	14.81
18,328	451	9.64	486	10.61	517	11.61	543	12.99	566	14.78	591	16.49
19,473	474	11.46	510	12.47	539	13.47	565	14.85	587	16.56	610	18.35
20,619	497	13.51	532	14.55	560	15.60	587	16.82	610	18.54	630	20.40
21,764	520	15.82	556	16.82	584	17.98	608	19.17	632	20.77	652	22.59

Source: Bayler Blower Co.

[a]Wheel style: backward inclined
[b]Wheel diameter: $50\frac{1}{2}$ inches
[c]Maximum fan speed: 1134 rpm
[d]Performances underlined are those at maximum efficiency
[e]Brake horsepower = bhp

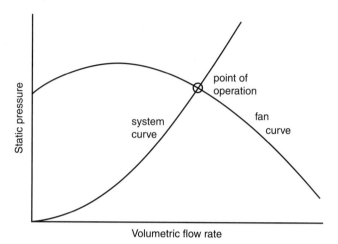

Figure 17.1 System and fan characteristic curves.

and the fan properly selected, the two performance curves will intersect at the design flow rate. If system pressure losses have not been accurately specified or if undesirable inlet and outlet conditions exist, design conditions will not be obtained. Dampers and fan speed changes can provide some variability on operating conditions.

There are a number of process and equipment variables that are classified as part of a fan specification. These include:

Flow rate (acfm)
Temperature
Density
Gas stream characteristics
Static pressure that needs to be developed
Motor type
Drive type
Materials of construction
Fan location
Noise controls (briefly discussed earlier)

With respect to drive type, belt drives are usually employed if the power is <200 hp. Direct drives are preferred for large horsepower systems. Direct drive units have lower maintenance costs and lower power transmission losses. However, the speed cannot be varied; if a speed change is required, the cost can be expensive. For materials of construction, mild carbon steel is commonly employed when treating dry air up to temperatures approaching 1000°F. Fiberglass may be used for corrosive conditions but the temperature should not exceed 250°F.

With respect to fan location, the fan may be located downstream or upstream of a particular piece of equipment in the process. Fans located downstream are referred to as *induced draft* or *negative pressure* fans. Leakage occurs into, rather than out of, the unit. Possible lower flow rates and lower temperatures may lead to a smaller fan and reduced operating costs. Equipment costs could be higher because of the need for heavier construction material when operating under negative (vacuum) pressure. Fans located upstream may have erosion problems due to possible particulate loading; in addition, there may be an accumulation of particulates. A larger volumetric flow rate is possible, which would require a larger fan and high horsepower costs.

With respect to noise controls, fans can create objectionable noise problems in the work area or neighboring residences. To minimize noise effects, the fan should, of course, be properly designed and properly operated. A fan should therefore be operated close to the point of maximum efficiency to reduce noise levels. Putting a fan in proper balance can also be an effective solution. If noise problems are expected or are already present, acoustical insulation should be applied to the fan housing. If the noise level is severe, insulation should also be added to the ductwork.[2]

Gas (or air) horsepower (ghp) and brake horsepower (bhp) are the two terms of interest. These may be calculated from Equations (17.7) and (17.8):

$$ghp = 0.0001575 q_a \Delta P = q_a \Delta P / 6356 \qquad (17.7)$$
$$bhp = 0.0001575 q_a \Delta P / \eta_f \qquad (17.8)$$

where q_a = volumetric flow rate (acfm)
   ΔP = SP = static pressure head developed (in. H_2O)
   η_f = fractional fan efficiency

For cold start-ups, the static pressure can be significantly higher than the normal operating pressure drop, and the fan must be able to handle the increased resistance. The correction for the cold static pressure may be calculated from Equation (17.9):

$$\Delta P_c = \Delta P(\rho_c / \rho) \qquad (17.9)$$

where ρ = gas density (lb/ft^3)
   c = subscript denoting cold gas

Illustrative Example 17.3   Calculate the hp required to process a 6500-acfm gas stream from a process. The pressure drop across various pieces of equipment has been estimated to be 6.4 in. H_2O. The pressure loss for duct work, elbows, valves, expansion–contraction losses, etc., are estimated at 4.4 in. H_2O. Assume an overall fan-motor efficiency of 63%.

Solution   The total pressure drop, ΔP (in. H_2O), is

$$\Delta P = (6.4 + 4.4) = 10.8 \text{ in. } H_2O$$

The brake horsepower required is calculated by Equation (17.7).

$$\text{bhp} = \frac{(1.575 \times 10^{-4})q_0 \Delta P}{\eta_f} = \frac{(1.575 \times 10^{-4})(6500)(10.8)}{(0.63)}$$

$$= 17.55\,\text{bhp}$$

Note that the term 1.575×10^{-4} is a conversion factor to obtain units of horsepower.

17.3 PUMPS

Pumps are required to transport liquids, liquid–solid mixtures such as slurries and sludges, auxiliary fuel, etc. Pumps are also needed to transport water to and/or from such peripheral devices as boilers, quenchers, scrubbers, and so on.

As indicated earlier, pumps may be classified as reciprocating, rotary, or centrifugal. The reciprocating and rotary types are referred to as *positive displacement* pumps because unlike the centrifugal type, the liquid or semiliquid flow is broken up into small portions as it passes through the pump. These three classes of units are described below.

Reciprocating pumps operate by the direct action of a piston on the liquid contained in a cylinder. As the liquid is compressed by the piston, the higher pressure forces it through discharge valves to the pump outlet. As the piston retracts, the next batch of low pressure liquid is drawn into the cylinder and the cycle is repeated. The piston may be either directly steam driven or moved by a rotating crankshaft through a crosshead. The rate of liquid delivery is a function of the volume swept out by the piston and the number of strokes per unit time. A fixed volume is delivered for each stroke but the actual delivery may be less because of both leakage past the piston and failure to fill the cylinder when the piston retracts. The volumetric efficiency of the pump is defined as the ratio of the actual volumetric discharge to the pump displacement. For well maintained pumps, the volumetric efficiency is at least 95%. Reciprocating pumps are used for some applications.

Reciprocating pumps can deliver the highest pressure of any type of pump (20,000 psig); however, their capacities are relatively small compared to the centrifugal pump. Also, because of the nature of the operation of the reciprocating pump, the discharge flow rate tends to be somewhat pulsating. Liquids containing abrasive solids can damage the machined surfaces of the piston and cylinder. Because of its positive displacement operation, reciprocating pumps can be used to measure liquid volumetric flow rates.

The *rotary* pump combines rotation of the liquid with positive displacement. The rotating elements mesh with elements of the stationary casing in much the same way that two gears mesh. As the rotating elements come together, a pocket is created that first enlarges, drawing in liquid from the inlet or suction line. As rotation continues, the pocket of liquid is trapped, reduced in volume, and then forced into the discharge line at a higher pressure.

The flow rate of liquid from a rotary pump is a function of size and speed of rotation and is slightly dependent on the discharge pressure. Unlike reciprocating pumps, rotary pumps deliver nearly constant flow rates. Rotary pumps are used on liquids of almost any viscosity as long as the liquids do not contain abrasive solids. For this reason, they are very effective with many high viscosity mixtures. They operate in moderate pressure ranges (5000 psig), have small-to-medium capacities, and like the reciprocating pump, can be used for metering liquids.

Centrifugal pumps are the most widely used in the process industry because of simplicity of design, low initial cost, low maintenance, and flexibility of application. Centrifugal pumps have been built to move as little as a few gallons per minute against a small head, and as much as several thousand gallons per minute against a pressure of several hundred pounds force per square inch (psi). In its simplest form, this type of pump consists of an impeller rotating within a casing. Fluid enters the pump near the center of the rotating impeller and is thrown outwards by centrifugal force. The kinetic energy of the fluid increases from the center of the impeller to the tips of the impeller vanes. This high velocity is converted to a high pressure as the fast-moving fluid leaves the impeller and is driven into slower moving fluid in the volute or diffuser.[3]

Not all centrifugal pumps produce the radial flow (directed away from the axis of rotation) described previously; many produce an axial flow (directed along the axis of rotation) and others a combination of the two. The turbine type of centrifugal pump has smaller, straighter vanes, is driven at high speed, and generates a highly radial flow to produce higher pressures at lower flow rates. The axial-flow type employs multibladed propellers that generate a highly axial flow, resulting in large flow rates at lower pressures. For a specific balance between flow rate and pressure, impellers can be shaped to provide results between that of the turbine and axial-flow type; these are referred to as *mixed-flow* impellers.

The impeller is the heart of the centrifugal pump. It consists of a number of curved vanes or blades that are shaped in such a way as to give smooth fluid flow between the blades. In the *straight-vane, single-suction closed* impeller, the surfaces of the vanes are defined by straight lines parallel to the axis of rotation. The *double-suction* impeller is, in fact, two single-suction impellers arranged back-to-back in a single casing. Centrifugal pump casings may be of several designs but their main function is to convert kinetic energy imparted to the fluid by the impeller into a higher pressure. In addition, the casing provides an inlet and an outlet for the pump and contains the fluid. Casings may be either the *volute* or the *diffuser* type. The *volute* type has a continuous flow area that allows the velocity to decrease gradually, thereby reducing eddy formation; this minimizes the loss of energy due to turbulence. The *diffuser*-type casing has stationary guides that offers the liquid a widening path from impeller to casing; this also keeps turbulence to a minimum.[3]

In pumping fluids at a facility, the operation of a pump may result in either the loss of liquid or the release of air contaminants, or both. The reciprocating and centrifugal pumps can also be sources of hazardous emissions. The opening in the cylinder through which the connecting rod drives the piston is the major source of contaminant release from a reciprocating pump. In centrifual pumps, normally the only potential source of leakage occurs where the drive shaft passes through the casing.

Several methods have been devised for sealing the clearance between the pump shaft and fluid casing. For most applications, packed seals and mechanical seals are widely used. Packed seals can be used on both positive displacement and centrifugal pumps. A typical packed seal consists of a stuffing box filled with a sealing material or packing that encases the moving shaft. The stuffing box is fitted with a take-up ring that compresses the packing and causes it to tighten around the shaft. Materials used for packing vary with the fluid's temperature, physical and chemical properties, pressure, and pump type. Some commonly used materials are metal, rubber, leather, and plastics. For cases where the use of mechanical seals is not feasible, specialized pumps such as canned-motor, diaphragm, or electromagnetic pumps are used. These specialty pumps are used where no leakage can be tolerated and are available in a limited range of sizes; most are for low flow rates and all are of single- or two-stage construction. They have been used to handle both high temperature and very low temperature liquids. These pumps follow the same hydraulic principles as the traditional centrifugal pump. Because of their small size, they operate with rather low efficiencies, but in dangerous applications, efficiency must be sacrificed for safety.[4]

Illustrative Example 17.4   A pump produces 25 kPa pressure when a valve is closed to shut down the flow completely, and 5 kPa when the valve is opened to allow a flow rate of $2\,\mathrm{m}^3/\mathrm{s}$. Assuming the pressure-flow rate curve is parabolic, what is the pressure when the flow rate is $1\,\mathrm{m}^3/\mathrm{s}$?

Solution   The parabolic pump pressure-flow rate curve can be written as:

$$P = a - bq^2$$

Enter the given data to obtain two equations for the constants a and b.

$$25 = a - b(0)$$
$$5 = a - b(2)^2$$

Solve the equations simultaneously for a and b.

$$a = 25$$
$$5 = a - 4b$$
$$b = \frac{a-5}{4} = \frac{25-5}{4} = 5$$

Write the equation for the pressure P with the values for the constants.

$$P = 25 - 5q^2$$

Substitute the known value for q and compute P.

$$P = 25 - 5(1)^2$$
$$= 20 \, \text{kPa}$$

Pumps are employed in industry for one and/or a combination of the following reasons:

1. The transportation of liquids from trucks or delivery vehicles to storage tanks.
2. The transportation of liquids from storage tanks to process units.
3. The transportation of slurries and/or sludges to/from storage tanks to/from process units.
4. The transportation of liquids from one process unit to another.
5. The delivery of water to the quench units and heat boilers (where applicable).
6. The delivery and circulation of water and/or caustic solutions in absorbers and scrubbers (where applicable).

Pump discharge pressures for (2) and (3) are roughly 75 and 50 psig, respectively. Actual discharge pressures will vary, however, with fluid viscosity and line pressure drop (which is a function of site-specific logistics).

A general relationship that may be used to determine pump hp requirements is[4]

$$\text{hp} = \frac{7.27 \times 10^{-5} \dot{m} \Delta P}{\rho \times \eta_p} \quad (17.10)$$

where $\dot{m}$ is the liquid flow rate, ΔP is the developed pressure drop, ρ is the liquid density and η_p is the pump-motor efficiency (on a fraction basis).

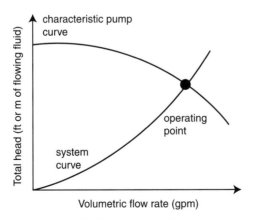

Figure 17.2 Pump characteristic curve.

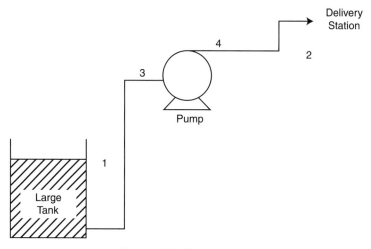

Figure 17.3 Pump system.

The pressure increase (head) developed by a centrifugal pump is generally a function of the discharge rate. In a real pump, as the flow rate becomes large, the total head delivered by the pump decreases. To obtain the discharge rate for a given system, the engineer has to solve, simultaneously, the system curve of head, h_p, versus discharge with the pump curve head, h_c. The system h_p is usually obtained by applying Bernoulli's equation (or the equivalent). The pump characteristic curve of h_c is usually supplied by the manufacturer. The point where the two curves intersect (as indicated earlier—see Fig. 17.1) yields the operating conditions for the system, as shown in Fig. 17.2.

Consider the flow of fluid in Fig. 17.3. The liquid is pumped from the large tank (point 1) to the delivery station (point 2). Applying a modified form of Bernoulli's equation to the liquid system between stations 1 and 2 gives:

$$\frac{P_1}{\rho} + \frac{v_1^2}{2g_c} + z_1 \frac{g}{g_c} = \frac{P_2}{\rho} + \frac{v_2^2}{2g_c} + z_2 \frac{g}{g_c} - h_s + h_f$$

or

$$h_s = \frac{P_2 - P_1}{\rho} + \frac{v_2^2 - v_1^2}{2g_c} + (z_2 - z_1)\frac{g}{g_c} + h_f \qquad (17.11)$$

where h_s is the input work (pump head) and h_f is the head loss due to fluid friction, with both treated as positive numbers. The units of each term as written in the above equation are energy/mass.

For any flow rate, the pump head required to maintain that flow is calculated from Equation (17.11). In this procedure, one constructs a head-versus-discharge curve. This curve of h_p vs q is termed the system curve. For a given centrifugal pump, the pump manufacturer supplies a characteristic head-versus-discharge curve. Once again, where the two curves intersect is the operating point of the pump.

When a centrifugal pump is operating at high flow rates, the high velocities occurring at certain points in the eye of the impeller or at the vane tips cause local pressures to fall below the vapor pressure of the liquid. This is predicted by the Bernoilli equation. Vaporization occurs at these points, forming bubbles that collapse violently upon moving to a region of higher pressure or lower velocity. This momentary vaporization and destructive collapse of the bubbles is called cavitation. This should be avoided if maximum capacity is to be obtained and damage to the pump prevented. The shock of bubble collapse causes severe pitting of the impeller and creates considerable noise and vibration. Cavitation may be reduced or eliminated by reducing the pumping rate or by slight alterations in the impeller design to give better streamlining. Note, however, that cavitation usually does not occur at low flow rates on any given pump.[5]

Both vapor binding and cavitation may be eliminated by maintaining a pressure at the pump inlet that is significantly higher than the vapor pressure of the liquid being pumped. The required margin of pressure is called the net positive suction head (NPSH). It is a function of the pump design and is usually specified by the manufacturer for pumps that handle liquids such as preheated boiler feed, steam condensate, or volatile liquids. Usually, the NPSH is of the order of 2 to 10 ft-lb$_f$/lb of fluid and increases with increasing throughput.[5]

From an engineering point-of-view, the question that needs answering is will the pump work or will it cavitate. The NPSH determines the answer. Cavitation will not occur if the sum of the pressure plus the velocity and dynamic heads in the suction line is greater than the vapor pressure head. The NPSH is therefore defined as:

$$\text{NPSH} = \frac{P_i g_c}{\rho g} + \frac{v_i^2}{2g} - \frac{p' g_c}{\rho g} \tag{17.12}$$

where P_i and v_i are the pressure and velocity at the pump inlet and p' is the liquid vapor pressure. If the pump inlet is at a height, z_i, and if the reservoir free surface is at pressure, P_{ac}, and height, z_a, one may use Bernoulli's equation to rewrite NPSH as:

$$\frac{P_a g_c}{\rho g} + z_a = \frac{P_i g_c}{\rho g} + \frac{v_i^2}{2g} + z_i + h_{fi}' \tag{17.13}$$

where h_{fi}' is the friction head loss between the reservoir and the pump inlet, with the prime representing units of height of flowing fluid. Substituting from Equation (17.13) into Equation (17.12) gives:

$$\text{NPSH} = \left(\frac{P_a - p'}{\rho}\right) \frac{g_c}{g} + (z_a - z_i) - h_{fi}' \tag{17.14}$$

Pump manufacturers will usually specify the NPSH requirement of the pump. The reader should note that the units change from energy/mass to height of flowing fluid by multiplying each of the terms in the energy equations by (g/g_c).

The sensitivity of a device to cavitation occurrence is measured by the dimensionless cavitation number, Ca. It is the ratio of available pressure above the vapor pressure to the *dynamic pressure* of the flow. When a fluid moves at a velocity, v, its dynamic pressure, P_{dyn}, is defined as the kinetic energy of flow per unit volumetric flow rate of the fluid, i.e.,

$$P_{dyn} = \frac{\rho v^2}{2g_c} \tag{17.15}$$

Therefore, the cavitation number, Ca, is expressed by the following equation:

$$Ca = \frac{P_a - p'}{P_{dyn}} = \frac{P_a - p'}{\rho v^2 / 2g_c} \tag{17.16}$$

where P_a is the ambient pressure, p' is the vapor pressure, v is the fluid velocity, and ρ is the fluid density. In some liquid flow applications, a pump is placed above the liquid being pumped. The pump has to draw in the liquid in order to pump it out. It is highly undesirable to have the pump drawing vapor rather than the liquid. An example is the use of a pump to draw water from a well. Clearly, the vapor pressure of a liquid limits the maximum height above a sump at which a pump may be located.

There are seven items of process data that must be specified for pumps.[6] The first four are basic, and can be obtained from the process flowsheet (which should include a heat and material balance) or from tables of physical properties. The basic four process items are:

1. Fluid being pumped
2. Operating temperature
3. Specific gravity, both at 60°F and at operating conditions
4. Viscosity at operating conditions—this is particularly important for positive displacement pumps

The next three items usually require calculation. These are:

5. Capacity
6. Head (or pressure)
7. Net Positive Suction Head (NPSH)

Centrifugal pumps are very widely used in industry. The use of a pump is required any time a liquid needs to overcome pressure or static constraints. Pump applications range from the small-scale every-day uses as those in household piping systems to large machinery, plants, and pipelines. Pumps can be found as integral parts of systems in laboratories, automobiles, and anything else that necessitates a liquid flow system.

Centrifugal pumps are dependable and long-lasting. They can tolerate internal corrosion and erosion without substantially decreasing performance. Flexible operation provides good flow-control characteristics over a wide capacity range at constant speed. Capacities can be conveniently varied by throttling the discharge. These pumps are usually quiet, need little attention, and operate pulsation free. Due to the long life of the pumps, maintenance costs are also low. They can be easily disassembled, few parts have close tolerances, and worn parts can be quickly replaced. Centrifugal pumps pay for themselves in a short time. The low investment cost is due to simple designs, direct-coupled motor arrangements, and a wide range of material, size, performance and operating characteristics. Such pumps occupy small space without shelter. Piping arrangements are usually simple.

Pumps rarely influence plant layout except where a common standby for two services might require the rearrangement of process equipment. Pumps are generally placed close to process vessels and arranged esthetically.

On the negative side, capacity, head, and efficiency rapidly decrease with increasing viscosity. In addition, the centrifugal pump normally cannot transfer liquid having vapor content. At low flows (below 15–20% of design capacity), the centrifugal pump can become unstable. Thus, a minimum flow is often specified.

Pumps are selected by specialists and the piping designer has little influence on the basic selection. However, the layout designer can request preferred orientation for suction and discharge and the NPSH limitations (to meet required equipment elevations).[7]

17.3.1 Parallel Pumps

Pumps can also be set up in parallel. This configuration is used to administer systems where higher capacity is desired. Pumps in parallel work as two single pumps with merging outlets. In addition to creating greater capacity, pumps in parallel can be helpful for systems that cannot be shut down for long periods of time, as each pump works independently of the other.

The calculations used for pumps in parallel are much like those used for pumps in series. The capacity and volumetric flow rate of the pair of pumps are found in the same way as that of the single pump and series pumps. The total head of the system is simply the greater head produced from each of the two pumps; these values are calculated as the head is calculated for a single pump. The power and total efficiency are calculated in the same way as those for two pumps in series.

Illustrative Example 17.5 It is necessary to deliver a liquid with density and viscosity similar to that of water to a process unit at a rate of 80 gal/min. The pump must operate against a pressure of 180 psi. An external gear pump with characteristics shown in Fig. 17.4 is available with variable speed drive. At what speed should the pump be operated? What horsepower is needed to maintain flow?

Solution Plotting this information on Fig. 17.4 indicates that the speed is between the 400 and 600 rpm lines (point A). Interpolation gives a speed of about 425 rpm. Interpolating again on the hp curve gives a value of about 17 hp (point B). Using this data with Equation (17.11), an efficiency of 50% results in a pump with a horsepower of approximately 15 hp.

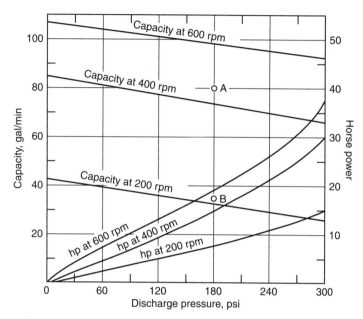

Figure 17.4 Performance characteristics of an external gear pump.

Illustrative Example 17.6 The characteristic curve, h'_c, of a centrifugal pump is approximated as

$$h'_c = 42 - 0.0047q^2$$

where h'_c is the total (or developed) head in ft of water and q is the liquid discharge in gpm. The pump efficiency, η, is about 60% and is independent of the discharge rate. The pump is to be used in a water flow system in which the pump head in feet of water is calculated as:

$$h'_p = 12 + 0.0198q^2$$

The density of water is 62.4 lb/ft³. Determine the water flow rate, the total pump head, the ideal (theoretical or fluid) pump power and the brake horsepower of the pump.

Solution The equation to be solved for the flow rate is obtained by equating h_c and h_p

$$h'_c = h'_p$$

$$42 - 0.0047q^2 = 12 + 0.0198q^2$$

$$30 = 0.0245q^2$$

$$q = 35 \text{ gpm} = 0.078 \text{ cfs}$$

Calculate the pump head

$$h'_c = 42 - 0.0047(35)^2$$
$$= 36.24 \text{ ft of water}$$

Calculate the mass flow rate of water

$$\dot{m} = \rho q = 62.4(0.078)$$
$$= 4.87 \text{ lb/s}$$

The fluid power requirement, $\dot{W}_{s,id}$, may now be calculated

$$\dot{W}_{s,id} = \dot{m} h'_c = 4.87(36.24) = 176 \text{ lbf} \cdot \text{ft/s}$$
$$= 0.32 \text{ hp}$$

Finally, the brake horsepower, $\dot{W}_s$, or bhp is

$$\dot{W}_s = \text{bhp} = \frac{\dot{W}_{s,id}}{\eta_p} = \frac{0.32}{0.6}$$
$$= 0.53 \text{ hp}$$

Illustrative Example 17.7   As indicated in pump flow, the lowest pressure usually occurs at the pump inlet. For a pump handling a known liquid at a specified temperature, determine the inlet absolute pressure (in Pa, lb_f/ft^2, and psi) that is liable to cause cavitation, given that the fluid is water at 40°C and 1 atm.

Solution   The vapor (or saturation) pressures of several common liquids at 20°C are listed in Table A.2 (in the Appendix). The vapor pressure of water at different temperatures is listed in Table A.4 in the Appendix.

Obtain the vapor pressure of water at 40°C from Table A.4.

$$p' = 7.375 \text{ kPa} = 0.0728 \text{ atm} = 1.069 \text{ psi} = 153.97 \text{ psf}$$
$$= 0.752 \text{ m of water} = 2.47 \text{ ft of water}$$

Noting that atmospheric pressure is equivalent to 33.91 ft of water, then the pump elevation should not exceed 33.91 ft. Since the water vapor pressure at 40°C is 2.47 ft of water, then the maximum elevation at which the pump may be placed is $(33.91 - 2.47) = 31.44$ ft (9.58 m); otherwise the pump will draw water vapor instead of liquid water. Therefore an inlet pressure greater than 31.44 ft of water is liable to cause cavitation.

17.4 COMPRESSORS

Compressors, unlike fans and pumps, find only limited and specialized application. They are primarily employed to increase the pressure of a fluid in some types of systems, e.g., when liquids are broken up into tiny droplets (atomized) before entering a unit. This can be accomplished through the use of a high pressure stream of air or steam that impinges on the liquid stream and atomizes it. The pressurizing of the air or steam is accomplished through the use of compressors. Compressors are also used for atomization on certain types of air pollution control devices such as venturi scrubbers, which depend on fine water droplets to remove particulates from the flue gas stream.

Compressors operate in a similar fashion to pumps and have the same classifications: rotary, reciprocating, and centrifugal. An obvious difference between the two operations is the large decrease in volume resulting from the compression of a gaseous stream compared to the negligible change in volume caused by the pumping of a liquid stream.

Centrifugal compressors are employed when large volumes of gases are to be handled at low to moderate pressure increases (0.5–50 psi). *Rotary* compressors have smaller capacities and can achieve discharge pressures up to 100 psi. *Reciprocating* compressors are the most common type used in industry and are capable of compressing small gas flows to as much as 3500 psig. With specially designed compressors, discharge pressures as high as 25,000 psig can be reached, but these devices are capable of handling only very small volumes, and do not work well for all gases. For the applications mentioned earlier, atomizing of liquids for combustion or of venturi scrubber water for gas cleaning, reciprocating compressors are normally used.

The following equation may be used to calculate compressor power requirements when the compressor operation is adiabatic and the gas (usually air) follows ideal gas behavior[2]

$$W_s = \left(\frac{\gamma RT}{\gamma - 1}\right)\left[\left(\frac{P_2}{P_1}\right)^{(\gamma-1)/\gamma} - 1\right] \qquad (17.17)$$

where W_s = compressor work required per lbmol of air
R = 1.987 Btu/(lbmol · °R)
T = air temperature at compressor inlet conditions (°R)
P_1, P_2 = air inlet and discharge pressures
γ = ratio of the heat capacity at constant pressure to that at constant volume—typically 1.3 for air.

Illustrative Example 17.8 Compressed air is to be employed in the nozzle to assist the atomization of a liquid. The air requirement for the nozzle is 7.5 lb/min at 40 psia. If atmospheric air is available at 60°F and 1.0 atm, calculate the power requirement.

Solution The ideal gas law will apply at these conditions. Set the coefficient γ for air equal to 1.3. The compressed energy requirement (delivered to the air) is given by Equation (17.17):

$$W_s = -\frac{\gamma R T_1}{\gamma - 1}\left[\left(\frac{P_2}{P_1}\right)^{(\gamma-1)/\gamma} - 1\right] = -\frac{(1.3)(1.987)(520)}{1.3 - 1}\left[\left(\frac{40}{14.7}\right)^{(1.3-1)/1.3} - 1\right]$$

$$= -1163\,\text{Btu/lbmol of air}$$

The power is

$$hp = -(1163)(7.5/29)(778)$$
$$= -234.000\ \text{ft-lb}_f/\text{min}$$

This result may be divided by 33,000 to yield a hp requirement of 7.1 hp. The power required has the opposite sign. The reader should also note that this represents the *minimum* power required to accomplish this job.

The temperature ratio across a compression stage is

$$T_2/T_1 = (P_2/P_1)^{(k-1)/k} \quad \text{adiabatic operation} \tag{17.18}$$

$$T_2/T_1 = (P_2/P_1)^{(N-1)/N} \quad \text{polytropic operation} \tag{17.19}$$

where $k =$ adiabatic exponent, C_p/C_v
$N =$ polytropic exponent, $(N - 1)/N = (K - 1)/KE_p$
$P_1, P_2 =$ suction, discharge pressures, psia
$T_1, T_2 =$ suction, discharge temperatures, °R
$E_p =$ polytropic efficiency, fraction

It is normally assumed that the usual centrifugal compressor is uncooled internally and thus follows a polytropic path. The temperature of the gas is often limited to protect against temperature excursions. Intercooling can be employed to retain temperatures at reasonable levels during high overall compression ratio applications.

REFERENCES

1. T. Shen, Y. Choi-McGuinn, and L. Theodore, "Hazardous Waste Incineration Manual," USEPA Air Pollution Training Institute, Research Triangle Park, NC, 1988.

2. J. Santoleri, J. Reynolds, and L. Theodore, "Introduction to Hazardous Waste Incineration," 2nd edition., John Wiley and Sons, Hoboken, NJ, 2000.

3. A. Foust, L. Wenzel, C. Chung, L. Maus, and L. Andrews, "Principles of Unit Operations," John Wiley & Sons, Hoboken, NJ, 1950.

4. USEPA, "Engineering Handbook for Hazardous Waste Incineration," RTP, NC Research Group, USEPA Contract No. 68-03-3025, 1981.

5. G. Brown and Associates, "Unit Operations," John Wiley & Sons, Hoboken, NJ, 1950.

6. A. Younger and J. Ruiter, "Selecting Pumps and Compressors," Chemical Engineering, New York, June 26, 1981.

7. R. Kern, "How to get the Best Process-plant Layouts for Pumps and Compressors," Chemical Engineering, New York, December 5, 1977.

NOTE: Additional problems are available for all readers at www.wiley.com. Follow links for this title.

18

VALVES AND FITTINGS

As indicated in Chapter 13, pipes, tubing and other conduits are used for the transportation of gases, liquids and slurries. These ducts are often connected and may also contain a variety of valves and fittings, including expansion and contraction joints. Types of connecting conduits include:

1. Threaded.
2. Bell-and-spigot.
3. Flanged.
4. Welded.

Extensive information on these classes of connections is available in the literature.[1]

Details on valves, fittings, and changes in cross-sectional area are provided below.

18.1 VALVES[2]

Because of the diversity of the types of systems, fluids, and environments in which valves must operate, a vast array of valve types have been developed. Examples of the common types are the globe valve, gate valve, ball valve, plug valve, pinch valve, butterfly valve, and check valve. Each type of valve has been designed to meet specific needs. Some valves are capable of throttling flow, other valve types can only stop flow, others work well in corrosive systems, and others handle high pressure fluids. Each valve type has advantages and disadvantages. Understanding these differences and how they effect the valve's application or operation is necessary for the successful operation of a facility.

Valves have two main functions in a pipeline: to control the amount of flow or to stop the flow completely. There are many different types of valves; the most commonly used are the gate valve and the globe valve. The gate valve contains a disk that slides at right angles to the flow direction. This type of valve is used primarily for on–off control of a liquid flow. Because small lateral adjustments of the disk can cause extreme changes in the flow cross-sectional area, this type of valve is not suitable for accurately adjusting flow rates. As the fluid passes through the gate valve, only a small amount of turbulence is generated; the direction of flow is not altered and the flow cross-sectional area inside the valve is only slightly smaller than that of the pipe. As a result, the valve causes only a minor pressure drop. Problems with abrasion and erosion of the disk arise when the valve is used in positions other than fully open or fully closed.

Unlike the gate valve, the globe valve—so called because of the spherical shape of the valve body—is designed for more sensitive flow control. In this type of valve, the liquid passes through the valve in a somewhat circuitous route. In one form, the seal is a horizontal ring into which a plug with a slightly beveled edge is inserted when the stem is closed. Good control of flow is achieved with this type of valve, but at the expense of a higher pressure loss than a gate valve.

Stop valves are used to shut off or, in some cases, partially shut off the flow of fluid. Stop valves are controlled by the movement of the valve stem. Stop valves can be divided into four general categories: globe, gate, butterfly, and ball valves. Plug valves and needle valves may also be considered stop valves.

Ball valves, as the name implies, are stop valves that use a ball to stop or start the flow of fluid. The ball performs the same function as the disk in the globe valve. When the valve handle is operated to open the valve, the ball rotates to a point where the hole through the ball is in line with the valve body inlet and outlet. When the valve is shut, which requires only a 90-degree rotation of the handwheel for most valves, the ball is rotated so the hole is perpendicular to the flow openings of the valve body and flow is stopped.

A plug valve is a rotational motion valve used to stop or start fluid flow. The name is derived from the shape of the disk, which resembles a plug. The simplest form of a plug valve is the petcock. The body of a plug valve is machined to receive a tapered or cylindrical plug. The disk is a solid plug with a bored passage at a right angle to the longitudinal axis of the plug. In the open position, the passage in the plug lines up with the inlet and outlet ports of the valve. When the plug is turned 90° from the open position, the solid part of the plug blocks the ports and stops fluid flow.

The relatively inexpensive pinch valve is the simplest in any valve design. Pinch valves are suitable for on–off and throttling services. However, the effective throttling range is usually between 10% and 95% of the rated flow capacity. Pinch valves are ideally suited for the handling of slurries, liquids with large amounts of suspended solids, and systems that convey solids pneumatically. Because the operating mechanism is completely isolated from the fluid, these valves also find application where corrosion or metal contamination of the fluid might be a problem.

The pinch control valve consists of a sleeve molded of rubber or other synthetic material and a pinching mechanism. All of the operating portions are completely external to the valve.

A butterfly valve is a rotary motion valve that is used to stop, regulate, and start fluid flow. Butterfly valves are easily and quickly operated because a 90° rotation of the handle moves the disk from a fully closed to a fully opened position. Larger butterfly valves are actuated by hand wheels connected to the stem through gears that provide mechanical advantage at the expense of speed. Butterfly valves possess many advantages over gate, globe, plug, and ball valves, especially for large valve applications. Savings in weight, space, and cost are the most obvious advantages. The maintenance costs are usually low because there are a minimal number of moving parts and there are no pockets to trap fluids.

Finally, check valves are designed to prevent the reversal of flow in a piping system. These valves are activated by the flowing material in the pipeline. The pressure of the fluid passing through the system opens the valve, while any reversal of flow will close the valve. Closure is accomplished by the weight of the check mechanism, by back pressure, by a spring, or by a combination of these means. The general types (classification) of check valves are swing, tilting-disk, piston, butterfly, and stop.

Valves are also sometimes classified according to the resistance they offer to flow. The low resistance class of valves includes the straight-through flow units; for example, gate, ball, plug, and butterfly valves. Valves having a change in direction are high resistance valves; examples include globe and angle valves.

18.2 FITTINGS[2]

A fitting is a piece of equipment that has for its function one or more of the following:

1. The joining of two pieces of straight pipe (e.g., couplings and unions).
2. The changing of pipeline direction (e.g., elbows and Ts).
3. The changing of pipeline diameter (e.g., reducers and bushings).
4. The terminating of a pipeline (e.g., plugs and caps).
5. The joining of two streams (e.g., Ts and Ys).

A coupling is a short piece of pipe threaded on the inside and used to connect straight sections of pipe with no change in direction or size. When a coupling is opened, a considerable amount of piping must usually be dismantled. A union is also used to connect two straight sections but differs from a coupling in that it can be opened conveniently without disturbing the rest of the pipeline. An elbow is an angle fitting used to change flow direction, usually by 90°, although 45° elbows are also available. In addition, a T (shaped like the letter T) can be used to change flow direction; this fitting is more often used to combine two streams into one; that is, when two branches of piping are to be connected at the same point. A reducer is a coupling for two pipe sections of different diameter. A bushing is also a connector for pipes of different diameter, but, unlike the reducer coupling, is threaded on both the inside and outside. The larger pipe screws onto the outside of the bushing and the smaller pipe screws into the inside of the bushing. Plugs, which are threaded on the outside, and caps, which are threaded on the inside, are used to terminate a pipeline. Finally, a Y (shaped like the letter Y) is similar to the T and is used to combine two streams.

Fittings may be classified as reducing, expanding, branching or deflecting. Reducing or expanding fittings are ones that change the area for flow; these include reducers, bushings, and sudden expansions and contractions. Branch fittings are Ts, crosses, or side outlet elbows. Deflecting fittings change the direction of flow, for example, elbows and bends.

18.3   EXPANSION AND CONTRACTION EFFECTS

If the cross-section of a conduit enlarges gradually so that the flowing fluid velocity does not undergo any disturbances, energy losses are minor and may be neglected. However, if the change is sudden, as in a rapid expansion, it can result in additional friction losses. For such sudden enlargement/expansion situations, the exit loss can be represented by

$$h_{f,e} = \frac{v_1{}^2 - v_2{}^2}{2g_c}; \quad e = \text{sudden expansion, SE} \tag{18.1}$$

where $h_{f,e}$ is the loss in head, v_2 is the velocity at the larger cross-section and v_1 is the velocity at the smaller cross-section. When the cross-section of the pipe is reduced suddenly (a contraction), the loss may be expressed by:

$$h_{f,c} = \frac{Kv_2{}^2}{2g_c}; \quad c = \text{sudden contraction, SC} \tag{18.2}$$

where v_2 is the velocity in the small cross-section and K is a dimensionless loss coefficient that is a function of the ratio of the two cross-sectional areas. Both of the above calculations receive additional treatment in the next section.

In addition to specifying the pipe, fitting and (where applicable) flange materials required, it is also necessary to specify the nuts and bolts and gasket material to be used in most joints. Information is available on steel nuts and bolts for various pressure and temperature services.[3] The choice of gaskets is wide and in general is determined by the material being handled in the pipe line; there are no generalized suggestions. The thickness of any insulation required must be considered in laying out pipe lines for accessibility, ease of maintenance, clearances, and supports. Anchors and supports may also have to be considered. Steam lines or hot process lines must be designed to either have inherent flexibility or to actually incorporate expansion loops or joints in their runs. Lines subject to appreciable expansion must also be adequately anchored at certain intervals so that the expansion joints function properly. Inherent flexibility can be designed in the piping systems by the inclusion of right angle bends in at least two directions. Care must also be taken that the stresses set up in a pipe system due to expansion are not transmitted to the nozzles on the pieces of equipment being connected by the piping. The magnitude of these stresses or forces is sufficient in many cases to distort or actually break these nozzles. In the case of pumps, the force may be sufficient to cause the rotating shaft to bind or cause the packing to wear out very rapidly and cause excessive maintenance.[3,4]

18.4 CALCULATING LOSSES OF VALVES AND FITTINGS

Pipe systems, as mentioned above, include inlets, outlets, bends, and other devices (e.g., valves, fittings) that create a pressure drop. This drop, which results in energy loss, may be greater than that due to flow in a straight pipe. Thus, these additional so-called minor losses may not be so minor in some applications. Pressure loss data for a wide variety of valves and fittings have been compiled in terms of either the loss (or resistance) coefficient, K, or the "equivalent-length," L_{eq}. Details on these two concepts follow.

The dimensionless loss coefficient is defined as:

$$K = \frac{h_{f,m}}{v^2/2g_c} \qquad (18.3)$$

where $h_{f,m}$ is the minor head loss due to the device and $v^2/2g_c$ is defined as the dynamic or velocity head. This equation may be rewritten as

$$h_{f,m} = K \frac{v^2}{2g_c} \qquad (18.4)$$

The units of $h_{f,m}$ must have units consistent with $v^2/2g_c$ (e.g., ft · lb$_f$/lb) for K to be dimensionless. A pipe system may have several valves and fittings. The minor losses of all of them can therefore be expressed in terms of the aforementioned velocity (dynamic) head. Then, they can all be summed, i.e.,

$$\sum h_{f,m} = \frac{v^2}{2g_c} \sum K \qquad (18.5)$$

The resistance coefficient, K, for open valves, elbows and tees are listed in Table 18.1. The listed K values depend on the nominal diameter, the valve or fitting type, and whether it is screwed or flanged. The tabulated valve loss coefficients are for *fully open* valves. To account for *partially open* valves, the following ratio is employed

$$\frac{K \text{ of partially open valve}}{K \text{ of fully open valve}} \qquad (18.6)$$

Table 18.1   Resistance coefficients K for open valves, elbows, and tees

Nominal diameter, inch	Screwed				Flanged				
	1/2	1	2	4	1	2	4	8	20
Valves (fully open):									
Globe	14	8.2	6.9	5.7	13	8.5	6.0	5.8	5.5
Gate	0.30	0.24	0.16	0.11	0.80	0.35	0.16	0.07	0.03
Swing check	5.1	2.9	2.1	2.0	2.0	2.0	2.0	2.0	2.0
Angle	9.0	4.7	2.0	1.0	4.5	2.4	2.0	2.0	2.0
Elbows:									
45° regular	0.39	0.32	0.30	0.29					
45° long radius					0.21	0.20	0.18	0.16	0.14
90° regular	2.0	1.5	0.95	0.64	0.50	0.39	0.30	0.26	0.21
90° long radius	1.0	0.72	0.41	0.23	0.40	0.30	0.19	0.15	0.10
180° regular	2.0	1.5	0.95	0.64	0.41	0.35	0.30	0.25	0.20
180° long radius					0.40	0.30	0.21	0.15	0.10
Tees:									
Line flow	0.90	0.90	0.90	0.90	0.24	0.19	0.14	0.10	0.07
Branch flow	2.4	1.8	1.4	1.1	1.0	0.80	0.64	0.58	0.41

Values for gate and globe valves are given in Table 18.2 for estimation purposes.

The loss coefficient K_{SE} due to a sudden expansion (SE) between two different sizes of pipes D_1 and D_2 ($D_1 < D_2$) for Equation (18.7)

$$h_{f,e} = K_{SE} \frac{v_1^2}{2g_c}$$ (18.7)

is given by (see Fig. 18.1)

$$K_{SE} = \left[1 - \left(\frac{D_1}{D_2} \right)^2 \right]^2$$ (18.8)

Combining Equations (18.7) and (18.8) gives

$$h_{f,e} = \left[1 - \left(\frac{D_1}{D_2} \right)^2 \right]^2 \frac{v_1^2}{2g_c}$$ (18.9)

Table 18.2   Increased losses of partially open valves

Condition		Ratio K/K (Fully Open Condition)	
		Gate Valve	Globe Valve
Open		1.0	1.0
Closed	25%	3.0–5.0	1.5–2.0
	50%	12.0–22.0	2.0–3.0
	75%	70–120.0	6.0–8.0

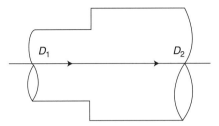

Figure 18.1   Sudden expansion.

Illustrative Example 18.1   Calculate K_{SE} if there is a sudden expansion in which the diameter D_1 doubles to D_2, that is, $D_2 = 2D_1$.

Solution   Apply Equation (18.8).

$$K_{SE} = \left[1 - \left(\frac{D_1}{D_2} \right)^2 \right]^2$$

Set

$$D_2 = 2D_1$$

Thus,

$$K_{SE} = \left[1 - \left(\frac{1}{2} \right)^2 \right]^2$$
$$= 9/16$$
$$= 0.562$$

Note that the loss coefficient K_{SE} is based on the velocity head of the smaller pipe.

A sudden contraction (see Fig. 18.2) generally leads to a higher pressure drop. The loss coefficient, K_{SC}, can be *approximated* by the equation:

$$K_{SC} = 0.42 \left[1 - \left(\frac{D_2}{D_1} \right)^2 \right] \tag{18.10}$$

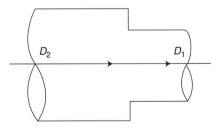

Figure 18.2   Sudden contraction.

Note that values of 0.50 rather than 0.42 have been reported in the literature; for this case,

$$K_{SC} = 0.50 \left[1 - \left(\frac{D_2}{D_1} \right)^2 \right] \tag{18.11}$$

Note that the loss coefficient, K_{SC}, is based on the velocity head of the smaller pipe at exit conditions.

The above information on the various minor losses is summarized in Table 18.3.

Another approach to describe these losses is through the equivalent length concept. L_{eq} is the length of straight pipe that produces the same head loss as the device in question. The term, L_{eq}, is related to the loss coefficient through Equation (18.12)

$$L_{eq} = K \frac{D}{4f} \tag{18.12}$$

This equation may be rewritten as

$$K = 4f \frac{L_{eq}}{D} \tag{18.13}$$

The L_{eq} in the above equation is not usually expressed as actual feet of straight pipe, but rather as a certain number of pipe diameters. Table 18.4 provides some of these values.[5]

Crane Co. provides information on the equivalent length of straight pipe that would have the same effect as the device. Their results, published by Bennett and Myers,[6] appear in the equivalent length chart in Fig. 18.3.

It should be noted that either the K or the L_{eq} approach assumes a "zero-length" fitting/device. The resistance of the fitting is therefore taken as the total resistance of the pipe/fitting system less the resistance of the length of straight pipe of the section of flow under study.

Using the above approach, the total friction can be expressed as

$$h_f = \left[4f \frac{L}{d} + \sum K_c + \sum K_e + \sum K_f \right] \frac{v^2}{2g_c} \tag{18.14}$$

The above equation may also be written in terms of equivalent lengths instead of the loss coefficients.

Table 18.3 Minor loss calculations

Configuration	Calculation of K
Entrance losses	Equation (18.9)
Exit losses	Equations (18.10) and (18.11)
Fittings: elbows, tees	Table 18.1
Valves: Fully open	Table 18.1
Partially open	Tables 18.1 and 18.2

Table 18.4   **Friction loss of screwed fittings, valves, etc.**

	Equivalent Lengths, Pipe Diameters
45° elbows	15
90° elbows, standard radius	32
90° elbows, medium radius	26
90° elbows, long sweep	20
90° square elbows	60
180° close return bends	75
180° medium-radius return bends	50
Tee (used as elbow, entering run)	60
Tee (used as elbow, entering branch)	90
Couplings	Negligible
Unions	Negligible
Gate valves, open	7
Globe valves, open	300
Angle valves, open	170
Water meters, disk	400
Water meters, piston	600
Water meters, impulse wheel	300

Illustrative Example 18.2   Calculate the equivalent length of pipe that would cause the same head loss for a gate and globe valve located in piping with a diameter of 3 inches.

Solution   The equivalent length of piping that will cause the same head loss for a particular component can be determined by multiplying the L/D for that component by the diameter of pipe. Refer to Table 18.4.

The L/D for a fully open gate valve is 7. The L/D for a fully open globe valve is 300. Based on the definition of the equivalent length

$$L_{eq} = (L/D)D$$

Thus

$$L_{eq}(\text{gate}) = (7)(3) = 21 \text{ in.}$$
$$L_{eq}(\text{globe}) = (300)(3) = 900 \text{ in.}$$

Illustrative Example 18.3   Water is flowing at room temperature through 30 ft of 3/8 in. pipe at a velocity of 10 ft/s. The pipe contains a flanged globe valve. What is the pressure drop along the length of the pipe? Assume that the friction factor is given by the equations provided in Illustrative Example 14.3, i.e.,

$$f = 0.0015 + 0.125/(\text{Re})^{0.30}$$

Solution   At room temperature,

$$\rho = 62.4 \text{ lb/ft}^3$$
$$\mu = 6.72 \times 10^{-4} \text{ lb/ft} \cdot \text{s}$$

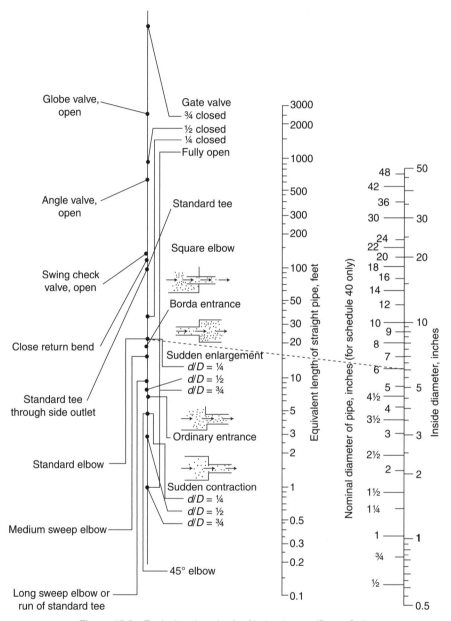

Figure 18.3   Equivalent lengths for friction losses (Crane Co.).

The Reynolds number is

$$\text{Re} = \frac{Dv\rho}{\mu}$$

$$= \frac{(0.375/12)(10)(62.4)}{6.72 \times 10^{-4}}$$

$$= 29{,}000$$

For

$$f = 0.0015 + 0.125/(\text{Re})^{0.30}$$

$$f = 0.0015 + 0.125/(29{,}000)^{0.30}$$

$$= 0.0015 + 0.0057$$

$$= 0.0072$$

The pressure drop is calculated by some modest of manipulation of either Equation (13.15) or (14.3).

$$\Delta P = \frac{2f\rho v^2 L}{Dg_c}$$

$$= \frac{(2)(0.0072)(62.44)(10)^2(30)}{(0.375/12)(32.2)}$$

$$= 2681 \, \text{lb}_f/\text{ft}^2$$

$$= 2681/62.4$$

$$= 43.0 \, \text{ft} \cdot \text{lb}_f/\text{lb}$$

$$= 516 \, \text{in. H}_2\text{O}$$

Illustrative Example 18.4   Refer to Illustrative Example 18.3. Determine the frictional loss for the fitting.

Solution   The frictional loss for the globe valve is calculated by employing Equation 18.4:

$$h_{f,\,\text{fitting}} = K_f \frac{v^2}{2g_c}$$

The loss coefficient K_f is obtained by linear extrapolation in Table 18.1.

$$K_f \cong 22$$

Substitution

$$h_f = \frac{(22)(10)^2}{(2)(32.2)}$$

$$= 34.16 \frac{\text{ft} \cdot \text{lb}_f}{\text{lb}}$$

Illustrative Example 18.5   Calculate the total pressure drop in lb_f/ft^2 (psf) for the 30 ft pipe system.

Solution   The total pressure drop is given by the sum of the results of the two previous problems. Thus,

$$\Delta P_T = (\Delta P + h_f)\rho$$

$$= (34.16 + 43.0)\,62.4$$

$$= 4815 \, \text{lb}_f/\text{ft}^2$$

Illustrative Example 18.6   Water flows from a large open tank (point 1) through a new cast iron pipe (10 in diameter and 5000 ft length). The discharge point (point 2) to the atmosphere is 260 ft below the tank level ($z_1 - z_2 = 260$ ft). The pipe entrance is sharp cornered (see Fig. 18.4). Water properties are $\rho = 62.4 \, \text{lb/ft}^3$ and $\mu/\rho = 1.082 \times 10^{-5} \, \text{ft}^2/\text{s}$. Calculate the water volumetric flow rate in ft^3/s.

Solution   Obtain the relative roughness of pipe from Table 14.1. For cast iron:

$$k = 0.00085 \, \text{ft}$$
$$D = 10 \, \text{in} = 0.833 \, \text{ft}$$
$$k/D = 0.001$$

Since the flow rate is unknown, the Reynolds number cannot be immediately calculated. Assume a Fanning friction factor in the 0.004–0.005 range. Select the upper limit, that is,

$$f = 0.005$$

The entrance loss coefficient is estimated from Equations (18.10) and (18.11).

$$K = 0.45 \, \text{(average value)}$$

Calculate the friction head loss in terms of the line velocity

$$h_f{}' = 4f \frac{L}{D} \frac{v_2^2}{2g} = (0.02)\frac{5000}{0.833}\frac{v_2^2}{2g} = 12\frac{v_2^2}{2g}$$

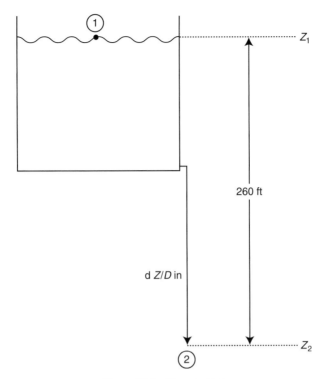

Figure 18.4 Flow system.

Apply Bernoulli's equation between points 1 and 2 to calculate v_2

$$P_1 = P_2 = 0\,\text{psig (both locations open to the atmosphere)}$$
$$v_1 = 0 \text{ (large tank)}$$
$$h_s = 0 \text{ (no shaft head)}$$

$$260 = \frac{v_2^2}{2g} + 12\frac{v_2^2}{2g} + 0.45\frac{v_2^2}{2g}$$

$$= 13.45\frac{v_2^2}{2g}$$

The positive root is:

$$v_2 = 11.75\,\text{ft/s}$$

Check on the flow condition

$$\text{Re} = \frac{Dv}{\mu/\rho} = \frac{(0.833)(11.75)}{(1.082 \times 10^{-5})}$$

$$= 904{,}600$$

Read f from Fig. 14.2 to check on the assumed f for $k/D = 0.001$.

$$f = 0.005; \quad \text{agrees}$$

Finally, calculate the volumetric flow rate.

$$q = vS = 11.75 \frac{\pi(0.833)^2}{4}$$
$$= 6.4 \, \text{ft}^3/\text{s}$$

Illustrative Example 18.7   Two large water reservoirs (see Fig. 18.5) are connected by 2000 ft of 3-in. schedule 40 pipe. Water at 20°C is to be pumped from one to the other at a rate of 200 gallon per minute. Both tanks are open to the atmosphere and the level in both reservoirs are the same. The pipes are commercial steel. Calculate the friction loss neglecting minor losses due to entrance and exit effects and assuming fully developed flow.

Solution   From Table A.5 in the Appendix, for 3 in. schedule 40 pipe:

$$D = 3.068 \, \text{in.} = 0.0779 \, \text{m}$$

Obtain the roughness of the pipe, k, from Table 14.1

$$k = 0.046 \, \text{mm}$$
$$k/D = 0.046 \times 10^{-3}/0.0779$$
$$= 0.0006$$

The flow velocity of water is ($q = 200 \, \text{gpm} = 0.0126 \, \text{m}^3/\text{s}$)

$$v = \frac{q}{S} = \frac{q}{\pi D^2/4} = \frac{0.0126}{\pi(0.0779)^2/4}$$
$$= 2.64 \, \text{m/s}$$

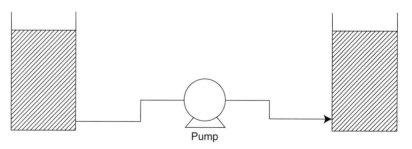

Pump

Figure 18.5   Water reservoir.

Check the flow regime

$$Re = \frac{vD}{\nu} = \frac{(2.64)(0.0779)}{10^{-6}} = 205{,}656$$

Therefore, the flow is turbulent.
From the Fanning chart, Fig. 14.2, at

$$Re = 205{,}656 \quad \text{and} \quad k/D = 0.0006$$

one obtains

$$f = 0.00345$$

Calculate the head loss

$$h_f' = 4f\frac{L}{D}\frac{v^2}{2g_c} = 4(0.00345)\frac{(2000)(0.3048)}{0.0779}\frac{2.64^2}{2} = 376.3\,\text{J/kg}$$

Apply Bernoulli's equation between stations 1 and 2. Note that $P_1 = P_2 = 1$ atm, $v_1 = v_2$ and $z_1 = z_2$.

$$\frac{\Delta P}{\rho} + \frac{\Delta v^2}{2g_c} + \Delta z\frac{g}{g_c} = h_s - h_f$$

The above equation reduces to

$$h_s = h_f$$

or

$$h_s' = h_f'$$

Therefore

$$h_s' = h_f' = 376.3\,\text{J/kg}$$

Illustrative Example 18.8   Refer to Illustrative Example 18.7. What are the ideal pumping requirements in kW and hp (i.e., calculate the fluid power)? What is the required pressure rise in the pump?

Solution   Calculate the pressure rise across the pump.

$$\Delta P = \rho g h_f' = (1000)(9.807)(38.39)$$
$$= 376{,}500\,\text{Pa} = 54.61\,\text{psi}$$

Finally, calculate the ideal pumping requirement (the fluid power).

$$\dot{W}_s = q\Delta P = 0.0126(475{,}000)$$
$$= 5985\,\text{kW} = 8.0\,\text{hp}$$

18.5 FLUID FLOW EXPERIMENT: DATA AND CALCULATIONS

One of the experiments conducted in the Chemical Engineering Laboratory at Manhattan College was concerned with flow through pipes and fittings. Students performed the experiment and later submitted a report. In addition to theory, experimental procedure, discussion of results, and so on, the report contained sample calculations. The following is an (edited) example of those calculations that cover a wide range of flow through pipes and fittings principles and applications.

A photograph of the flow through pipes and fittings experimental setup is provided in Fig. 18.7. A line diagram of the system follows in Figure 18.8. (The system was designed by one of the authors approximately 40 years ago.) Water enters the system at different flow rates and flows along the outer or inner loop and exits the system. As the water flows through the system, the pressure drops along the various pipe lengths and fittings are read from two mercury manometers. One manometer is solely dedicated to read the pressure drop across the orifice meter. Pressure taps are located at the start and end of each section of pipe, fitting and valve to assist in taking the pressure drop across the appropriate section.

Referring to Fig. 18.7, the outer loop consists of a/an:

1. length of 3/8 inch diameter brass pipe,
2. straight portion of tee,
3. gate valve,

Figure 18.7 Flow through pipes and fittings.

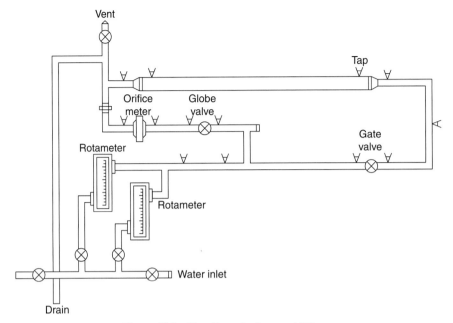

Figure 18.8 Flow through pipes and fittings.

4. 90° elbow plus junction,
5. 90° elbow only,
6. expansion from 3/8 to 1/2 inch diameter pipe,
7. length of 1/2 inch diameter brass pipe, and
8. contraction from 1/2 to 3/8 inch diameter pipe

The inner pipe consists of a/an:

1. length of 3/8 inch diameter brass pipe,
2. pair of right angle elbow portion of tees,
3. globe valve,
4. length of 3/8 inch diameter brass pipe, and
5. orifice meter

The following sample calculations represent some of the calculations that were performed in the analysis of this experiment. The sample calculations are for both outer and inner loops for 80% of the maximum flow rate in the 6 gpm rotameter. Therefore,

$$q = (0.8)(6\,\text{gpm}) = 4.8\,\text{gpm}$$

$$= 0.01069\,\text{ft}^3/\text{s}$$

The cross-section area of the outer pipe is

$$S = \frac{\pi D^2}{4}$$

$$= \frac{\pi (0.0413^2)\,\text{ft}^2}{4}$$

$$= 0.00134\,\text{ft}^2$$

The velocity is therefore,

$$v = \frac{0.01069\,\text{ft}^3/\text{s}}{0.00134\,\text{ft}^2} = 7.98\,\text{ft/s}$$

The Reynolds number for this flow rate is

$$\text{Re} = \frac{Dv\rho}{\mu}$$

$$= \frac{(62.4)(7.98\,\text{ft/s})(0.0413\,\text{ft})}{7.29 \times 10^{-4}}$$

$$= 28{,}210$$

This Re number indicates that the flow rate is turbulent.

Calculate the pressure drop across the length of 3/8 inch diameter pipe. The equation used to calculate the pressure drop is:

$$\Delta P = P_1 - P_2 = \left(\frac{g}{g_c}\right)(R_m)(\rho_{\text{Hg}} - \rho_{\text{H}_2\text{O}})$$

where ρ_{Hg} = density of mercury, 999.2 kg/m³
   $\rho_{\text{H}_2\text{O}}$ = density of water, 13,600 kg/m³
   R_m = vertical distance between two meniscuses, (7.0 cm) 0.070 m

Substituting

$$\Delta P = \frac{(9.8)(0.070)(13{,}600 - 999)(14.7)(144)}{1.01325 \times 10^5}$$

$$= 180.59\,\text{lb}_\text{f}/\text{ft}^2$$

After obtaining the pressure drop for the length of 3/8 inch diameter pipe between pressure taps 1 and 2 (3.979 ft), the pressure drop per unit length is determined as follows:

$$\Delta P_{L\frac{3}{8}} = \frac{\Delta P}{L} = \frac{180.59}{3.979}$$

$$= 45.38\,\text{lb}_\text{f}/\text{ft}^3$$

The Fanning friction factor is given

$$f = \frac{D\Delta P g_c}{2L\rho v^2}$$

where ΔP = pressure drop calculated above, 180.59 lb_f/ft^2
    D = diameter of the pipe, 0.0413 ft
    g_c = Newton's law constant 32.2 ft $\cdot$ lb/lb$_f$ $\cdot$ s^2
    L = length of pipe, 3.979 ft
    ρ = density of water, 62.4 lb/ft^3
    v = average velocity of the fluid in the pipe, 7.98 ft/s

Substituting,

$$f = \frac{(0.0413)(180.59)(32.2)}{(2)(3.979)(62.4)(7.98)}$$

$$= 0.0076$$

This is in reasonable agreement with the Fanning friction plot.
    Calculation of friction losses are determined from the Bernoulli equation:

$$\frac{P_2 - P_1}{\rho} + \frac{g}{g_c}(z_2 - z_1) + \frac{\alpha_2 v_2^2 - \alpha_1 v_1^2}{2g_c} = h_f$$

where $P_1 - P_2$ = pressure drop at points 1 and 2, lb/ft^2
    ρ = fluid density, lb/ft^3
    g = acceleration due to gravity, ft/s^2
    g_c = Newton's conversion constant, ft $\cdot$ lb/lb$_f$ $\cdot$ s^2
    z_1, z_2 = vertical distances at points 1 and 2, ft
    α = kinetic energy correction factor
    v_1, v_2 = velocities at points 1 and 2, ft/s
    h_f = friction generated per pound mass of fluid, ft $\cdot$ lb$_f$/lb

The Bernoulli equation above simplifies to the equation below for the length of 3/8 inch diameter brass pipe since there is no height difference along the pipe, and the velocity of the water along the pipe is assumed to be constant. Also assume $\alpha_1 = \alpha_2 = 1.05$ for turbulent flow.

$$h_f = \frac{P_2 - P_1}{\rho}$$

$$h_f = \frac{180.59}{62.4}$$

$$= 2.894 \text{ ft} \cdot lb_f/lb$$

To calculate the frictional losses due to the straight portion of tee, the total frictional loss between taps 2 and 3 was found and the frictional losses due to the length of the pipe between the two taps subtracted from it

$$h_{ft} = \frac{P_2 - P_1}{\rho} - \frac{\left(\Delta P_{L\frac{3}{8}}\right)(L)}{\rho} = \left(\frac{110.94}{62.4}\right) - \frac{(45.38)(2.02)}{62.4}$$

$$h_{ft} = 0.308 \text{ ft} \cdot \text{lb}_f/\text{lb}$$

The frictional losses due to the gate valve are calculated in a manner similar to the frictional loss from the straight portion of tee.

$$h_{fg} = \frac{P_2 - P_1}{\rho} - \frac{(\Delta P_L)L}{\rho} = \left(\frac{108.36}{62.4}\right) - \frac{(45.38)(0.66)}{62.4}$$

$$h_{fg} = 1.26 \text{ ft} \cdot \text{lb}_f/\text{lb}$$

The frictional losses due to the two 90° elbows plus junction (union) were determined as above. The velocity is again assumed to be constant as the water moves up the elevation

$$h_f = \frac{P_2 - P_1}{\rho} - \frac{\left(\Delta P_{L\frac{3}{8}}\right)(L)}{\rho} - \frac{g(z_2 - z_1)}{g_c}$$

$$= \left(\frac{394.75}{62.4}\right) - \frac{(45.38)(2.35)}{62.4} - \frac{(32.2)(2.001)}{32.2}$$

$$h_f = 2.62 \text{ ft} \cdot \text{lb}_f/\text{lb}$$

The frictional losses due to the elbow alone are also determined as above. The velocity is again assumed to be constant as the water moves up the elevation

$$h_f = \frac{P_2 - P_1}{\rho} - \frac{(\Delta P_L)L}{\rho} - \frac{g(z_2 - z_1)}{g_c}$$

$$= \left(\frac{608.89}{62.4}\right) - \frac{(45.38)(1.283)}{62.4} - \frac{(32.2)(1.083)}{32.2}$$

$$h_f = 7.74 \text{ ft} \cdot \text{lb}_f/\text{lb}$$

The frictional losses due to the union are determined by subtracting the losses due to the two elbows and the length of the pipe between taps 4 and 6 from the total frictional loss determined in above. Thus,

$$h_f = (2.62 - 7.74)$$
$$= -5.12 \text{ ft} \cdot \text{lb}_f/\text{lb}$$

The frictional loss cannot be negative since this would imply energy is generated by the system due to the flow of the fluid. This result is attributed to experimental error.

The frictional loss in the length of $1/2$ inch diameter brass pipe is determined the same way as the loss in the length of $3/8$ inch diameter brass pipe:

$$h_f = \frac{P_2 - P_1}{\rho}$$

$$h_f = \frac{92.87}{62.4}$$

$$= 1.488 \text{ ft} \cdot \text{lb}_f/\text{lb}$$

After obtaining the pressure drop for the length of $1/2$ inch diameter pipe between pressure taps 7 and 8, the pressure drop per unit length (6.33 ft) in this section is determined as follows:

$$\Delta P_{L\frac{1}{2}} = \frac{\Delta P}{L} = \frac{92.87}{6.33}$$

$$= 14.67 \text{ lb}_f/\text{ft}^3$$

The frictional losses due to expansion and contraction are calculated by including the change in velocity, which is factored into the calculation. For the expansion,

$$h_f = \frac{P_2 - P_1}{\rho} - \frac{(\Delta P_{L1})(L_1) + (\Delta P_{L2})(L_2)}{\rho} - \frac{\alpha_2 v_2{}^2 - \alpha_1 v_1{}^2}{2g_c}$$

$$= \left(\frac{5.16}{62.4}\right) - \frac{(45.38)(0.2625 \text{ ft}) + (14.67)(0.4921)}{62.4} - \frac{(1.05)(5.00^2 - 7.98^2)}{(2)(32.2)}$$

$$h_f = 0.41 \text{ ft} \cdot \text{lb}_f/\text{lb}$$

The frictional losses due to the contraction from $1/2$ to $3/8$ inch pipe are similarly calculated. The only difference is v_2 and v_1 that are the velocities in the $3/8$ inch and $1/2$ inch pipe, respectively.

$$h_f = \frac{P_2 - P_1}{\rho} - \frac{(\Delta P_{L1})(L_1) + (\Delta P_{L2})(L_2)}{\rho} - \frac{\alpha_2 v_2{}^2 - \alpha_1 v_1{}^2}{2g_c}$$

$$= \left(\frac{110.94}{62.4}\right) - \frac{(45.38)(1.00) + (14.67)(1.01)}{62.4} - \frac{(1.05)(7.98^2 - 5.00^2) \text{ ft/s}}{(2)(32.2)}$$

$$h_f = 0.18 \text{ ft} \cdot \text{lb}_f/\text{lb}$$

The globe valve calculation is performed the same way as the determination of the frictional losses across the gate valve in the outer loop. Only the result will be provided here

$$h_{fg} = \frac{P_2 - P_1}{\rho} - \frac{(\Delta P_L)(L)}{\rho}$$

$$= 6.30 \, \text{ft} \cdot \text{lb}_f/\text{lb}$$

Sample calculations for the determination of the frictional loss coefficient, K, for the straight through tee, orifice meter, and expansion from 3/8 inch to 1/2 inch diameter pipe are shown below. The determination of the frictional loss coefficient for all the other fittings is similar to the one for the straight through tee. Therefore, those calculations are not shown here. The frictional losses due to the straight through tee in the system are given by the equation below:

$$h_{f, \text{fitting}} = K_f \frac{v_1^2}{2g_c}$$

where v_1 = average velocity in pipe leading to fitting, ft/s
K_f = frictional loss coefficient for fitting

Re-arranging the equation above gives

$$\log_{10}(h_{f, \text{fitting}}) = 2 \log_{10} V_1 + \log_{10}(K_f) - \log_{10}(2g_c)$$

The equation above is similar to the equation of a straight line

$$y = mx + b$$

Therefore,

$$b = \log_{10}(K_f) - \log_{10}(2g_c)$$

The value of b is obtained from the equation of a trend line in the graph of log h_f versus log v for the straight through tee fitting. From the graph (not shown here)

$$b = -2.2148$$

so that

$$K_f = 10^{-2.2148 + \log_{10}(2*32.2)}$$

$$= 0.39$$

REFERENCES

1. A. Foust, L. Wenzel, C. Chung, L. Maus, and L. Andrews, "Principles of Unit Operations," John Wiley & Sons, Hoboken, NJ, 1950.

2. J. Santoleri, J. Reynolds, and L. Theodore, "Introduction to Hazardous Waste Incineration," 2nd edition, John Wiley and Sons, Hoboken, NJ, 2000.

3. J. Walker and S. Crocker, "Piping Handbook," McGraw-Hill, New York, 1939.

4. C. E. Lapple, "Fluid and Particle Dynamics," University of Delaware, Newark, Delaware, 1951.

5. W. Badger and J. Banchero, "Introduction to Chemical Engineering," McGraw-Hill, New York, 1955.

6. C. Bennett and J. Meyers, "Momentum, Heat, and Mass Transfer," McGraw-Hill, New York, 1962.

NOTE: Additional problems are available for all readers at www.wiley.com. Follow links for this title.

19

FLOW MEASUREMENT

19.1 INTRODUCTION

Measurement of a flowing fluid can be difficult since it requires that the mass or volume of material be quantified as it moves through a pipe or conduit. Problems may arise due to the complexity of the dynamics of flow. Further, flow measurements draw on a host of physical parameters that are also often difficult to quantify.

This chapter serves to review standard industrial methods that are employed to measure fluid *flow rates*. Information provided can include the velocity or the amount of fluid that passes through a given cross-section of a pipe or conduit per unit time. Local velocity variations across the cross-section or short-time fluctuations (e.g., turbulence) are not considered. These concerns can be important, particularly the former. For example, in air pollution applications, it is often necessary to traverse a stack to obtain local velocity variations with position.

Hydrodynamic methods are primarily used by industry in the measuring the flow of fluids. These methods include the use of the following equipment:

1. Pitot tube
2. Venturi meter
3. Orifice meter

Other approaches include weighing, direct displacement, and dilution. Weighing involves mass or gravitational approaches which, as one might suppose, cannot be

Fluid Flow for the Practicing Chemical Engineer. By J. Patrick Abulencia and Louis Theodore
Copyright © 2009 John Wiley & Sons, Inc.

used for gases. Direct displacement can be applied to liquids and is based on a displacement of either a moving part of the unit or the moving fluid. Dilution methods involve adding a second fluid of a known rate to the stream of fluid to be measured and determining the concentration of this second fluid at some displaced point. In addition, there is the vane anemometer that is in effect a windmill consisting of a number of light blades mounted on radial arms attached to a common spindle rotating in two jeweled bearings; when placed parallel to a moving gas stream, the forces on the blades cause the spindle to rotate at a rate depending mainly on the gas velocity. An extension of this unit is the hot-wire anemometer that essentially consists of a fine, electrically-heated wire exposed to the gas stream in which the velocity is being measured; the velocity of the gas determines the cooling effect upon the wire, which in turn affects the electrical resistance. Finally, the rotameter is the most widely used form of area meter that is essentially a vertical tapered glass tube inserted into a pipe line by means of special end connections and containing a float that moves up and down as the flow increases or decreases; graduations are etched onto the side of the tube to indicate the rate of flow.

This chapter will primarily key on the five hydrodynamic methods listed above, introducing the subject with pressure measurement and the general topic of manometry.

19.2 MANOMETRY AND PRESSURE MEASUREMENTS

Pressure is usually measured by allowing it to act across some area and opposing it with some type of force (e.g., gravity, compressed spring, electrical, and so on). If the force is gravity, the device is usually a manometer.

A very common device to measure pressure is the Bourdon-tube pressure gauge. It is a reliable and inexpensive direct displacement device. It is made of a stiff metal tube bent in a circular shape. One end is fixed and the other is free to deflect when pressurized. This deflection is measured by a linkage attached to a calibrated dial (see Fig. 19.1). Bourdon gauges are available with an accuracy of $\pm 0.1\%$ of the full scale.

Other pressure gauges measure the pressure by the displacement of the sensing element electrically. Among the common methods are capacitance, resistive and inductive. However, the interest in this section is primarily with the manometer.

Consider the open manometer shown in Fig. 19.2. P_1 is unknown and P_a is the known atmospheric pressure. The heights z_a, z_1, and z_2 are also known. Applying Bernoulli's hydrostatic equation at points 1 and 2, and again at points a and 2 yields:

$$P_1 - P_2 = -\rho_1 \frac{g}{g_c}(z_1 - z_2) \tag{19.1}$$

$$P_2 - P_a = -\rho_2 \frac{g}{g_c}(z_2 - z_a) \tag{19.2}$$

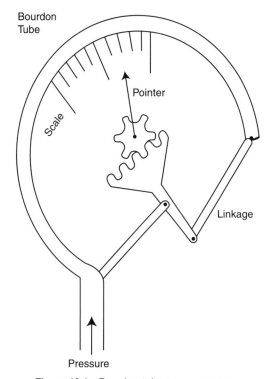

Figure 19.1   Bourdon-tube pressure gauge.

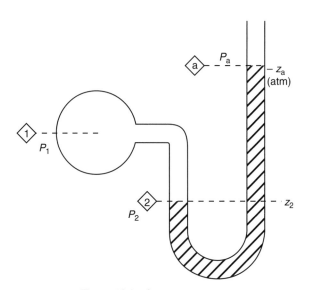

Figure 19.2   Open manometer.

If the two equations are added, one obtains:

$$P_1 - P_a = -\left[\rho_1 \frac{g}{g_c}(z_1 - z_2) - \rho_2 \frac{g}{g_c}(z_2 - z_a)\right] \qquad (19.3)$$

The reader should refer to Chapter 10 for more details on manometry.

Illustrative Example 19.1   Find the air pressure (P_1) in the oil tank pictured in Fig. 19.3, given the heights and densities of the fluids in the manometer. The oil has a specific gravity of 0.8, the specific gravity of mercury is 13.6, and the density of air is $1.2 \, \text{kg/m}^3$. Employ SI units.

Solution   Apply the manometer equation between points 1 and 2:

$$P_1 + \rho_1 \frac{g}{g_c} z_1 = P_2 + \rho_2 \frac{g}{g_c} z_2$$

Since $z_1 = z_2$ and $\rho_1 = \rho_2$

$$P_1 = P_2$$

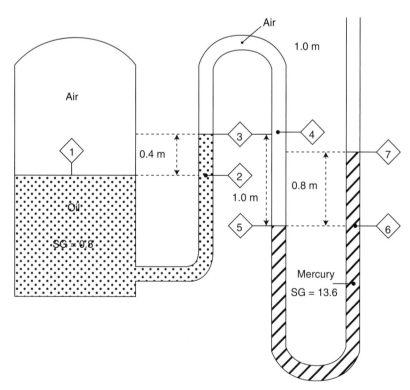

Figure 19.3   Air pressure in a tank.

Also apply the manometer equation between points 2 and 3:

$$P_2 + \rho_2 \frac{g}{g_c} z_2 = P_3 + \rho_3 \frac{g}{g_c} z_3$$

Since $\rho_2 = \rho_3 = \rho_{oil}$ and $z_3 - z_2 = 0.4$ m of oil

$$P_2 = P_3 + \rho_{oil} \frac{g}{g_c} (z_3 - z_2)$$
$$= P_3 + (0.8)(1000)(9.807)(0.4)$$
$$= P_3 + 3138$$

Consider points 4 and 5:

$$P_3 = P_4 = P_5 + \rho_{air} \frac{g}{g_c} (z_5 - z_4)$$
$$= P_5 - (1.2)(9.807)(1)$$
$$= P_5 - 11.77$$

Consider points 6 and 7:

$$P_5 = P_6 = P_7 + \rho_{Hg} \frac{g}{g_c} (z_7 - z_6)$$
$$= P_7 + (13,600)(9.807)(0.8)$$
$$= P_7 + 106,700$$

Since $P_7 = 0$ (gauge basis),

$$P_5 = P_6 = 106,700 \, \text{Pag}$$

Back substitute to obtain the desired pressures.

$$P_3 = P_4 = P_5 - 11.77 = 106,700 - 11.77 = 106,688 \, \text{Pag} = 2.05 \, \text{atm}$$
$$P_2 = P_1 = 106,688 + 3138 = 109,826 \, \text{Pag} = 211,151 \, \text{Pa} = 2.08 \, \text{atm}$$

Note:

1. The contribution of the gas section of the manometer to the answer is negligible.
2. Manometers that are open to the atmosphere are gauge-pressure devices.
3. The cross-sectional area of the manometer tube does not affect the calculation.

Liquid depths in tanks are commonly measured by the scheme shown in Fig. 19.4. Compressed air (or nitrogen) bubbles slowly through a dip tube in the liquid. The flow of the air is so slow that it may be considered static. The tank is vented to the atmosphere. The gauge pressure reading at the top of the dip tube is then primarily due to the liquid depth in the tank.

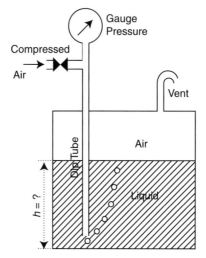

Figure 19.4 Liquid depth measurement.

19.3 PITOT TUBE

Bernoulli's equation provides the basis to analyze some devices for fluid flow measurement. A common device is the Pitot tube shown in Fig. 19.5. It essentially consists of one tube with an opening normal to the direction of flow and a second tube in which the opening is parallel to the flow. It measures both the static pressure (through the side holes at station 2) and the stagnation, or impact, pressure (through the hole in the front at station 1). Applying Bernoulli's equation between stations 1 and 2, one obtains (after neglecting frictional effects):

$$P_1 + \frac{\rho v_1{}^2}{2g_c} + \rho \frac{g}{g_c} z_1 = P_2 + \frac{\rho v_2{}^2}{2g_c} + \rho \frac{g}{g_c} z_2 \qquad (19.4)$$

Since

$$v_1 = 0 \text{ (stagnation)}$$
$$z_1 = z_2 \text{ (horizontal)}$$
$$v_2 = v = \text{fluid velocity}$$

$$v = \sqrt{\frac{2(P_1 - P_2)g_c}{\rho}} \qquad (19.5)$$

This is the Pitot tube formula.

The pressure difference $(P_1 - P_2)$ is often measured by connecting the ends of the Pitot tube to a manometer. The manometer liquid (density ρ_M) develops a differential height, h, due to the flowing fluid. Applying Bernoulli's equation at the manometer

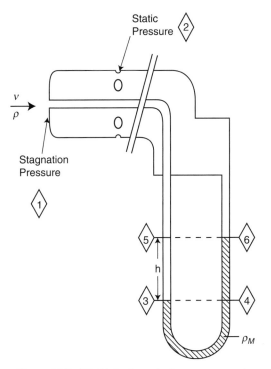

Figure 19.5   Pitot tube for velocity measurement.

(as presented in Fig. 19.5 yields):

$$P_3 = P_4$$

and

$$P_5 + \rho \frac{g}{g_c} h = P_6 + \rho_M \frac{g}{g_c} h \tag{19.6}$$

In addition

$$P_5 - P_6 = P_1 - P_2 = \frac{g}{g_c} h(\rho_M - \rho) \tag{19.7}$$

This is a modified form of Equation (19.3). Substituting Equation (19.7) into the Pitot tube formula, Equation (19.5) gives

$$v = \sqrt{\frac{2gh(\rho_M - \rho)}{\rho}} \tag{19.8}$$

This equation has also been written as

$$v = C\sqrt{\frac{2gh(\rho_M - \rho)}{\rho}} \qquad (19.9)$$

The term C is included to account for the assumption of negligible frictional effects. However, for most Pitot tubes, C is approximately unity.

Illustrative Example 19.2   A Pitot tube is located at the center line of a horizontal 12-inch ID pipe transporting dry air at 70°F and at atmospheric pressure. The horizontal deflection on a U-tube (inclined 10 inch horizontal to 1 inch vertical and connected to the impact and static openings) shows 2 inch of water. Calculate the actual velocity of air at the point where the reading is taken and the average velocity through this cross-section if the average velocity is 81.5% of the maximum velocity.

Solution   The density of the gas at the point of reading is approximately

$$\rho = 0.075 \, \text{lb/ft}^3$$

Owing to the 10-to-1 inclination, the actual difference in levels is only 0.2 inch of water.

The velocity is calculated directly from Equation (19.8):

$$v = \sqrt{\frac{2gh(\rho_M - \rho)}{\rho}}; \quad \rho_M - \rho \approx \rho_M$$

$$= \sqrt{\frac{(2)(32.2)(0.2/12)(62.4)}{0.075}}$$

$$= 29.9 \, \text{ft/s}$$

This represents the velocity at the point where the reading was taken; that is, the centerline of the pipe. Thus,

$$v_{\text{max}} = 29.9 \, \text{ft/s}$$

Since the flowing fluid is air at a high velocity, the flow has a high probability of being turbulent. For this condition, assume (see Chapter 14)

$$\bar{v}/v_{\text{max}} = 0.815$$

so that

$$\bar{v} = (0.815)(29.9)$$
$$= 24.4 \, \text{ft/s}$$

Illustrative Example 19.3   Refer to Illustrative Example 19.3. Calculate the mass flow rate of the air.

Solution   Since the area is 0.785 ft²

$$q = (24.4)(0.785)(60)$$
$$= 1150 \, \text{ft}^3 \, \text{min}$$

And,

$$\dot{m} = (1150)(0.075)(60)$$
$$= 5175 \, \text{lb/hr}$$

Illustrative Example 19.4   Water ($\rho = 1000 \, \text{kg/m}^3$, $\mu = 0.001 \, \text{kg/m} \cdot \text{s}$) flows in a circular pipe. The pipe is a 3 inch schedule 40 steel pipe. A Pitot tube is used to measure the water velocity. The liquid in the manometer is mercury ($\rho_{Hg} = 13,600 \, \text{kg/m}^3$). The manometer height, h ($z_6 - z_4$ in Fig. 19.5), is 7 cm. Determine the water velocity (in m/s and fps), volumetric flow rate (in m³/s and gpm), and flow regime.

Solution   Calculate the water velocity from the Pitot tube equation. Note

$$h = 0.07 \, \text{m}$$

and

$$(\rho_M - \rho)/\rho = 12.6$$

Employing Equation (19.8),

$$v = \sqrt{(2)(9.807)(0.07)(12.6)} = 4.2 \, \text{m/s}$$

Obtain the pipe inside diameter. Use Table A.5 in the Appendix. For a 3 inch schedule 40 pipe:

OD $= 3.5$ in
Wall thickness $= 0.216$ in
ID $= 3.068$ in $= 0.0779$ m
Pipe weight $= 7.58$ lb/ft

Calculate volumetric flow rate

$$S = \frac{\pi}{4}(0.0779)^2 = 0.00477\,\mathrm{m}^2$$

$$q = vS = (4.2)(0.00477) = 0.02\,\mathrm{m}^3/\mathrm{s} = 317\,\mathrm{gpm}$$

The flow is turbulent since

$$\mathrm{Re} = \frac{1000(4.2)(0.0779)}{0.001} = 327{,}180$$

Note that a Pitot tube measures the local velocity at only one point. To obtain the average velocity over the cross-section, it is necessary to read the velocity at a number of specific locations in the cross-section of the pipe. Also note that when the Pitot tube is used for measuring low-pressure gases, the pressure difference reading is usually extremely small, and can lead to large errors.

19.4 VENTURI METER

The Venturi meter is also a device for measuring a fluid flow rate. As shown in Fig. 19.6, it consists of three sections: a converging section to accelerate the flow, a short cylindrical section (called the throat), and a diverging section to increase the cross-sectional area to its original (upstream) value. There is a change in pressure between the upstream (point 1) and the throat (point 2). This pressure difference is measured (often with a manometer). The Venturi meter can determine the volumetric flow rate from either the pressure difference $(P_1 - P_2)$ or the manometer head (h). The development of pertinent equations is presented below.

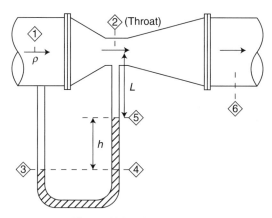

Figure 19.6 Venturi meter.

Refer to Fig. 19.6. From the conservation law of mass $\dot{m}_1 = \dot{m}_2$ at steady state. If the fluid is assumed incompressible, then:

$$\rho_1 = \rho_2 = \rho$$
$$q_1 = q_2$$

so that

$$\frac{\pi D_1^2}{4} v_1 = \frac{\pi D_2^2}{4} v_2 \tag{19.10}$$

and

$$v_1 = \left(\frac{D_2^2}{D_1^2}\right) v_2 \tag{19.11}$$

Applying Bernoulli's equation between points 1 and 2, and assuming no frictional losses (see Eq. 19.4)

$$P_1 + \frac{\rho v_1^2}{2g_c} + \rho \frac{g}{g_c} z_1 = P_2 + \frac{\rho v_2^2}{2g_c} + \rho \frac{g}{g_c} z_2 \tag{19.12}$$

For a horizontal Venturi meter, $z_1 = z_2$. Therefore, the above equation simplifies to:

$$P_1 + \frac{\rho v_1^2}{2g_c} = P_2 + \frac{\rho v_2^2}{2g_c} \tag{19.13}$$

Rearranging Equation (19.13) and substituting for v_1 from Equation (19.11) leads to:

$$v_2 = \sqrt{\frac{2g_c(P_1 - P_2)}{\rho[1 - (D_2^4/D_1^4)]}} \tag{19.14}$$

Substituting the manometer equation for $(P_1 - P_2)$, for example see Equation (19.7), into Equation (19.14) yields

$$v_2 = \sqrt{\frac{2gh(\rho_M - \rho)}{\rho[1 - (D_2^4/D_1^4)]}} \tag{19.15}$$

where once again ρ_M is the manometer fluid density, and ρ the flowing fluid density. The volumetric flow rate, q, is

$$q = \left(\frac{\pi D_2^2}{4}\right) v_2 \tag{19.16}$$

Equation (19.15) is often referred to as the Venturi formula. It applies to frictionless flow. To account for the small friction loss between points (1) and (2), a Venturi discharge coefficient, C_v, is introduced in the above equation, that is,

$$v_2 = C_v \sqrt{\frac{2g_c(P_1 - P_2)}{\rho[1 - (D_2^4/D_1^4)]}} = C_v \sqrt{\frac{2gh(\rho_M - \rho)}{\rho[1 - (D_2^4/D_1^4)]}} \qquad (19.17)$$

For well-designed Venturi meters, C_v, is approximately 0.96.

There is a permanent pressure loss, ΔP_L, in the Venturi of about 10% of $(P_1 - P_2)$. This means that 90% of the $(P_1 - P_2)$ is recovered in the divergent section of the Venturi. This pressure loss causes an overall loss of energy. The power requirement to operate a Venturi meter (or the power loss) is calculated from the volumetric flow rate and the pressure loss, that is,

$$\dot{W}_L = q(\Delta P_L) \qquad (19.18)$$

where

$$\Delta P_L = 0.1(P_1 - P_2) \qquad (19.19)$$

Illustrative Example 19.5   A Venturi meter has gasoline flowing through it. The upstream diameter, D_1, is 0.06 m and the throat diameter, D_2, is 0.02 m. The manometer fluid is mercury, with a height difference, h, of 35 mm Hg. Gasoline properties are $\rho = 680 \, \text{kg/m}^3$ and $\mu = 2.9 \times 10^{-4} \, \text{kg/m} \cdot \text{s}$. The density of mercury is $13,600 \, \text{kg/m}^3$. Find the flow rate of gasoline in m^3/s and gpm. If P_1, the upstream pressure is 1 atm, what is the pressure, P_2, at the throat of the Venturi meter? If the pressure loss is 10% of $(P_1 - P_2)$, calculate the power loss.

Solution   Calculate the velocity of gasoline at the throat using the following data

$$h = 0.035 \, \text{m}$$
$$D_2/D_1 = 1/3$$
$$\rho_M/\rho = 13,600/680 = 20$$

Assume $C_v = 1.0$ and substitute into Equation (19.15)

$$v_2 = \sqrt{\frac{2(9.807)(0.035)(20 - 1)}{1 - (1/3)^4}}$$

$$= 3.63 \, \text{m/s}$$

Calculate the volumetric flow rate.

$$q = \frac{\pi(0.02)^2}{4} \quad (3.63)$$
$$= 0.00114 \, \text{m}^3/\text{s}$$
$$= 18.1 \, \text{gpm}$$

Calculate P_2 from the manometer equation and the corresponding pressure loss, ΔP_L.

$$P_2 = 101{,}325 - (9.807)(0.035)(13{,}600 - 680)$$
$$= 96{,}900 \, \text{Pa}$$
$$\Delta P = P_1 - P_2$$
$$= 101{,}325 - 96{,}900 = 4425 \, \text{Pa}$$

For a 10% loss,

$$\Delta P_L = 0.1(4425)$$
$$= 442.5 \, \text{Pa}$$

Calculate the power loss:

$$\dot{W}_L = (0.00114)(442.5)$$
$$= 0.5 \, \text{W} = 6.71 \times 10^{-4} \, \text{hp}$$

Illustrative Example 19.6   Refer to Illustrative Example 19.6. If gasoline has a vapor pressure of 50,000 Pa, what flow rate will cause cavitation to occur?

Solution   Set $P_2 = p' = 50{,}000$ Pa and use Equation (19.14):

$$v_2 = \sqrt{\frac{2g_c(P_1 - P_2)}{\rho[1 - (D_2{}^4/D_1{}^4)]}} = \sqrt{\frac{2(101{,}325 - 50{,}000)}{680[1 - 1^4/(1/3)^4]}}$$
$$= 12.36 \, \text{m/s}$$

Also note that

$$q = \frac{\pi 0.02^2}{4} \, 12.36$$
$$= 0.0388 \, \text{m}^3/\text{s}$$

19.5 ORIFICE METER

Another device used for flow measurement is the orifice meter (see Fig. 19.7). The pressure difference is measured (often with a manometer) between the upstream (point 1) and the orifice (point 2). Although it operates on the same principles as a Venturi meter, orifice plates can be easily changed to accommodate a wide range of flow rates.

The orifice can be employed to determine either the volumetric flow rate from the pressure difference, $(P_1 - P_2)$, or from the manometer head, h. For a horizontal orifice meter, the velocity equation is the same as the Venturi meter, that is,

$$v_2 = C_o\sqrt{\frac{2g_c(P_1 - P_2)}{\rho[1 - (D_2{}^4/D_1{}^4)]}} = C_o\sqrt{\frac{2gh(\rho_M - \rho)}{\rho[1 - (D_2{}^4/D_1{}^4)]}} \qquad (19.20)$$

The volumetric flow rate is once again:

$$q = \left(\frac{\pi D_2{}^2}{4}\right)v_2 \qquad (19.21)$$

The orifice meter is simpler in construction and less expensive than a Venturi meter, and occupies less space. However, it has a lower pressure recovery (around 70%). The discharge coefficient, C_o, for drilled-plate orifices is shown in Fig. 19.8 where C_o is a function of D_2/D_1 and the Reynolds number at the throat, Re_2. At Re_2 values greater than 20,000, the discharge coefficient, C_o, is approximately constant at 0.61–0.62.

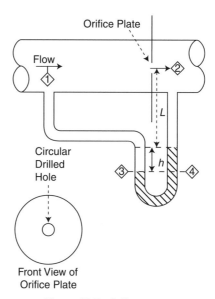

Figure 19.7 Orifice meter.

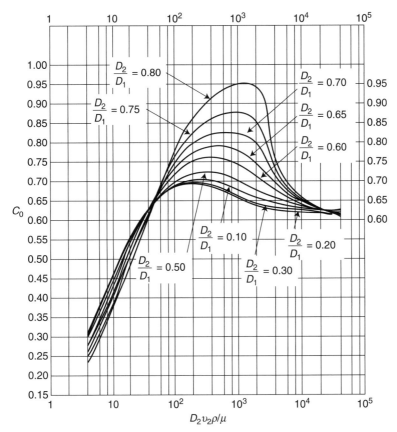

Figure 19.8 Discharge coefficients for drilled plate orifices. D_2 is the orifice diameter, and D_1 is the pipe diameter. The abscissa is the Reynolds number based on the orifice conditions.

An approximate equation that can be used to estimate the pressure recovery in an orifice meter is[1]:

$$\text{Percentage Pressure Recovery } \Delta P_{\text{rec}} = 14\frac{D_2}{D_1} + 80\left(\frac{D_2}{D_1}\right)^2 \tag{19.22}$$

Since

$$\text{Percentage Pressure Loss} = 100 - \Delta P_{\text{rec}}$$

the pressure loss in the orifice meter is:

$$\Delta P_L = \left(1 - \frac{\text{Percentage Pressure Recovery}}{100}\right)(P_1 - P_2) \tag{19.23}$$

The power loss (or power consumption) due to the orifice meter is (in consistent units):

$$\dot{W}_L = q \Delta P_L \tag{19.24}$$

Illustrative Example 19.7   An orifice meter equipped with flange taps is installed to measure the flow rate of air in a circular duct of diameter, $D_1 = 0.25$ m. The orifice diameter, $D_2 = 0.19$ m. The air is flowing at a rate of $1.0 \, \text{m}^3/\text{s}$ at 1 atm. Under these conditions, the air density, ρ, is $1.23 \, \text{kg/m}^3$ and the absolute viscosity, μ, is $1.8 \times 10^{-5} \, \text{kg/(m} \cdot \text{s)}$. Water is used as the manometer fluid. Calculate the pressure drop if the orifice discharge coefficient is 1.0, the actual pressure drop and the manometer head.

Solution   Calculate the velocity through the orifice.

$$v_2 = \frac{4(1)}{\pi (0.19)^2}$$
$$= 35.3 \, \text{m/s}$$

Assuming $C_o = 1$, calculate ΔP using Equation (19.20):

$$\Delta P = \frac{(1.23)(35.3)^2 [1 - (0.19/0.25)^4]}{2}$$
$$= 510 \, \text{Pa} = 52 \, \text{mm H}_2\text{O}$$

The orifice Reynolds number is

$$\text{Re} = \frac{(1.23)(35.3)(0.19)}{1.8 \times 10^{-5}} = 458{,}310$$

And

$$\frac{D_2}{D_1} = \frac{0.19}{0.25} = 0.76$$

The actual discharge coefficient, C_0, from Fig. 19.8 is

$$C_0 = 0.62$$

The actual pressure drop is calculated by once again rearranging Equation (19.20):

$$\Delta P = \frac{510}{(0.62)^2} = 1327 \, \text{Pa} = 135 \, \text{mm H}_2\text{O}$$

The percent pressure recovery and pressure loss may now be calculated.

$$\text{Percentage Pressure Recovery} = 14(0.76) + 80(0.76)^2 = 56.8\%$$
$$\text{Percentage Pressure Loss} = 43.2\%$$

The actual pressure loss, ΔP_L, in the orifice meter is then

$$\Delta P_L = 0.432(1326.2) = 573 \text{ Pa}$$

Illustrative Example 19.8   Refer to Illustrative Example 19.8. Calculate the actual power consumption of the orifice meter.

Solution   Calculate the power consumption using Equation (19.18)

$$\dot{W}_L = q(\Delta P_L) = (1.0)(\Delta P_L) = 1.0(573) = 573 \text{ W} = 0.77 \text{ hp}$$

The reader should note that if the pressure drop or the manometer head is given, the calculation of the volumetric flow rate will usually involve trial-and-error.

Illustrative Example 19.9   Air at ambient condition is flowing at the rate of 0.50 lb/s in a 4-in ID pipe. What sized orifice would produce an orifice pressure drop of 10 in H_2O?

Solution   This requires the simultaneous solution of Equations (19.6) and (19.7). At ambient conditions

$$\rho = 0.075 \text{ lb/ft}^3$$

The volumetric flowrate is therefore

$$q = \frac{(0.5)}{(0.075)}$$
$$= 6.67 \text{ ft}^3/\text{s}$$

From Equation (19.16)

$$v_2 = q/(\pi D_2^2/4)$$

From Equation (19.17)

$$v_2 = C_v \sqrt{\frac{2gh(\rho_M - \rho)}{\rho[1 - (D_2/D_1)^4]}}$$

Equating the above two equations, and solving by trial-and-error (assuming $C_v = 0.61$) gives $D_2 \cong 2.825$ in.

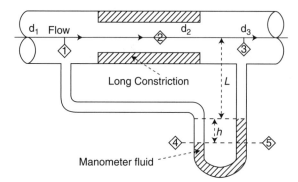

Figure 19.9   Constriction meter.

Is the assumption of $C_v = 0.61$ reasonable? This is left as an exercise for the reader. [Hint: Check to see if Re > 20,000 at the throat.]

Another way to measure the fluid flow rate in a pipe is to insert a long constriction (of smaller diameter than the pipe) inside the pipe and measure the pressure drop (or head loss) across the constriction (as shown in Fig. 19.9). The flow goes through a sudden contraction when it enters the constriction and through a sudden expansion as it exits the constriction. "Major" loss due to friction may be neglected. Only "minor" losses due to sudden expansion and sudden contraction are considered in the calculation of the flow rate. The calculation of the head loss follows the method outlined earlier for a sudden contraction and expansion.

The term *flowmeter* is sometimes used to designate any restricted opening or tube through which the rate of flow has been determined by calibration. For example, a 6-inch pipe may be tapered down to 2 inches and then enlarged back to 6 inches. The pressure drop through this "opening" provides a measure of the rate of flow, but this relation should be determined by calibration.

19.6   SELECTION PROCESS

Several factors should be considered in selecting a flow measurement device. Engineering decisions on the selection process should consider the following:

1. Is the fluid phase a gas or liquid?
2. The range (or capacity) of the device
3. Accuracy
4. Desired readout
5. Fluid properties
6. Internal environment

7. External environment
8. Capital cost
9. Operating cost
10. Reliability

Considering the complexity and diversity of flow meter measuring devices and the wide range of flow conditions encountered in industrial applications, one should carefully compare the different options that are available before purchasing a device. It should be noted that it is possible that a number of devices may be suitable for a given application.

REFERENCE

1. I. Farag, "Fluid Flow," A Theodore Tutorial, Theodore Tutorials, East Williston, NY, 1996.

NOTE: Additional problems are available for all readers at www.wiley.com. Follow links for this title.

20

VENTILATION

20.1 INTRODUCTION

Indoor air pollution is rapidly becoming a major health issue in the United States. Indoor pollutant levels are quite often higher than outdoors, particularly where buildings are tightly constructed to save energy. Since most people spend nearly 90% of their time indoors, exposure to unhealthy concentrations of indoor air pollutants is often inevitable. The degree of risk associated with exposure to indoor pollutants depends on how well buildings are ventilated and the type, mixture, and amounts of pollutants in the building.[1]

Industrial ventilation is the field of applied science concerned with controlling airborne contaminants to produce healthy conditions for workers and a clean environment for the manufacture of products. However, to claim that industrial ventilation will prevent contaminants from entering the workplace is naive and unachievable. More to the point, and within the realm of achievement, is the goal of controlling contaminant exposure within prescribed limits. To accomplish this goal, one must be able to describe the movement of contaminants in quantitative terms that take into account:

1. The spatial and temporal rate at which contaminants are generated;
2. The velocity field of the air in the workplace;
3. The spatial relationship between source, workers, and openings through which air is withdrawn or added;
4. Exposure limits (time-concentration relationships) that define unhealthy conditions.[1]

Fluid Flow for the Practicing Chemical Engineer. By J. Patrick Abulencia and Louis Theodore
Copyright © 2009 John Wiley & Sons, Inc.

In general, most control/recovery equipment are more economically efficient when handling higher concentrations of contaminants, all else being equal. Therefore, the gas handling system should be designed to concentrate contaminants in the smallest possible volume of air. This is important since, exclusive of the fan, the cost of the control equipment is based principally upon the volume of gas to be handled and not on the quantity of contaminant to be removed. The reduction of emissions by process and system control is an important adjunct to any cleaning technology.

Although ventilation does not remove the contaminants from the workplace, it does provide an opportunity to either recover/control or dilute (into the atmosphere) any problem emissions. Regarding manufacture and production, some unions and European nations either require or recommend that closed operations be employed.

The reader is referred to the work of Heinsohn[2] for an excellent treatment of ventilation. This topic is also addressed in Chapter 29. In addition, it should be noted that several of the Illustrative Examples at the end of the chapter were excerpted and/or adapted from publications[3,4] resulting from NSF sponsored faculty workshops.

20.2 INDOOR AIR QUALITY

Indoor air quality (IAQ) is a major concern because indoor air pollution may present a greater risk of illness than exposure to outdoor pollutants. People spend nearly 90% of their time indoors. This situation is compounded as sensitive populations the very young, the very old, and sick people—who are potentially more vulnerable to disease, spend many more hours indoors than the average population.

Indoor air quality problems have become more serious and of greater concern now than in the past because of a number of developments that are believed to have resulted in increased levels of harmful chemicals in indoor air. Some of those developments are the construction of more tightly sealed buildings to save on energy costs, the reduction of the ventilation rate standards to save still more energy, the increased use of synthetic building materials and synthetics in furniture and carpeting that can release harmful chemicals, and the widespread use of new pesticides, paints, and cleaners.

Some of the immediate health effects of indoor air quality problems are irritation of the eyes, nose and throat, headaches, dizziness and fatigue, asthma, pneumonitis, and "humidifier fever." Some of the long-term health effects of indoor air quality problems are respiratory diseases and cancer. These are most often associated with radon, asbestos, and second-hand tobacco smoke. The U.S. EPA (EPA), in a report to Congress in 1989, estimated that the costs of IAQ problems were in the tens of billions of dollar per year. The major types of costs from IAQ problems are direct medical costs, lost productivity due to absence from the job because of illness, decreased efficiency on the job, and damage to materials and equipment.

20.3 INDOOR AIR/AMBIENT AIR COMPARISON

Outdoor air pollution and indoor air pollution share many of the same pollutants, concerns, and problems. Both can have serious negative impacts on the health of the population. Not too many years ago, it was a common practice to advise people with respiratory problems to stay indoors on days when pollution outdoors was particularly bad. The assumption was that the indoor environment provided protection against outdoor pollutants. Recent studies conducted by the EPA have found, however, that the indoor levels of many pollutants are often two to five times, and occasionally more than 100 times, higher than corresponding outdoor levels. Such high levels of pollutants indoors are of even greater concern than outdoors because most people spend more time indoors than out. Indoors is where most people work, attend school, eat, sleep, and even where much of their recreational activity takes place.

Among the consumer and commercial products that release pollutants into the indoor air are pesticides, adhesives, cosmetics, cleaners, waxes, paints, automotive products, paper products, printed materials, air fresheners, dry cleaned fabrics, and furniture. In addition to the "active" ingredient in all the products mentioned, many products contain so-called inert ingredients that are also contaminants when released into indoor air. Examples include solvents, propellants, dyes, curing agents, flame retardants, mineral spirits, plasticizers, perfumes, hardeners, resins, binders, stabilizers, and preservatives. Aerosol products produce droplets, which remain in the air long enough to be inhaled. This allows the inhalation of some chemicals that would not be volatile enough to be inhaled otherwise. One consumer product that produces indoor air contaminants and merits special mention is tobacco smoke.

The single most important building or structural source of contaminant is formaldehyde contained in building materials such as plywood, adhesives, insulating materials such as urea formaldehyde foam, floor, and wall coverings. Depending on the individual type of construction and maintenance practices, there can be many other building sources of IAQ problems. Damp or wet wood, insulation, walls, and ceilings can be breeding places for allergens and pathogens that can become airborne. Allergens and pathogens can also originate from poorly maintained humidifiers, dehumidifiers, and air conditioners. If the building has openings to the soil, radon can enter the building in those areas where radon occurs. The building's heating plant can be the source of contaminants such as CO and NO_x. Automobile exhaust from attached garages is another source of carbon monoxide and nitrogen oxides. Particulates such as asbestos from crumbling insulation and lead form the sanding of lead-based paints are additional contaminants that can become part of the indoor air.

Personal activities can be sources of indoor air contaminants such as pathogenic viruses and bacteria, and a number of harmful chemicals such as products of human and animal (pet) respiration. Houseplants can release allergenic spores. Pet products, such as flea powder, can be sources of pesticides and pets produce allergenic dander when they lick themselves.

Outdoor sources of indoor air contaminants are widely varied. Polluted outdoor air can enter the indoor space through open windows, doors, and ventilation intakes. Most outdoor air is less contaminated than indoor air. In some cases, however, such as with a nearby smokestack, a parking lot, heavy street traffic, or an underground garage, outdoor air can be a significant source of indoor contaminants.

Outdoor pesticide applications, barbecue grills, and garbage storage areas can also bring outdoor contaminants into the building if placed close to a window or door, or the intake of a ventilation system. Improper placement of the intake of a ventilation system near a loading dock, parking lots, the exhaust from restrooms, laboratories, manufacturing spaces, and other exhausts of contaminated air is a major source of indoor air pollution. Other outdoor sources of indoor air pollutants are hazardous chemicals entering the structure from the soil. Examples are the aforementioned radon gas, methane and other gases from sanitary landfills, and vapors from leaking underground storage tanks of gasoline, oil, and other chemicals penetrating into basements. Polluted water can give off substantial quantities of harmful chemicals during showering, dishwashing, and similar activities.

20.4 INDUSTRIAL VENTILATION SYSTEMS

The major components of an industrial ventilation system include the following:

1. Exhaust hood
2. Ductwork
3. Contaminant control device
4. Exhaust fan
5. Exhaust vent or stack

Several types of hoods are available. One must select the appropriate hood for a specific operation to effectively remove contaminants from a work area and transport them into the ductwork. The ductwork must be sized such that the contaminant is transported without being deposited within the duct; adequate velocity must be maintained in the duct to accomplish this. Selecting a control device that is appropriate for the contaminant removal is important to meet certain pollution control removal efficiency requirements. The exhaust fan is the workhorse of the ventilation system. The fan must provide the volumetric flow at the required static pressure, and must be capable of handling contaminated air characteristics such as dustiness, corrosivity, and moisture in the air stream. Properly venting the exhaust out of the building is equally necessary to avoid contaminant recirculation into the air intake or into the building through other openings. Such problems can be minimized by properly locating the vent pipe in relation to the aerodynamic characteristics of the building. In addition, all or a portion of the cleaned air may be recirculated to the workplace. Primary (outside) air may be added to the workplace and is referred to as makeup air; the temperature and humidity of the makeup air may have to be controlled. It also may be necessary to exhaust a portion of the room air.

A line diagram of a typical industrial ventilation system is provided in Fig. 20.1. Note that either the control device or the fan (or both) can be located in the room/workplace.

Exposure to contaminants in a workplace can be reduced by proper ventilation. Ventilation can be provided either by *dilution ventilation* or by a *local exhaust*

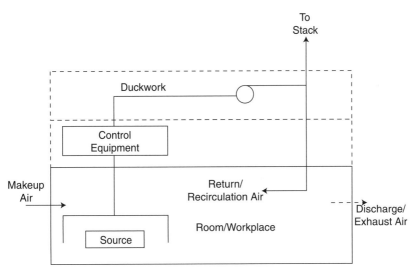

Figure 20.1   Industrial ventilation system components.

system. In dilution ventilation, air is brought into the work area to dilute the contaminant sufficiently to minimize its concentration and subsequently reduce worker exposure. In a local exhaust system, the contaminant itself is removed from the source through hoods.

A local exhaust is generally preferred over a dilution ventilation system for health hazard control because a local exhaust system removes the contaminants directly from the source, whereas dilution ventilation merely mixes the contaminant with uncontaminated air to reduce the contaminant concentration. Dilution ventilation may be acceptable when the contaminant concentration has a low toxicity, and the rate of contaminant emission is constant and low enough that the quantity of required dilution air is not prohibitively large. However, dilution ventilation is generally not acceptable when the acceptable concentration is less than 100 ppm.

In determining the quantity of dilution air required, one must also consider mixing characteristics of the work area in addition to the quantity (mass or volume) of contaminant to be diluted. Thus, the amount of air required in a dilution ventilation system is much higher than the amount required in a local exhaust system. In addition, if the replacement air requires heating or cooling to maintain an acceptable workplace temperature, then the operating cost of a dilution ventilation system may further exceed the cost of a local exhaust system.

The amount of dilution air required in a dilution ventilation system can be estimated using the following expression:

$$q = K(q_c/C_a) \tag{20.1}$$

where q = dilution air flowrate; K = dimensionless mixing factor; q_c = flowrate of pure contaminant vapor; and C_a = acceptable contaminant concentration. For more details, see Illustrative Example 20.2.

Illustrative Example 20.1 Discuss in some detail the term "air exchange rate." The answer should explain what is meant by the term "air exchange rate" and then discuss some of the variables that affect it. The discussion should address the following terms: infiltration, exfiltration, wind effects, stack effects, combustion effects, natural ventilation, and forced ventilation.[5]

Solution The term "air exchange rate" is the rate at which indoor air is replaced with outdoor air. The units of the air exchange rate are "air changes per hour" or "ach." If the volume of air in a building is replaced twice in 1 hour, the air exchange rate would be two. If the volume of air in a building is replaced once in 2 hours, the air exchange rate would be 0.5. The air exchange rate can be calculated by dividing the rate at which outdoor air enters the building in m^3/hr (or ft^3/hr) by the volume of the building in m^3 (or ft^3). If the air exchange rate were 1 ach, it would not mean that every molecule of indoor air would have been replaced at the end of 1 hour. Just which molecules were replaced would depend on a number of factors. Some of those factors are infiltration, exfiltration, wind effects, stack effects, combustion effects, natural ventilation, and forced ventilation.

Infiltration and exfiltration refer to the uncontrolled leakage of air into or out of the building through cracks and other unintended openings in the outer shell of the building. In addition to leakage around windows and doors, infiltration and exfiltration can occur at points such as openings for pipes, wires, and ducts. The rate of infiltration and exfiltration can vary greatly depending on such factors as wind temperature differences between indoors and outdoors, as well as the operation of stacks and exhaust fans.

Wind effects result from wind striking one side of a building causing positive pressure on that side and lower pressure on the opposite side (the leeward side). Air is forced into the building on the windward side and out the leeward side. Some buildings may be somewhat protected from wind effects by terrain, trees, and other buildings.

The tendency of warm air to rise in a room or through the levels of a multilevel building result in what is known as stack effects. In winter, when there is a large temperature difference between indoor and outdoor air, rising warm air escapes through openings at the top of the building and outdoor air is drawn in at the bottom of the building. The effect is usually less pronounced in the summer because of smaller temperature differences and the direction of the flow may be reversed.

Combustion effects often arise from fires in fireplaces, stoves, and heating systems. The combustion uses up indoor air (oxygen), which causes pressure in the building to drop. Outdoor air is then drawn in. This effect can double infiltration rates. Use of outdoor air in a heating system or fireplace substantially reduces this effect.

Natural ventilation is air that is drawn into a building through windows, doors, and other controlled openings. Natural ventilation results from wind striking the building and/or temperature differences between the outdoor air and the indoor air.

Forced ventilation refers to drawing air into a building through fans and ducts. The effectiveness in removing contaminants from indoor air by the use of forced ventilation can vary widely. Fans used to exhaust specific sources of pollutants (such as the kitchen stove) can be very effective. Most forced ventilation systems are used to circulate air-conditioned air. Whole house fans and the forced ducted ventilation systems of large buildings must be carefully balanced by the air supply to prevent backdrafts of contaminants from stacks and heating plants.

Illustrative Example 20.2:   Estimate the dilution ventilation required in an indoor work area where a toluene-containing adhesive in a nanotechnology process is used at a rate of 3 gal/8 h workday. Assume that the specific gravity of toluene (C_7H_8) is 0.87, that the adhesive contains 4 vol% toluene, and that 100% of the toluene is evaporated into the room air at 20°C. The plant manager has specified that the concentration of toluene must not exceed 80% of its threshold limit value (TLV) of 100 ppm.[7]

As described earlier, the following equation can be used to estimate the dilution air requirement:

$$q = K(q_c/C_a) \tag{20.1}$$

where q = dilution air flowrate; K = dimensionless mixing factor that accounts for less than complete mixing characteristics of the contaminant in the room, the contaminant toxicity level and the number of potentially exposed workers; usually, the value of K varies from 3 to 10, where 10 is used under poor mixing conditions and when the contaminant is relatively toxic; q_c = volumetric flowrate of pure contaminant vapor, c; C_a = acceptable contaminant concentration in the room, volume or mole fraction (ppm $\times$ 10^{-6}).

Solution   The dilution air can be estimated from (see Problem statement)

$$q = K(q_c/C_a) \tag{20.1}$$

Since the TLV for toluene is 100 ppm and C_a is 80% of the TLV,

$$C_a = [0.80(100)] \times 10^{-6} = 80 \times 10^{-6} \quad \text{(volume fraction)}$$

The mass flowrate of toluene is

$$\dot{m}_{tol} = \left(\frac{3 \text{ gal}_{adhesive}}{8 \text{ h}}\right) \left(0.4 \frac{\text{gal}_{toluene}}{1 \text{ gal}_{adhesive}}\right) \left[\frac{(0.87)(8.34 \text{ lb})}{1 \text{ gal}_{toluene}}\right]$$

$$= 1.09 \text{ lb/h}$$

$$= \left(\frac{1.09 \text{ lb}}{1 \text{ h}}\right) \left(\frac{454 \text{ g}}{1 \text{ lb}}\right) \left(\frac{1 \text{ h}}{60 \text{ min}}\right)$$

$$= 8.24 \text{ g/min}$$

Since the molecular weight of toluene is 92,

$$\dot{n}_{tol} = 8.24/92$$
$$= 0.0896 \, gmol/min$$

The resultant toluene vapor volumetric flowrate q_{tol} is calculated directly from the ideal gas law:

$$q_{tol} = \frac{(0.0896 \, gmol/min)[0.08206 \, atm \cdot L/(gmol \cdot K)](293K)}{1 \, atm}$$
$$= 2.15 \, L/min$$

Therefore, the required diluent volumetric flowrate is (with $K = 5$)

$$q = \frac{(5)(2.15 \, L/min)}{80 \times 10^{-6}}$$
$$= 134{,}375 \, L/min$$
$$= \left(134{,}375 \, \frac{L}{min}\right)\left(\frac{1 \, ft^3}{28.36 \, L}\right)$$
$$= 4748 \, ft^3/min$$

Illustrative Example 20.3   A certain poorly-ventilated chemical storage room (10 ft × 20 ft × 8 ft) has a ceiling fan but no air conditioner. The air in the room is at 51°F and 1.0 atm pressure. Inside this room, a 1 lb bottle of iron (III) sulfide (Fe_2S_3) sits next to a bottle of sulfuric acid containing 1 lb H_2SO_4 in water. An earthquake (or perhaps the elbow of a passing technician) sends the bottles on the shelf crashing to the floor where the bottles break, and their contents mix and react to form iron (III) sulfate [$Fe_2(SO_3)$] and hydrogen sulfide (H_2S).

Calculate the maximum H_2S concentration that could be reached in the room assuming rapid complete mixing by the ceiling fan with no addition of outside air.[6]

Solution   Balance the chemical equation:

amount	Fe_2S_3	+	$3H_2SO_4$	$\longrightarrow$	$Fe_2(SO_4)_3$	+	$3H_2S$
before	1 lb		1 lb		0		0
reaction	0.0048 lbmol		0.010 lbmol		0		0

The molecular weights of Fe_2S_3 and H_2SO_4 are 208 and 98, respectively.

The terms *limiting reactant* and *excess reactant* refer to the actual number of moles present in relation to the stoichiometric proportion required for the reaction to proceed to completion. From the stoichiometry of the reaction, 3 lbmol of H_2SO_4 are required to react with each lbmol of Fe_2S_3. Therefore, sulfuric acid is the limiting reactant and the iron (III) sulfide is the excess reactant. In other words, 0.0144 lbmol of H_2SO_4 is required to react with 0.0048 lbmol of Fe_2S_3, or 0.030 lbmol of Fe_2S_3 is required to react with 0.010 lbmol of H_2SO_4.

Calculate the moles of H_2S generated n_{H_2S}:

$$n_{H_2S} = (0.010 \text{ lbmol } H_2SO_4)(3H_2S/3H_2SO_4)$$
$$= 0.010 \text{ lbmol}$$

Next, convert the moles to mass:

$$m_{H_2S} = (0.010 \text{ lbmol } H_2S)(34 \text{ lb/lbmol } H_2S)$$
$$= 0.34 \text{ lb}$$

The final H_2S concentration in the room in ppm, C_{H_2S}, can now be calculated. At $32°$ and 1 atm, 1 lbmol of an ideal gas occupies 359 ft^3; at $51°$ 1 lbmol occupies

$$V = 359 \left(\frac{460 + 51}{460 + 32} \right) = 373 \text{ ft}^3$$

Therefore,

$$C_{H_2S} = \frac{(0.34 \text{ lb}) \left(\dfrac{373 \text{ ft}^3}{\text{lbmol air}} \right) \left(\dfrac{\text{lbmol air}}{29 \text{ lb}} \right) (10^6)}{1600 \text{ ft}^3}$$
$$= 2733 \text{ ppm}$$

This concentration of H_2S far exceeds an acceptable value.

Illustrative Example 20.4: Vinyl Chloride Application   The vinyl chloride fugitive emission rate in a process was estimated to be 10 g/min by a series of bag tests conducted for the major pieces of connections (i.e., flanges and valves) and pump seals. Determine the flowrate of air ($25°C$) necessary to maintain a level of 1.0 ppm by dilution ventilation. Correct for incomplete mixing by employing a safety factor of 10. Also consider partially enclosing the process and using local exhaust ventilation. Assume that the process can be carried out in a hood with an opening 30 inches wide by 25 inches high with a face velocity greater than 100 ft/ min to ensure high capture efficiency. What will be the flowrate of air required for local exhaust ventilation? Which ventilation method seems better?

Solution   Convert the mass flowrate of the vinyl chloride (VC) to volumetric flow-rate q, in cm^3/min and acfm. First, use the ideal gas law to calculate the density.

$$\rho = \frac{P(MW)}{RT}$$
$$= \frac{(1 \text{ atm})(78 \text{ g/gmol})}{\left(82.06 \dfrac{cm^3 \cdot atm}{mol \cdot K} \right)(298K)}$$
$$= 0.00319 \text{ g/cm}^3$$

$$q = \text{(mass flowrate)}/\text{(density)}$$
$$= (10\,\text{g/min})/(0.00319\,\text{g/cm}^3)$$
$$= 3135\,\text{cm}^3/\text{min}$$
$$= 0.1107\,\text{acfm}$$

Calculate the air flowrate in acfm, q_{air}, required to meet the 1.0 ppm constraint with the equation

$$q_{air} = (0.1107\,\text{acfm})/10^{-6}$$
$$= 1.107 \times 10^5\,\text{acfm}$$

Apply the safety factor to calculate the actual air flowrate for dilution ventilation:

$$q_{air,dil} = (10)(1.107 \times 10^5\,\text{acfm})$$
$$= 1.107 \times 10^6\,\text{acfm}$$

Now consider the local exhaust ventilation by first calculating the face area of the hood S, in square feet:

$$S = \text{(Height)(Width)}$$
$$= (30\,\text{in.})(25\,\text{in.})(\text{ft}^2/144\,\text{in.}^2)$$
$$= 5.21\,\text{ft}^2$$

The air flowrate in acfm $q_{air,exh}$, required for a face velocity of 100 ft/min is then

$$q_{air,exh} = (5.21\,\text{ft}^2)/(100\,\text{ft/min})$$
$$= 521\,\text{acfm}$$

Since the air flowrate for dilution ventilation is approximately 2000 times higher than the local ventilation air flowrate requirement, and considering the high cost of large blowers to handle high air flowrates, the local ventilation method appears to be the better method for this case.

Illustrative Example 20.5[8,9] Your consulting firm has received a contract to develop, as part of an emergency preparation plan, mathematical models describing the concentration of a chemical in a medium-sized ventilated laboratory room. The following information/data (SI units) is provided:

$V =$ volume of room, m^3
$q_0 =$ volumetric flow rate of ventilation air, m^3/min
$c_0 =$ concentration of the chemical in ventilation air, gmol/m^3

c = concentration of the chemical leaving ventilated room, gmol/m^3
c_1 = concentration of the chemical initially present in ventilated room, gmol/m^3
r = rate of disappearance of the chemical in the room due to reaction and/or other effects, $\text{gmol}/\text{m}^3 \cdot \text{min}$.

As an authority in the field (having taken several Theodore courses), you have been requested to:

1. Obtain the equation describing the concentration in the room as a function of time if there are no "reaction" effects, that is, $r = 0$.
2. Obtain the equation describing the concentration in the room as a function of time if $r = -k$. Note once again that the minus sign is carried since the agent is disappearing.
3. Obtain the equation describing the concentration in the room as a function of time if $r = -kc$. Note once again that the minus sign is carried since the chemical is disappearing.
4. For Part 2, discuss the effect on the resultant equation if k is extremely small, that is, $k \rightarrow 0$.

Solution   Use the laboratory room as the control volume. Apply the conservation law for mass to the chemical

$$\left\{ \begin{array}{c} \text{rate of mass} \\ \text{in} \end{array} \right\} - \left\{ \begin{array}{c} \text{rate of mass} \\ \text{out} \end{array} \right\} + \left\{ \begin{array}{c} \text{rate of mass} \\ \text{generated} \end{array} \right\} = \left\{ \begin{array}{c} \text{rate of mass} \\ \text{accumulated} \end{array} \right\}$$

Employing the notation specified in the problem statement gives:

$$\{\text{rate of mass in}\} = q_0 c_0$$
$$\{\text{rate of mass out}\} = q_0 c$$
$$\{\text{rate of mass generated}\} = rV$$
$$\{\text{rate of mass accumulated}\} = \frac{dV}{dt}$$

Substituting above gives

$$q_0 c_0 - q_0 c + rV = \frac{dV}{dt}$$

Since the laboratory room is constant, V may be taken out of the derivative term. This leads to

$$\frac{q_0}{V}(c_0 - c) + r = \frac{dc}{dt}$$

The term V/q_0 represents the average residence time that the chemicals reside in the room and is usually designated as τ. The above equation may then be rewritten as

$$\frac{dc}{dt} = \frac{c_0 - c}{\tau} + r$$

1. If $r = 0$,

$$\frac{dc}{dt} = \frac{c_0 - c}{\tau}$$

separating variables

$$\frac{dc}{c_0 - c} = \frac{dt}{\tau}$$

$$\int_{c_i}^{c} \frac{dc}{c_0 - c} = \int_{0}^{t} \frac{dt}{\tau}$$

$$\ln\left(\frac{c_0 - c}{c_0 - c_i}\right) = \frac{t}{\tau}$$

$$\left(\frac{c_0 - c}{c_0 - c_i}\right) = e^{-t/\tau}$$

$$c = c_0 - (c_0 - c_i)e^{-t/\tau} = c_0 + (c_i - c_0)e^{-t/\tau}$$

2. If $r = -k$

$$\frac{dc}{dt} = \frac{c_0 - c}{\tau} - k = \frac{c_0}{\tau} - k - \frac{c}{\tau} = \left(\frac{c_0 - k\tau}{\tau}\right) - \frac{c}{\tau}$$

$$\frac{dc}{\left[\left(\dfrac{c_0 - k\tau}{\tau}\right) - \dfrac{c}{\tau}\right]} = dt$$

$$\frac{dc}{[(c_0 - k\tau) - c]} = \frac{dt}{\tau}$$

$$\int_{c_i}^{c} \frac{dc}{[(c_0 - k\tau) - c]} = \int_{0}^{t} \frac{dt}{\tau}$$

$$-\ln\left[\frac{(c_0 - k\tau) - c}{(c_0 - k\tau) - c_i}\right] = \tau$$

$$\frac{(c_0 - k\tau) - c}{(c_0 - k\tau) - c_i} = e^{-(t/\tau)}$$

$$c = c_i e^{-t/\tau} + (c_0 - k\tau)\left[1 - e^{-(t/\tau)}\right]$$

3. If $r = -kc$

$$\frac{dc}{dt} = \frac{c_0 - c}{\tau} - kc = \frac{c_0}{\tau} - \frac{c}{\tau} - kc$$

$$= \frac{c_0}{\tau} - c\left(\frac{k\tau + 1}{\tau}\right)$$

$$\frac{dc}{\left[\frac{c_0}{\tau} - \left(\frac{1 + k\tau}{\tau}\right)c\right]} = dt$$

$$\int_{c_i}^{c} \frac{dc}{[c_0 - (1 + k\tau)c]} = \int_{0}^{t} \frac{dt}{\tau}$$

$$-\left(\frac{1}{1 + k\tau}\right)\ln\left[\frac{c_0 - (1 + k\tau)c}{c_0(1 + k\tau)c_i}\right] = \frac{t}{\tau}$$

$$\left(\frac{c_0 - (1 + k\tau)c}{c_0 - (1 + k\tau)c_i}\right) = e^{-(t/\tau)(1+k\tau)}$$

$$c = c_i e^{-(t/\tau)(1+k\tau)} + \left(\frac{c_0}{1 + k\tau}\right)\left[1 - e^{-(\frac{t}{\tau})(1+k\tau)}\right]$$

4. If $k = 0$, see solution 2.

$$c = c_i e^{-t/\tau} + (c_0 - k\tau)[1 - e^{-t/\tau}]$$

$$= c_i e^{-t/\tau} + c_0 - c_0 e^{-t/\tau}$$

$$= c_0 + (c_i - c_0)e^{-t/\tau}$$

Illustrative Example 20.6   Refer to Illustrative Example 20.5. For Part 1, qualitatively discuss the effect on the final equation if the volumetric flow rate, v_0, varies sinusoidally. Also, qualitatively discuss the effect on the final equation if the inlet concentration, c_0, varies sinusoidally.

Solution If v_0 varies and τ varies, solving the equation becomes more complex. Variations need to be included in the describing equation

$$\frac{dc}{dt} = \frac{c_0 - c}{\tau} + r; \quad \tau = \tau(t)$$

This may require numerical solution. If both $c_0 = c_0(t)$ and $v_0 = v\ d(t)$, the solution again applies.

Illustrative Example 20.7 Refer to Illustrative Example 20.5.

1. Calculate minimum air ventilation flow rate into the room containing $10\ \mathrm{ng/m^3}$ (ng = nanograms) of a toxic chemical to assure that the chemical concentration does not exceed $35.0\ \mathrm{ng/m^3}$. The chemical is being generated in the laboratory at a rate of $250\ \mathrm{ng/min}$. Assume steady-state conditions.
2. Calculate the steady-state concentration in the laboratory: the initial concentration of the chemical is $500\ \mathrm{ng/m^3}$. There is no additional source of chemical generated and the ventilation air is essentially pure, i.e., there is no background chemical concentration.

Solution

1. The applicable model for this case is:

$$q_0(c_0 - c) + rV = V\frac{dc}{dt}$$

Under steady-state, $dc/dt = 0$. Pertinent information includes

$$rV = 250\ \mathrm{ng/min}$$
$$c_0 = 10\ \mathrm{ng/m^3}$$
$$c = 35\ \mathrm{ng/m^3}$$

Substituting gives

$$
\begin{aligned}
q_0 &= \frac{-rV}{c_0 - c} \\
&= \frac{rv}{c - c_0} \\
&= \frac{250}{35 - 10} \\
&= 10\ \mathrm{m^3/min} = 353\ \mathrm{ft^3/min}
\end{aligned}
$$

2. The applicable model is:

$$q_0(c_0 - c) + rV = V\frac{dc}{dt}$$

Once again, for steady-state condition, $dc/dt = 0$. In addition, based on the information provided, $r = 0$ and $c_0 = 0$. Therefore, and as expected,

$$c = 0$$

Illustrative Example 20.8   Refer to Illustrative Examples 20.5 and 20.7. If the room volume is $142 \, m^3$, the flowrate of the $10 \, ng/m^3$ ventilation air is $12.1 \, m^3/min$, and the chemical is being generated at a steady rate of $30 \, ng/min$, calculate how long it would take for the concentration to reach $20.7 \, ng/m^3$. The initial concentration in the laboratory is $85 \, ng/m^3$. How long would it take to reach $12.2 \, ng/m^3$?

Solution   First note that

$$\tau = 142/12.1 = 11.73 \text{ min}$$

$$k = r/V = 30/142 = 0.211 \text{ ng}/(m^3 \cdot min)$$

The applicable describing equation is:

$$c = c_i e^{-t/\tau} + (c_0 + k\tau)[1 - e^{-(t/\tau)}]$$

Substituting gives

$$20.7 = 85e^{-(t/11.73)} + (10 + 2.48)[1 - e^{-(t/11.73)}]$$

Solving by trial-and-error gives (approximately)

$$t = 29 \text{ min}$$

Calculate the steady-state concentration for this condition. The applicable model is obtained from Part 2 of Illustrative Example 20.5, after setting $dc/dt = 0$:

$$\frac{c_0 + k\tau}{\tau} - \frac{c}{\tau} = 0$$

Solving and substituting gives

$$c = c_0 + k\tau$$
$$= 10 + (0.211)(11.73)$$
$$= 12.48 \text{ ng}/m^3$$

Since this is the steady-state concentration, it will take an infinite period of times to reach this value. This steady-state concentration represents the minimum concentration achievable in the laboratory based on the conditions specified. Therefore, the concentration will never reach a value of $12.2\,\text{ng/m}^3$.

REFERENCES

1. L. Theodore and R. Kunz, "Nanotechnology: Environmental Implications and Solutions," John Wiley & Sons, Hoboken, NJ, 2005.

2. R. Heinsohn, "Industrial Ventilation: Engineering Principle," John Wiley & Sons, Hoboken, NJ, 1991.

3. R. Dupont, T. Baxter, and L. Theodore, "Environmental Management: Problems and Solutions," CRC-Lewis Publishers, Boca Raton, FL, 1998.

4. K. Ganesan, L. Theodore, and R. Dupont, "Air Toxins: Problems and Solutions," Gordon and Beach, New York, 1996.

5. L. Theodore, "Nanotechnology: Basic Calculations for Engineers and Scientists," John Wiley & Sons, Hoboken, NJ, 2007.

6. J. Santoleri, J. Reynolds, and L. Theodore, "Introduction to Hazardous Waste Incineration," 2nd edition, John Wiley & Sons, Hoboken, NJ, 2000.

7. L. Theodore, Personal notes—final exam problem, Accident & Emergency Management, 2000.

8. J. Reynolds, J. Jeris, and L. Theodore, "Handbook of Chemical and Environmental Engineering Calculations," John Wiley & Sons, Hoboken, NJ, 2004.

9. L. Theodore, "Chemical Reaction Kinetics," A Theodore Tutorial, East Williston, NY, 1992.

NOTE: Additional problems are available for all readers at www.wiley.com. Follow links for this title.

21

ACADEMIC APPLICATIONS

The illustrative examples provided in this chapter pertain to academic applications. Those readers desiring more technical and industry-oriented calculations should bypass these examples and proceed directly to the next chapter. There are 18 Illustrative Examples in this chapter; several of the earlier examples are qualitative in nature.

Illustrative Example 21.1   Qualitatively explain pipe schedule number.

Solution   The wall thickness of a pipe is specified by a schedule number, which is a function of the internal pressure and allowable stress. The describing equation is:

$$\text{Schedule number} = 1000 \, P/S$$

where P = internal working pressure (lb_f/in^2 gauge)
        S = allowable stress (lb_f/in^2)
There are 10 schedule numbers in use and these range from 10 to 160; the thickness of the pipe wall increases with the schedule number. Schedule 40 is the most commonly used pipe thickness for normal temperature and pressure applications.

Illustrative Example 21.2   Selecting the appropriate pipe diameter to handle a particular liquid flow application is a function of many varieties. Provide information on suggested pipe diameters for various flow ranges.[1]

Fluid Flow for the Practicing Chemical Engineer. By J. Patrick Abulencia and Louis Theodore
Copyright © 2009 John Wiley & Sons, Inc.

Table 21.1   Nominal pipe diameters for liquid flows

Capacity (gpm)	Nominal Pipe Diameter (in.)
0–15	1
15–70	2
70–150	3
150–250	4

Solution   Typical pipe diameters for various liquid flow capacities are given in Table 21.1.

Illustrative Example 21.3   Provide information on how one would estimate the minimum pipe thickness required for a particular application.

Solution   In calculating the minimum thickness of pipe wall (t) required for any specific pressure, temperature, and corrosive condition, the following formula may be employed for estimation purposes

$$t = \frac{PD_o}{2S} + C$$

where t = minimum pipe wall thickness allowable on inspection, in.
P = maximum internal service pressure, $lb/in.^2$
D_o = outside diameter of the pipe, in.
S = allowable bursting stress in the pipe material, $lb/in.^2$
C = allowance for threading, mechanical strength and corrosion.

Appropriate values for S and C are available in the literature.[2]

Illustrative Example 21.4   Provide a qualitative discussion of pipes and tubing.

Solution   The manufacturing of various classes of pipe uses many materials including ceramic, metal, and plastic. Pipes and tubing are used as conduits for transporting liquids and gases. In the past, piping materials have included wood and lead. However, current materials include ceramic, metal, and plastic. Metal pipes are commonly made from unfinished black (lacquer) or galvanized steel, brass, and ductile iron. Other sources of metal pipes include copper, which is popular for plumbing systems. Plastic tubing is widely used for its light weight, chemical resistance, non-corrosive properties, and ease of making connections. Raw materials include polyvinyl chloride (PVC), polyvinyl dichloride (CPVC), poly-ethylene (PE), polybutylene (PB), and acrylonitrile butadiene styrene (ABS). Ceramic pipes are usually used for low pressure applications such as gravity flow or drainage.

Commonly used steel pipe ratings are Schedule 40 (standard) and Schedule 80 (extra strong). In most cases in the U.S., Schedule 40 piping is used for heating applications, while Schedule 80 is employed for high pressure applications or cases where higher than normal corrosion rates are expected.

Illustrative Example 21.5   Briefly describe the following four classes of fittings.

1. Fittings that extend or terminate pipe runs
2. Fittings that change a pipe's direction
3. Fittings that connect two or more pipes
4. Fittings that change pipe size

Solution

1. Couplings extend a run by connecting two lengths of pipe. They are available in all standard pipe sizes and nearly all varieties of pipe. They are called reducing couplings if they are connecting differently-sized pipe.

   Caps and plugs end a run of pipe by closing it off with a watertight seal.

2. Elbows change the direction of pipes. The most commonly used are 90° and 45° elbows, but they are also available in other sizes. They are identified by their angle but they are ordinarily referred to by number only. An elbow may be female at both ends, or in the case of a street elbow, may be male on one end and female on the other.

3. Tees offer the most varieties of any type of fitting. Tees are fittings in the shape of a "T" where the top of the "T" is the continuous pipe run, and the vertical section is a branch connected to it. They may be reducing tees, where the branch and/or one end of the through section is a smaller diameter than the inlet. They may have side-inlets, which allow for a fourth pipe to join them. These may be left- or right-handed depending on which side the inlet enters.

4. Reducers can be couplings, tees, or elbows, where one end is smaller than the other. This reduces the pipe's diameter between the inlet and outlet. In the case of fittings that connect more than two pipes, one of the outlets is of a smaller diameter, (not counting side-inlets, which are always smaller). Some reduce pipe only one size; others can reduce several sizes. Both ends are female.

   Bushings serve the same purpose as reducers except that they have one male and one female end. In steel pipe, they are threaded inside and out; instead of screwing directly onto pipe threads, they screw into a coupling and pipe is threaded into them. They are virtually invisible once installed. In PVC and copper, they are not threaded but work the same way.

   Expanders serve the opposite purpose of reducers. They increase the pipes diameter between inlet and outlet.

Illustrative Example 21.6   List the typical uses and applications for each of the following six valves, plus their advantages and disadvantages.

1. Gate
2. Globe
3. Ball
4. Butterfly
5. Pinch
6. Plug

Solution

1. Gate valve

   Recommended uses:
      1. Fully open/closed, non-throttling
      2. Infrequent operation
      3. Minimal fluid trapping in line

   Applications: Oil, gas, air, slurries, heavy liquids, steam, noncondensing gases, and corrosive liquids

   Advantages:
      1. High capacity
      2. Tight shutoff
      3. Low cost
      4. Little resistance to flow

   Disadvantages:
      1. Poor control
      2. Cavitate at low pressure drops
      3. Cannot be used for throttling

2. Globe valve

   Recommended uses:
      1. Throttling service/flow regulation
      2. Frequent operation

   Applications: Liquids, vapors, gases, corrosive substances, slurries

   Advantages:
      1. Efficient throttling
      2. Accurate flow control
      3. Available in multiple ports

   Disadvantages:
      1. High pressure drop
      2. More expensive than other valves

3. Ball valve

   Recommended uses:

   1. Fully open/closed, limited-throttling
   2. Higher temperature fluids

   Applications: Most liquids, high temperatures, slurries

   Advantages:

   1. Low cost
   2. High capacity
   3. Low leakage and maintenance
   4. Tight sealing with low torque

   Disadvantages:

   1. Poor throttling characteristics
   2. Prone to cavitation

4. Butterfly valve

   Recommended uses:

   1. Fully open/closed or throttling services
   2. Frequent operation
   3. Minimal fluid trapping in line

   Applications: Liquids, gases, slurries, liquids with suspended solids

   Advantages:

   1. Low cost and maintenance
   2. High capacity
   3. Good flow control
   4. Low pressure drop

   Disadvantages:

   1. High torque required for control
   2. Prone to cavitation at lower flows

5. Pinch valve

   Recommended uses:

   1. Fully open/closed, or throttling services
   2. Abrasives and corrosives

   Applications: Medical, pharmaceutical, wastewater, slurries, pulp, powder and pellets

   Advantages:

   1. Streamlined flow
   2. High coefficient of flow

   Disadvantages:

   1. Limited materials
   2. Low shut-off capabilities
   3. Low pressure limits

6. Plug valve

   Recommended uses:

   1. Fully open/closed, non-throttling
   2. Maintain flow

   Applications: Sewage, sludge, and wastewater

   Advantages:

   1. Easy operation
   2. Medium to high flow
   3. Good shut off

   Disadvantages:

   1. Low cleanliness
   2. Inability to handle slurry

Illustrative Example 21.7   If water at 70°F is flowing through a 3/8 in schedule 40 brass pipe at a volumetric flow rate of 2.0 gpm, calculate the Reynolds number. Also determine whether the flow is in the laminar or turbulent region.

Solution   From Table A.5 in the Appendix,

D_i (3/8 inch pipe) = 0.493 in = 0.0411 ft
S_i (3/8 inch pipe) = 0.00133 ft^2

From Table A.4 in the Appendix,

μ (at 70°F) = 0.982 cP = 6.598 × 10^{-4} lb/ft · s
ρ = 62.4 lb/ft^3

The volumetric flow rate is converted to ft^3/s by noting that 1 gpm = 0.00228 ft^3/s.

$$q = (2.0)(0.00228)\,\text{ft}^3/\text{s}$$

$$= 0.00456\,\text{ft}^3/\text{s}$$

The velocity of the fluid is calculated by dividing the volumetric flow rate by the cross-sectional area of the 3/8 inch pipe:

$$v = \frac{q}{S}$$

$$= \frac{0.00456}{0.00133}$$

$$= 3.43\,\text{ft}/\text{s}$$

The Reynolds number is then calculated by employing Equation (12.1).

$$Re = \frac{D v \rho}{\mu}$$

$$= \frac{(0.0411)(3.43)(62.4)}{0.0006598}$$

$$= 13,330$$

Since the Reynolds number is above 2100, the flow of the water is in the turbulent region.

Illustrative Example 21.8   Part of a Fluid Flow Unit Operations experiment at Manhattan College requires the following calculation. Determine the Reynolds numbers for water flows of 6.0 gpm and 1.5 gpm through 3/8 in and 1/2 in schedule 40 pipes.

Solution   For water,

$$\rho = 62.4 \, lb/ft^3$$

and

$$\mu = 6.72 \times 10^{-4} \, lb/ft \cdot s$$

For 3/8 inch and 1/2 inch schedule 40 pipe, the outside and inside diameters can again be found in the Appendix and provided below.

Pipe size	D_o (in)	D_i (in)	D_o (ft)	D_i (ft)
3/8 in.	0.675	0.493	0.0563	0.041
1/2 in.	0.84	0.622	0.07	0.0518

Use the following equation to calculate the average velocity:

$$v = q/S = (gpm)\left(\frac{35.3}{264}\right)\left(\frac{1}{60}\right)\left(\frac{(4)(144)}{\pi D_i^2}\right)$$

$$= (0.409)(gpm)/(D_i)^2$$

where $v = ft/s$
    $q = gal/min$
    $D_i = inches$
For a 3/8 in pipe with an inside diameter of 0.493 inches, the results are

q	v
1.5	2.52
6.0	10.1

For a 1/2 in. pipe with an inside diameter of 0.622 inches, the results are

q	v
1.5	1.58
6.0	6.33

Use Equation (12.1) for the calculation of the Reynolds number

$$\text{Re} = Dv\rho/\mu$$

and the properties of water at room temperature provided above. The following values result:

For a 3/8 in. pipe and 1.5 gpm, Re = 9594
For a 3/8 in. pipe and 6 gpm, Re = 38,452
For a 1/2 in. pipe and 1.5 gpm, Re = 7600
For a 1/2 in. pipe and 6 gpm, Re = 30,447

This indicates that all of the flows are turbulent although for smaller pipes the result could approach the "intermediate" regime.

Illustrative Example 21.9   If water is flowing in an upward 15 ft vertical pipe in a 3/8 in. schedule 40 brass pipe, calculate the frictional loss using the Bernoulli equation if the pressure drop of the flowing fluid from bottom to top is 4.5 lb_f/ft^2.

Solution   The Bernoulli equation is employed. The velocity head is neglected since the water is flowing through a straight vertical pipe and the velocity can be assumed not to change. From Equation (17.11), with location 1 at the bottom of the pipe

$$\frac{P_2 - P_1}{\rho} + \frac{g}{g_c}(z_2 - z_1) + \frac{v_2^2 - v_1^2}{2g_c}^{\,0} = h_f$$

Substituting the information provided,

$$h_f = (-4.5)/(62.4) + (32.2/32.2)(15 - 0)$$
$$= -0.0721 + 15$$
$$= 14.9 \text{ ft} \cdot lb_f/lb$$

The frictional loss of the vertical pipe is therefore 14.9 ft · lb_f/lb. Note that the contribution of the "pressure" loss is negligible in comparison to potential or hydrostatic head.

Illustrative Example 21.10   A centrifugal pump is needed to transport water from sea level to 10,000 feet above sea level. Atmospheric pressure at 10,000 feet is 10.2 psi. Using a mass flow rate of 50 lb/s and an efficiency of no less than 65%, determine the actual pump work in ft · lb$_f$/s and hp. Neglect frictional losses.

Solution   Using Equations (13.4) or (17.11) and neglecting kinetic energy effects and frictional losses, one may write

$$\frac{\Delta P}{\rho} + \frac{\Delta v^2}{2g_c} + \Delta z\left(\frac{g}{g_c}\right) - h_s + h_f = 0$$

$$h_s = \frac{P_2 - P_1}{\rho} + \frac{v_2^2 - \cancel{v_1^2}^{\,0}}{\cancel{2g_c}} + (z_2 - z_1)\frac{g}{g_c} + \cancel{h_f}^{\,0}/\eta_p$$

where $P_2 =$ pressure at 10,000 ft ($z_2 = 10,000$ ft)
$\quad\quad\;\; P_1 =$ atmospheric pressure at sea level conditions ($z_1 = 0$ ft)
$\quad\quad\;\; \rho =$ density of water
$\quad\quad\;\; \eta_p =$ frictional efficency of pump

Plugging in for these values yields

$$h_s = \frac{(10.2 - 14.7)(144)}{62.4} + 10,000 - 0$$

$$= -10 + 10,000$$
$$= 9990 \text{ ft} \cdot \text{lb}_f/\text{lb}$$
$$= (9990)(50)$$
$$= 499,500 \text{ ft} \cdot \text{lb}_f/\text{s}$$
$$= 499,500/550$$
$$= 908 \text{ hp}$$

The above value represents the work delivered by the pump to the fluid (water). The actual pump work is calculated by dividing the above terms by the frictional efficiency. For example,

$$W_p = 908/0.65$$
$$= 1397 \text{ hp}$$

Finally the reader should note that there is a credit of "pressure" energy since the discharge pressure is lower than the input (inlet) pressure.

Illustrative Example 21.11   What is the maximum water velocity allowed in a pipe length of 150 ft, given a required pressure drop of no more than 5 lb$_f$/in^2.

Solution   First a calculation of the Reynolds number is required. However, the velocity is not known, so a trial-and-error method will be employed using the

Table 21.2   Trial-and-error calculations for Illustrative Example 21.11

v (ft/s)	Re	f	ΔP (lb/ft^2)	ΔP (lb/in^2)
3	5060	0.00956	965	7.00
2	3370	0.0107	480	3.33
1	1690	0.0130	146	1.01
1.5	2530	0.0116	292	2.0

following three equations:

$$\text{Re} = \frac{Dv\rho}{\mu} \tag{12.1}$$

$$\Delta P = \frac{4f\rho vL}{2D_i g_c} \tag{14.3}$$

$$f = 0.0014 + \frac{0.125}{\text{Re}^{0.32}}$$

The trial-and-error calculations are provided in the Table 21.2.

As can be seen from Table 21.2, a maximum velocity between 2 and 3 ft/s will allow for a pressure drop of no more than 5.0 lb$_f$/in^2 (psi). The exact maximum allowable velocity can be shown to be 2.53 ft/s with a corresponding Re and ΔP of 4257 and 5.00 psi, respectively.

Illustrative Example 21.12   Refer to Illustrative Example 21.4. If the pipe contains two globe valves and one straight through tee, what is the friction loss?

Solution   One must now include the friction losses for each fitting in the calculation. Employ Equation (18.14).

$$h_f = \left(4f\frac{L}{D} + 2K_{f,\text{globe}} + K_{f,\text{tee}}\right)\frac{v^2}{2g_c}$$

where $K_{f,\text{globe}} = 6.0$
$K_{f,\text{tee}} = 0.4$

The first term represents the frictional loss associated with the pipe; this value was previously "calculated" to be 5.0 psia. Subtracting the above equation gives:

$$h_f = (5.0)(144/62.4) + [(2)(6.0) + 0.4](2.53)^2/(2)(32.2)$$
$$= 11.5 + 1.23$$
$$= 12.73 \text{ ft} \cdot \text{lb}_f/\text{lb}$$

Illustrative Example 21.13   A Pitot tube is inserted in a 55 mm diameter circular pipe to measure the flow velocity. The tube is inserted so that it points upstream into the flow and the pressure sensed by the probe is the stagnation pressure. The static pressure is measured at the same location in the flow using a wall pressure tap. The change in elevation between the tip of the Pitot tube and the wall pressure tap is negligible. The flowing fluid is soybean oil at 20°C and the fluid in the manometer tube is mercury. Refer to Fig. 21.1. Is the height, h, correct or should the manometer fluid be higher on the left side? If the magnitude of h is 40 mm, determine the flow speed. If it is assumed that the velocity is uniform across the cross-section of the 55 mm pipe, what is the mass flow rate of the fluid? What is the flow type (laminar or turbulent)? Assume $\rho_{oil} = 919 \, kg/m^3$, $\mu_{oil} = 0.04 \, kg/m \cdot s$ and $\rho_M = 13,600 \, kg/m^3$.

Solution   Note that point 2 is a stagnation point. Thus, $P_2 > P_1$, and the manometer fluid should be higher on the left side ($h < 0$).

Calculate the flow velocity, given that $h = 40$ mm of mercury. Use the Pitot tube equation. See Equation (19.9) and assume $C = 1.0$.

$$v = \sqrt{2gh\left(\frac{\rho_M}{\rho} - 1\right)} = \sqrt{2(9.804)(0.04)\left(\frac{13,600}{919}\right) - 1}$$

$$= 3.29 \, m/s$$

Assuming a uniform velocity, calculate the mass flow rate

$$\dot{m} = \rho q = \rho v S = (919)(3.29)\pi(0.055)^2/4$$
$$= 7.18 \, kg/s$$

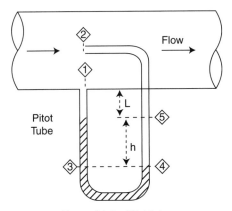

Figure 21.1   Pitot tube.

Calculate the Reynolds number

$$\mathrm{Re} = \frac{Dv\rho}{\mu} = \frac{(0.055)(3.29)(919)}{0.04}$$

$$= 4160$$

The flow is therefore turbulent.

Illustrative Example 21.14   Given a 50 ft pipe with flowing water, determine the flow rate if there is an expansion from 3/8 inch to 1/2 inch and immediately back to 3/8 inch with an overall pressure loss no greater than 2 lb$_f$/ft^2.

Solution   The flow rate will need to be determined using trial-and-error. Use the equation:

$$h_f = \left(4f\frac{L}{D} + K_e + K_c\right)\frac{v^2}{2g_c} \qquad (18.14)$$

with

$$K_e = \left[1 - \left(\frac{D_1}{D_2}\right)^2\right]^2 = \left(1 - \frac{S_1}{S_2}\right)^2 \qquad (18.8)$$

$$K_c = 0.4\left[1 - \left(\frac{D_2}{D_1}\right)^2\right] = 0.4\left(1 - \frac{S_2}{S_1}\right) \qquad (18.10)$$

Since (see Table A.5 in the Appendix) $S_{3/8} = 0.00133$ ft^2 and $S_{1/2} = 0.00211$ ft^2. A trial-and-error calculation is again required for the equation:

$$h_f = \left(4f\frac{50}{0.03125} + 0.06224 - 0.13334\right)\frac{v^2}{2(32.2)}$$

Results are provided in Table 21.3.

Table 21.3   Friction loss calculation for Illustrative Example 21.14

v	Re	f	h_f
10	14083.64	0.007279	72.22989
8	11266.91	0.007714	48.99488
6	8450.186	0.008323	29.73801
4	5633.457	0.009282	14.74203
2	2816.729	0.01124	4.463628
0.5	704.1821	0.016734	0.415472

The velocity for this piping system is estimated to be 1.93 ft/s by linear interpolation.

Illustrative Example 21.15   Refer to Illustrative Example 21.14. How would the velocity differ if an orifice meter were used instead?

Solution   If using an orifice meter instead of the expansion and contraction, the following equation would be used:

$$v_2 = C_o \sqrt{\frac{2gh(\rho_M - \rho)}{\rho[1 - (D_2/D_1)^4]}} \tag{19.17}$$

Obviously, the calculation cannot be performed since the diameter of the orifice has not been specified. One would expect the velocity to be lower since the orifice offers more resistance to flow.

Illustrative Example 21.16   Water flows at a velocity of 0.02 m/s in a concrete pipe (diameter, $D_p = 1.5$ m; length, $L_p = 20$ m; roughness, $k_p = 0.003$ m). This prototype is to be modeled in a lab using a 1/30th scale pipe (diameter, D_m; length, L_m; roughness, k_m). The fluid in the model is castor oil ($\rho_m = 961.3$ kg/m³, $\mu_m = 0.0721$ kg/m · s). Determine the model dimensions, the velocity of castor oil (v_m), and, if the pressure drop in the model (ΔP_m) is measured to be 100 kPa, what is the pressure drop (ΔP_p) in the prototype?

Solution   Refer to Chapter 3. Achieve geometric similarity using a circular pipe of 1/30 the scale:

$$\frac{D_m}{D_p} = \frac{D_m}{1.5} = \frac{1}{30}$$

$$D_m = 0.05 \text{ m} = 5 \text{ cm}$$

Achieve dynamic similarity using the dimensionless ratios k/D and L/D, plus the Reynolds number:

$$k_m = k_p \frac{D_m}{D_p} = 0.0001 \text{ m}$$

$$L_m = L_p \frac{D_m}{D_p} = 0.667 \text{ m}$$

$$\text{Re} = \frac{\rho_m v_m D_m}{\mu_m} = \frac{\rho_p v_p D_p}{\mu_p}$$

For water, $\rho_p = 1000 \text{ kg/m}^3$ and $\mu_p = 0.001 \text{ kg/m} \cdot \text{s}$. Therefore,

$$\frac{(961.3)v_m(0.05)}{0.0721} = \frac{(1000)(0.02)(1.5)}{0.001}$$

$$v_m = 45 \text{ m/s}$$

Finally, calculate ΔP in the prototype noting that $\text{Eu}_m = \text{Eu}_p$. See Illustrative Example (2.2).

$$\frac{\Delta P_m}{0.5\rho_m v_m{}^2} = \frac{\Delta P_p}{0.5\rho_p v_p{}^2}$$

$$\frac{10^5}{0.5(961.3)(45)^2} = \frac{\Delta P_p}{0.5(1000)(0.02)^2}$$

$$\Delta P_p = 0.0206 \text{ Pa}$$

Illustrative Example 21.17   Air at 75°F and 1 atm (kinematic viscosity = 0.00016 ft²/s) flows at a rate of 4800 cfm in an inclined (1 ft × 2 ft) commercial steel rectangular duct (Fig. 21.2). The duct is 1000 ft long and is inclined upward at 5° to the horizontal. Neglect "minor losses" and assume fully developed flow.

Calculate the equivalent diameter, the Reynolds number, the pressure drop in psf and psi, the ideal power requirement to move the air through the duct, and the percent of the total power due to friction. Also determine the brake (or actual) horsepower (bhp) of the blower if the efficiency is 60%.

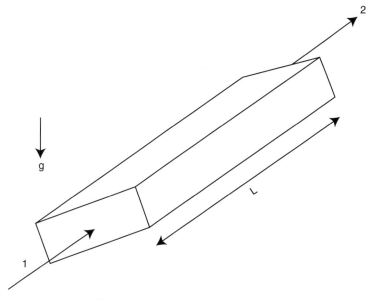

Figure 21.2 Inclined rectangular duct.

Solution   Calculate the equivalent diameter.

$$D_{eq} = \frac{4S}{l_p} = \frac{4(1)(2)}{2(1+2)} = 1.333 \text{ ft}$$

Obtain the air density from the ideal gas law.

$$\rho = \frac{P(MW)}{RT} = \frac{(1)(28.9)}{(0.7302)(460+75)} = 0.074 \text{ lb/ft}^3$$

Calculate flow velocity.

$$v = \frac{q}{S} = \frac{4800/60}{(1)(2)} = 40 \text{ ft/s}$$

The Reynolds number is:

$$Re = \frac{D_{eq}v}{\nu} = \frac{(1.333)(40)}{0.00016} = 3.33 \times 10^5$$

Therefore, the flow is in the turbulent regime.
   From Table 14.1, $k = 0.00015$ ft. Thus,

$$k/D = 0.00015/1.333 = 0.000113$$

Obtain the Fanning friction factor from the Chart in Fig. 14.2

$$f = 0.00375$$

Calculate the head loss using the Hagen–Poiseuille (Darcy–Weisbach) equation (see Eq. (14.3)) expressing the friction/head loss in feet of flowing fluid rather than energy/mass.

$$h_f' = 4f\frac{L}{D_{eq}}\frac{v^2}{2g} = 4(0.00375)\left(\frac{1000}{1.333}\right)\left(\frac{40^2}{2(32.174)}\right)$$

$$= 280 \text{ ft of air}$$

Apply Bernoulli's equation at the entrance and exit of the conduit. Note that the pipe is inclined, there is no shaft work, no (minor) head loss, and $v_1 = v_2$. Therefore,

$$\frac{P_1 g_c}{\rho g} = \frac{P_2 g_c}{\rho g} + h_f' + (z_2 - z_1)$$

$$P_1 - P_2 = \rho\frac{g}{g_c}(h_f' + z_2 - z_1)$$

Since

$$z_2 - z_1 = 1000\sin(5°) = 87.2 \text{ ft}$$

ΔP can now be calculated

$$\Delta P = 0.074(1)(280 + 87.2) = 27.17 \, \text{psf} = 0.19 \, \text{psi}$$

The fluid power requirement is

$$\dot{W}_s = q\Delta P = 80(27.17) = 2173.6 \, \frac{\text{ft} \cdot \text{lbf}}{\text{s}} = 3.95 \, \text{hp}$$

Note that

$$h_f' = 280 \, \text{ft of air}$$

Illustrative Example 21.18   Water is drawn from a reservoir and pumped through an equivalent length of 2 miles of a horizontal, circular concrete duct of 10 inch ID. At the end of the duct, the flow is divided into a network consisting of a 4 inch and a 3 inch ID pipe. The 4 inch line has an equivalent length of 200 ft and rises to a point 50 ft above the surface of the water in the reservoir, where the flow discharges to the atmosphere. This flow must be maintained at a rate of 1000 gal/min. The 3 inch line discharges to the atmosphere at a point 700 ft from the junction at the level of the surface of the water in the reservoir. Outline how to calculate the horsepower input to the pump, which has an efficiency of 70%.

Solution   A diagram of the system is shown in Fig. 21.3. The solution requires resolving five equations—three mechanical energy balances, one continuity and one pressure drop relationship that applies to the piping network. The three mechanical energy balance equations are

$$\frac{\Delta P_2}{\rho} + 50 + \left(\frac{4fLv^2}{2g_cD}\right)_2 = 0 \tag{1}$$

$$\frac{\Delta P_3}{\rho} + \left(\frac{4fLv^2}{2g_cD}\right)_3 = 0 \tag{2}$$

$$\frac{\Delta P_1}{\rho} + \left(\frac{4fLv^2}{2g_cD}\right)_3 = \eta_p W_p = h_s \tag{3}$$

The continuity equation,

$$\dot{m}_1 = \dot{m}_2 + \dot{m}_3 \tag{4}$$

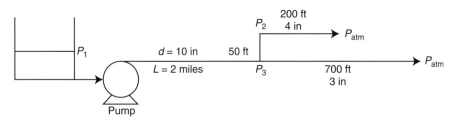

Figure 21.3   Application 21.18.

The pressure relationship is,

$$\Delta P_2 = \Delta P_3 \qquad (5)$$

where

$$\Delta P_2 = P_2 - P_{atm}$$
$$\Delta P_3 = P_3 - P_{atm}$$
$$\Delta P = P_2 - P_1$$
$$P_1 = P_{atm}$$

There are five unknowns—P_2, P_3, v_2, v_3 and W_p. The outline of the solution follows.

1. Calculate v_2 from the continuity equation
2. Calculate ΔP_2 from Equation (1).
3. Set $\Delta P_3 = \Delta P_2$.
4. Calculate v_3 from Equation (2). This will require a trial-and-error procedure since both v_3 and Re_3 (and f_3) are not known.
5. Calculate v_1 from the continuity equation since both v_2 and v_3 are known.
6. Calculate $\dot{W}_s$ in Equation (3)

$$(z_2 - z_1) = 87.2 \, \text{ft of air}$$

so that

$$h_f' + (z_2 - z_1) = 367.2 \, \text{ft of air}$$

Therefore, the percent of the energy loss due to friction is

$$280/367.2 = 0.762 = 76.2\%$$

Finally,

$$bhp = \frac{\dot{W}_s}{E} = \frac{3.93}{0.6}$$
$$= 6.56 \, \text{hp}$$

REFERENCES

1. J. Santoleri, J. Reynolds, and L. Theodore, "Introduction to Hazardous Waste Incineration," 2nd edition, John Wiley and Sons, Hoboken, NJ, 2000.
2. C. E. Lapple, "Fluid and Particle Dynamics," University of Delaware, Newark, Delaware, 1951.

NOTE: Additional problems are available for all readers at www.wiley.com. Follow links for this title.

22

INDUSTRIAL APPLICATIONS

The illustrative examples provided in this chapter primarily pertain to modulated applications. There are a total of 17 illustrative examples; the first three examples are qualitative in nature. Those readers preferring academic calculations should definitely consider reviewing the illustrative examples in the previous chapter.

Illustrative Example 22.1   List the various classifications of industrial piping.

Solution   Industrial piping can be divided into several major classifications. These are as follows:[1]

1. Outside overhead main lines.
2. Outside underground main lines.
3. Outside overhead lateral or distribution lines.
4. Outside underground lateral or distribution lines.
5. Process headers within buildings or tank farms.
6. Lateral distribution of process lines within buildings or tank farms.
7. Service headers in process or manufacturing buildings or tank farms.
8. Sewers, plumbing, and drain lines.

Illustrative Example 22.2   List the various services that employ piping.

Fluid Flow for the Practicing Chemical Engineer. By J. Patrick Abulencia and Louis Theodore
Copyright © 2009 John Wiley & Sons, Inc.

Solution   Piping may be classified by the services which they perform. Included in this list are piping to transport water, steam, compressed air, gas, sewage, drains, sanitary plumbing, process and instrument lines.

Illustrative Example 22.3   Outline how to determine the optimum economic pipe diameter for a flow system.[2]

Solution   The investment for piping can amount to an important part of the total cost for a chemical process. It is usually necessary to select pipe sizes that provide the minimum total cost of both capital and operating charges. For any given set of flow conditions, the use of an increased pipe diameter will result in an increase in the capital cost for the piping system and a decrease in the operating costs. (The operating cost is generally the energy costs associated with moving the fluid; that is, pumping the fluid of concern.) Thus, an optimum economic pipe diameter can be found by minimizing the sum of pumping (or energy) costs and capital charges of the piping system.

The usual calculational procedure is as follows:

1. Select a pipe diameter.
2. Obtain the annual operating cost.
3. Obtain the capital equipment cost.
4. Convert the capital cost to an annual basis.
5. Sum the two annual costs in steps 2 and 4.
6. Return to step 1.

The only variable that will appear in the resulting total-cost expressions is the pipe diameter. The optimum economic pipe diameter can be generated by taking the derivative of the total annual cost with respect to pipe diameter, setting the result equal to zero, and solving for the diameter. The derivative operation can be replaced by a trial-and-error procedure that involves calculating the total cost for various diameters and simply selecting the minimum.

Generally, the initial/capital cost (see Chapter 32 for more details) of the pipe and valves fittings is directly proportional to the diameter, as are the other factors of pipe operating cost, depreciation and maintenance, which are a constant percentage of the initial pipeline cost. The cost of pressure drop (i.e., cost of pumping or blowing), however, is inversely proportional to the diameter.

Illustrative Example 22.4   For a centrifugal pump operating at 1800 rpm, find the impeller diameter needed to develop a head of 200 ft.

Solution   Calculate the velocity needed to develop 200 ft of head. Use the equation

$$v^2 = 2gh$$

Substituting gives

$$v^2 = (2)(32.2\,\text{ft/s}^2)(200\,\text{ft})$$
$$= 12,880$$
$$v = 113.5\,\text{ft/s}$$

Next, calculate the number of feet that the impeller travels in one rotation:

$$(113\,\text{ft/s})/(1800\,\text{rpm}/60\,\text{s}) = 3.77\,\text{ft/rotation}$$

This represents the circumference of the impeller since it is equal to one rotation. The diameter of the impeller may now be calculated:

$$D = \text{circumference}/\pi$$
$$D = 3.77\,\text{ft}/\pi$$
$$D = 1.2\,\text{ft}$$

An impeller diameter of approximately 1.2 ft will therefore develop a head of 200 ft.

Illustrative Example 22.5   Water for a processing plant is required to be stored in a reservoir. It is believed that a constant supply of 1.2 m³/min pumped to the reservoir, which is 22 m above the water intake, would be sufficient. The length of the pipe is about 120 m and there is 15 cm diameter galvanized iron piping available. The line would need to include eight regular elbows. Calculate the total energy required, the theoretical power, and the head to accomplish the above task.

Solution   Assume the properties of water at 20°C are:

$$\rho = 998\,\text{kg/m}^3$$
$$\mu = 0.001\,\text{N} \cdot \text{s/m}^2$$

Cross-sectional area of pipe:

$$S = (\pi/4)D^2$$
$$= (0.785)(0.15)^2$$
$$= 0.0177\,\text{m}^2$$

Volume flow rate:

$$q = 1.2\,\text{m}^3/\text{min}$$
$$= (1.2)/(60)$$
$$= 0.02\,\text{m}^3/\text{s}$$

Velocity in the pipe:

$$v = (1.2/60)/0.0177$$
$$= 1.13 \, \text{m/s}$$

Calculate the Reynolds number.

$$\text{Re} = Dv\rho/\mu$$
$$= (0.15)(1.13)(998)/0.001$$
$$= 1.7 \times 10^5$$

The flow is clearly turbulent.

There are three contributions to the energy load. These may be calculated individually or from the Bernoulli equation. Individual calculations are provided below.

From Table 14.1, the roughness factor k is 0.0005 for galvanized iron so that

$$\text{roughness ratio, } k/D = 0.0005/0.15 = 0.003$$

From Fig. 14.2, the friction factor is:

$$f = 0.0053$$

Therefore, the friction loss of energy from Equation (14.3) is

$$h_f = 4fLv^2/2g_cD$$
$$= (4)(0.0053)(120)(1.13)^2/(2)(0.15)$$
$$= 10.8 \, \text{J}$$

For the eight elbows (from Table 18.1), the estimated value of K for one regular $90°$ elbow is 0.5.

$$K = 8(0.5)$$
$$= 4.0$$

The velocity head is

$$v^2/2g_c$$
$$\text{VH} = (1.13)^2/2$$
$$= 0.64 \, \text{J/kg}$$

The total loss from the elbows is therefore

$$= (4)(0.64)$$
$$- 2.56 \, \text{J/kg}$$

The energy to move 1 kg of water against a head of 22 m of water is

$$\Delta(PE) = \Delta z(g/g_c); \quad \Delta z = 22\text{ m}$$
$$= (22)(9.81)$$
$$= 215.8\text{ J/kg}$$

Total energy requirement per kg:

$$E_{tot} = 10.8 + 2.56 + 215.8$$
$$= 229.2\text{ J/kg}$$

The theoretical power requirement is

$$\dot{W}_s = (E_{tot})(q)(\rho)$$
$$= (229.2)(0.02)(998)$$
$$= 4574\text{ J/s}$$

The head (height of liquid) equivalent to the energy requirement is then

$$h = E_{tot}\left(\frac{g_c}{g}\right)$$
$$= 229.2/9.81$$
$$= 23.4\text{ m of water}$$

Illustrative Example 22.6   Oil is flowing through a standard $1\frac{1}{2}$-inch steel pipe containing a 1.00-inch square-edged orifice. The pressure differential across the orifice is indicated by two parallel vertical open tubes into which the oil rises from the two pressure taps. The oil is at 100°F; its specific gravity is 0.87 and its viscosity 20.6 cP. Calculate the reading on the gauge described, when the oil is flowing at a rate of 400 gal/hr.

Solution   Applying appropriate conversion factors, the orifice velocity is:

$$v_0 = (400)(144)/(0.785)(3600)(7.48)$$
$$= 2.72\text{ ft/s}$$

The Reynolds number is

$$Re_0 = D_0 v_0 \rho/\mu$$
$$= (1/12)(2.72)(0.87 \times 62.4)/(20.6 \times 0.000672)$$
$$= 889$$

Since

$$\frac{D_0}{D_1} = 1.00/1.61$$
$$= 0.62$$

proceed to Fig. 19.8 and note

$$C_o = 0.76$$

Equation (19.17) must be rearranged to solved for h.

$$h = \frac{v_0^2}{2gC_o^2}\left[1 - \left(\frac{D_0}{D_1}\right)^4\right]$$

Substituting,

$$= \frac{(2.72)^2}{(64.4)(0.76)^2}[1 - (0.62)^4]$$

$$= 0.170 \text{ ft of oil}$$

$$= 2.04 \text{ in of oil}$$

Illustrative Example 22.7 Natural gas consisting of essentially pure methane flows through a long straight standard 10-inch steel pipe into which is inserted a square-edged orifice 2.50 inches in diameter, with pressure taps, each 5.0 inches from the orifice plate. A manometer attached across the orifice reads 1.60 in H_2O. What is the mass rate of flow of gas through this line if the gas density is 0.054 lb/ft.

Solution   The ratio of orifice to pipe diameter is:

$$D_0/D_1 = 2.50/10.15 = 0.245$$

Assuming the Reynolds number in the orifice to be over 30,000. The coefficient C_o is therefore 0.61 from Fig. 19.8. Then, by Equation (19.19) with (D_0/D_1) assumed approximately zero,

$$v_0 = 0.61\sqrt{\frac{(64.4)(1.60)(62.4)}{(12)(0.054)}}$$

$$= 60.8 \text{ ft/s}$$

Using this result, the Reynolds number in the orifice is

$$\text{Re}_0 = \frac{D_0 v_0 \rho}{\mu} = \frac{(2.50)(60.7)(0.0540)}{(12)(0.011)(0.000672)} = 92,800$$

The assumption of $C_o = 0.61$ is permissible.
    The mass rate of flow is

$$\dot{m} = (60.7)(0.785)(2.50)^2(0.0540)(3600)/144$$

$$= 403 \text{ lb/hr}$$

Illustrative Example 22.8 In the gradual contraction pictured in Fig. 22.1, the upstream diameter (at station 1) is 10 cm and the downstream diameter (at station 6) is 6 cm. The flowing fluid is air at 20°C, which has a specific weight of $12\,\text{N/m}^3$ and $\mu = 0.0018$ cP. The manometer fluid is Meriam red oil (SG = 0.827); it indicates a manometer head, h, of 8 cm. Assume steady-state operation, constant properties, and no head losses.

Is there a stagnation point in the flow? Where? Compute the flow rate. Is the air flow incompressible? If the static pressure of the upstream air is 130,000 Pa absolute, calculate the static pressure of the air in the 6 cm pipe.

Solution   Calculate the density of air from the ideal gas law

$$\rho = 1.22\,\text{kg/m}^3$$

Is there a stagnation point? Yes, at station number 6.
Express the velocity at point 1 in terms of the velocity at point 2 by applying the continuity equation

$$\left(\pi D_1{}^2/4\right)v_1 = \left(\pi D_2{}^2/4\right)v_2$$

$$v_1 = v_2\left(\frac{D_2}{D_1}\right)^2$$

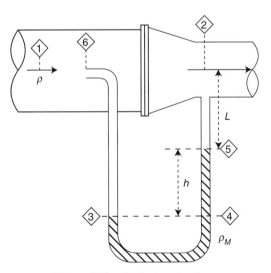

Figure 22.1 Gradual contraction.

Apply Bernoulli's equation between points 1 and 2, noting that $z_1 = z_2$:

$$P_1 + \frac{\rho v_1^2}{2g_c} = P_2 + \frac{\rho v_2^2}{2g_c}$$

$$P_1 - P_2 = \frac{\rho(v_2^2 - v_1^2)}{2g_c}$$

Substituting for the velocity, v_1, leads to

$$P_1 - P_2 = \frac{\rho v_2^2[1 - (D_2/D_1)^4]}{2g_c}$$

Apply Bernoulli's equation between points 1 and 6, noting that $z_1 = z_6$:

$$\frac{P_6 - P_1}{\rho} = \frac{v_1^2}{2g_c} + 0$$

Replacing v_1 by v_2 gives

$$\frac{P_6 - P_1}{\rho} = \frac{v_2^2}{2g_c}\left(\frac{D_2}{D_1}\right)^4$$

Combining the above two equations gives

$$P_6 - P_2 = \frac{\rho v_2^2}{2g_c}$$

Equate the pressure on both sides of the manometer.

$$P_3 = P_4$$

$$P_3 = P_6 + \rho\frac{g}{g_c}(L + h) = P_4 = P_2 + \rho\frac{g}{g_c}L + \rho_M\frac{g}{g_c}h$$

Therefore,

$$P_6 - P_2 = (\rho_M - \rho)\frac{g}{g_c}h$$

This is essentially the manometer equation. Equating $P_6 - P_2$ from the last two equations yields

$$\frac{v_2^2}{2g_c} = h\left(\frac{\rho_M}{\rho} - 1\right)\frac{g}{g_c}$$

or

$$v_2 = \sqrt{2gh\left(\frac{\rho_M}{\rho} - 1\right)} = \sqrt{2(0.08)(9.8)\left(\frac{827}{1.22} - 1\right)} = 32.58 \text{ m/s}$$

$$v_1 = v_2\left(\frac{D_2}{D_1}\right)^2 = 32.58(0.6)^2 = 11.73 \text{ m/s}$$

$$q = v_2 S_2 = (32.58)(\pi)(0.06)^2/4 = 0.092 \text{ m}^3/\text{s}$$

Calculate the Mach number from Equation (15.1)

$$c = 20\sqrt{T(^\circ\text{K})} = 20\sqrt{293} = 342.4 \text{ m/s}$$

$$Ma = \frac{v_2}{c} = \frac{32.58}{342.4} = 0.095$$

Noting that $0.095 < 0.3$, one can conclude that the flow is incompressible.

Given that $P_1 = 130,000$ Pa absolute, calculate P_2. Use Bernoulli's equation once again

$$P_1 - P_2 = \frac{\rho v_2^2[1 - (D_2/D_1)^4]}{2g_c} = \frac{(1.22)(32.54)^2[1 - (0.6)^4]}{2} = 562.2 \text{ Pa}$$

$$P_2 = 130,000 - 562.2 = 129,438 \text{ Pa absolute}$$

Illustrative Example 22.9   Water is flowing from an elevated reservoir through a conduit to a turbine at a lower level and out of the turbine through a similar conduit. At a point in the conduit 300 ft above the turbine, the pressure is 30 psia; at a point in the conduit 10 ft below the turbine, the pressure is 18 psia. The water is flowing at 3600 tons/hr, and the output at the shaft of the turbine is 1000 hp. If the efficiency of the turbine is known to be 90%, calculate the friction loss in the conduit in ft · lb$_f$/lb.

Solution   A pictorial representation of the system is given in Fig. 22.2.

Since the diameter of the conduit is the same at location 1 and 2, kinetic energy effects can be neglected and Bernoulli's equation takes the form of

$$\frac{\Delta P}{\rho} + \Delta z\frac{g}{g_c} - h_s + h_f = 0$$

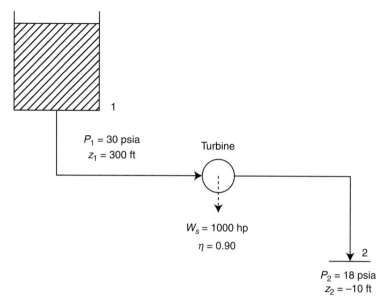

$P_1 = 30$ psia
$z_1 = 300$ ft

Turbine

$W_s = 1000$ hp
$\eta = 0.90$

2

$P_2 = 18$ psia
$z_2 = -10$ ft

Figure 22.2 Turbine problem.

The term h_s is calculated using the following equation

$$h_s = \frac{\dot{W}_s}{\eta \, \dot{m}}$$

$$E = \frac{1000}{(0.9)(3600)} \left(\frac{550 \text{ ft} \cdot \text{lb}_f/\text{s}}{\text{hp}}\right) \left(\frac{3600 \text{ s}}{\text{hr}}\right) \left(\frac{\text{tons}}{2000 \text{ lb}}\right)$$

$$= 305.6 \text{ ft} \cdot \text{lb}_f/\text{lb}$$

Substituting back into the modified Bernoulli's equation yields:

$$\frac{(18 - 30)(144)}{62.4} - 310(1) - 305.6 + h_f = 0$$

Solving for the friction loss, h_f, gives

$$h_f = -643.29 \text{ ft} \cdot \text{lb}_f/\text{lb}$$

Illustrative Example 22.10 Benzene is pumped from a large tank to a delivery station at a rate of 0.003 m³/s. The tank is at atmospheric pressure. The pressure at the delivery station is 350 kPa gauge. The pump station is 1.8 m above the level in the tank. The delivery station is 3.8 m above the benzene level in the tank. The diameter of the suction and discharge line is 0.03 m. The head loss in the system is estimated to be 8 m of benzene. The density of benzene is $865 \, \text{kg/m}^3$ and its vapor pressure is 26,200 Pa.

Determine the discharge, the lowest pressure in the system, and the NPSH based on the data given. If the pump manufacturer requires an NPSH of 8 m of benzene, is the height (1.8 m) adequate? If not, determine the desired height.

Solution   Refer to Fig. 22.3. Calculate the discharge velocity, v_2

$$v_2{}^2 = \frac{\dot{m}}{\pi D^2/4} = \frac{0.003}{\pi (0.03)^2/4} = 4.24 \, \text{m/s}$$

Calculate the lowest pressure in the system. Note that the lowest pressure occurs at the pump suction point (station 3). Since all the line diameters are the same $(D_3 = D_4 = D_2)$, the velocities are likewise the same. First calculate the dynamic or velocity head at station 3:

$$\text{dynamic head} \quad \frac{v_3{}^2}{2g} = \frac{(4.24)^2}{2(9.807)} = 0.917 \, \text{m};$$

Set $(z_1 = 0)$ so that

$$z_3 = 1.8 \, \text{m}$$

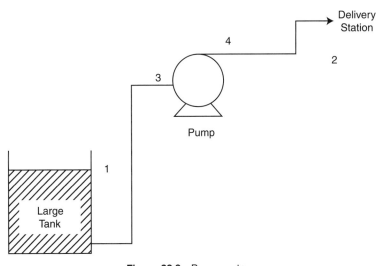

Figure 22.3 Pump system.

Apply Bernoulli's equation between the top of the tank (open to the atmosphere) and the inlet to the pump (station 3)

$$0 + 0 + 0 = \frac{P_3}{\rho g} + 0.917 + 1.8$$

$$\frac{P_3}{\rho g} = -2.717 \, \text{m of benzene}$$

$$P_3 = 101{,}325 - (2.717)(865)(9.807) = 78{,}277 \, \text{Pa}$$

Calculate the NPSH employing Equation (17.12)

$$\text{NPSH} = \frac{P g_c}{\rho g} + \frac{v^2}{2g} - \frac{p' g_c}{\rho g} = \frac{P - p' g_c}{\rho g} + \frac{v^2}{2g}$$

$$= \frac{78{,}277 - 26{,}200}{(865)(9.807)} + 0.917 = 7.06 \, \text{m benzene}$$

The manufacturer NPSH is 8 m, which is greater than the calculated NPSH of 7.06 m. Therefore, the suction point of the pump must be lowered.
    Calculate the new pressure.

$$\text{NPSH} = 8 = \frac{P g_c}{\rho g} + 0.917 - 3.09$$

$$\frac{P g_c}{\rho g} = 10.173 \, \text{m of benzene}$$

$$P = 10.173(865)(9.807) = 86{,}300 \, \text{Pa absolute} = -15{,}025 \, \text{Pag} = -1.77 \, \text{m of benzene}$$

Apply Bernoulli's equation to determine the height z.

$$0 = -1.77 + 0.917 + z_3$$

$$z_3 = 0.853 \, \text{m}$$

Illustrative Example 22.11   A storage tank on top of a building pumps 60°F water through an open pipe to it from a reservoir. The reservoir's water level is 10 ft above the pipe outlet, and 200 feet below the water level in the tank. Both tanks are open to the atmosphere. A 4 in. ID piping system ($k = 0.0018$ in.) contains two gate valves, five regular 90° elbows, and is 525 ft long. A flow rate of 610 gal/min is desired. Calculate the pump requirement (in hp) if it is rated as 60% efficient. Also, provide the pump requirements in units of kW, W, and N · m/s. Assume the density and viscosity of the water to be 62.37 lb/ft^3 and 1.129 cP, respectively.

Solution   A line drawing of the system is shown in Fig. 22.4.

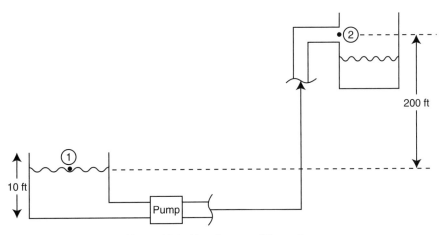

Figure 22.4 Line diagram of the system.

The pipe velocity is first calculated

$$q = 610\,\text{gal/min} = 1.36\,\text{ft}^3/\text{s}$$

$$v = \frac{q}{S} = \frac{1.36}{\pi(4/12)^2/4} = 15.6\,\text{ft/s}$$

The Reynolds number is

$$\text{Re} = \frac{Dv\rho}{\mu} = \frac{(0.333)(15.6)(62.37)}{1.129(6.72 \times 10^{-4})} = 427{,}480; \quad \text{turbulent flow}$$

In addition,

$$k/D = 0.0018/4$$
$$= 0.00045$$

From Fig. 14.2

$$f = 0.0046$$

The friction loss due to the length of pipe is:

$$h_{fp} = 4f\frac{L}{D}\frac{v^2}{2g_c} = 4(0.0046)\frac{525}{(4/12)}\frac{(15.6)^2}{2(32.174)}$$

$$= 110\,\text{ft} \cdot \text{lb}_f/\text{lb}$$

The friction due to the fittings (see Table 18.1) is:

$$K_{ff}(\text{gate}) = 2(0.11) = 0.22 \text{ ft} \cdot \text{lb}_f/\text{lb}$$
$$K_{ff}(\text{elbows}) = 5(0.64) = 3.2 \text{ ft} \cdot \text{lb}_f/\text{lb}$$

The friction due to the sudden contraction is obtained from Equation (18.10). Note that $D_1/D_2 = 0$, since the upstream diameter is significantly larger than the downstream diameter.

$$K_c = 0.42 \left[1 - \left(\frac{D_1}{D_2} \right) \right] = 0.42 \left[1 - \frac{S_1}{S_2} \right]^2$$

$$= 0.42$$

The friction due to a sudden expansion to the atmosphere (from Equation (18.8)) is

$$K_e = \left[1 - \left(\frac{S_1}{S_2} \right) \right]^2 \left[1 - \left(\frac{D_1}{D_2} \right)^2 \right]^2$$

$$= 1.0$$

The sum of the loss coefficients, $\sum K$, may now be calculated

$$\sum K = 0.22 + 3.2 + 0.42 + 1.0 = 4.84$$

These friction losses are therefore

$$h_f = \sum K \frac{v^2}{2g_c} = 4.84 \frac{(15.6)^2}{2(32.174)} = 18.3 \text{ ft} \cdot \text{lb}_f/\text{lb}$$

and

$$h_{f,\text{total}} = 110.2 + 18.3 = 128.5 \text{ ft-lb}_f/\text{lb}$$

Applying Bernoulli's equation:

$$W_s = \frac{\Delta P}{\rho} + \frac{v_2^2 - v_1^2}{2g_c} + (z_2 - z_1)\frac{g}{g_c} + h_{f,\text{total}}$$

Since both tanks are open to the atmosphere, $\Delta P = 0$ and $v_2 = 15.6 \text{ ft/s}$ (with $v_1 = 0$),

$$W_s = 0 + \frac{15.6^2}{2(32.174)} + 200 + 128.5$$

$$= 332.2 \text{ ft} \cdot \text{lb}_f/\text{lb}$$

The mass flow rate is

$$\dot{m} = q\rho = 1.36(62.37)$$
$$= 84.82 \text{ lb/s}$$

and the actual horsepower requirement is

$$\dot{W}_s = \frac{\dot{m}W_s}{\eta} = \frac{332.3(84.82)}{550(0.6)}$$

$$= 85.4 \text{ hp}$$

In other units:

$$\dot{W}_s = 85.4(0.7457) = 63.7 \text{ kW}$$

$$= 63{,}700 \text{ W} = 63{,}700 \text{ N} \cdot \text{m/s}$$

Illustrative Example 22.12 Turpentine is being moved from a large storage tank to a blender through a 700 ft pipeline. The temperature of the turpentine is 50°F and its specific gravity is 0.872. The top surface of the turpentine in the storage tank is 20 ft above floor-level and the discharge end of the pipe (directly over the blender) is 90 ft above floor level. Both the tank and pipe discharge end are open to the atmosphere. The line contains five 90° elbows, six wide open gate valves and one return bend. The average energy delivered by the pump is 401.9 ft · lb_f/lb of turpentine, the efficiency of the pump is 74%, and the average velocity of the turpentine in the line is 12.66 ft/s. The friction loss coefficients of contraction, elbows, bends and valves are to be assumed equal to 0.9, 2.2 and 0.2, respectively. Draw a diagram of this system and clearly show the location of the beginning point and the end point to be used in the solution of the problem. Determine the inside diameter of the pipeline, the volumetric flow rate in gal/min, and the brake horsepower of the pump.

Solution A line diagram of the system is provided in Fig. 22.5.
   Write the modified Bernoulli equation. See Equation (13.4) or (17.11).

$$\frac{\Delta P}{\rho} + \frac{\Delta(v^2)}{2g_c} + \Delta z \frac{g}{g_c} = \eta_p W_s - h_f$$

Calculate the friction loss in ft · lb_f/lb, noting that there is no pressure drop in the system.

$$0 + \frac{(12.66)^2}{2(32.174)} + \frac{(32.174)(90 - 20)}{32.174} = 0.74(401.9) - h_f$$

$$h_f = 224.9 \text{ ft} \cdot \text{lb}_f/\text{lb}$$

The friction can be determined from the friction loss coefficient due to the fittings. The friction loss is expressed in terms of Fanning friction factor and the diameter of the tube.
   The equation for the friction loss is first written.

$$h_f = \left[4f\frac{L}{D} + \sum K_c + \sum K_e + \sum K_f \right] \frac{v^2}{2g_c} \tag{8.14}$$

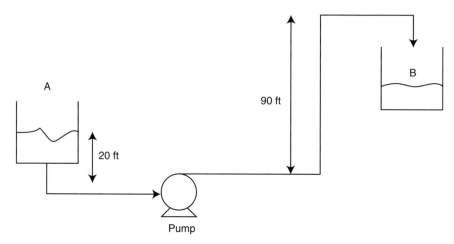

Figure 22.5   Line diagram of the system.

Substitution leads to

$$h_f = \left[4f\left(\frac{700}{D}\right) + 0.4 + 5(0.9) + 6(0.2) \right] \frac{(12.66)^2}{2(32.174)} = 20.68 + 6975 \left(\frac{f}{D}\right)$$

Employing the friction loss calculated from the Bernoulli equation and substituting into the friction loss equation above leads to

$$224.9 = 20.68 + 6975 \left(\frac{f}{D}\right)$$

so that

$$f = 0.0293D$$

The Reynolds number can also be expressed in terms of the tube diameter:

$$\text{Re} = \frac{\rho v D}{\mu} = \frac{0.872(62.4)(12.66)D}{1.76(6.72 \times 10^{-4})} = 582{,}250\,D$$

The tube diameter is determined by trial-and-error. First, guess the diameter in ft and then determine the Reynolds number from the equation above. Use the Reynolds number to determine the Fanning friction factor from the friction factor chart and then recalculate the diameter of the tube from the previous equation and compare the new diameter of the tube to the initial guess. Repeat the procedure until the assumed diameter and the calculated diameter are essentially the same. Note that the diameter calculated in each step may be used as the guess for the next step. The following table shows the trial-and-error calculation starting with a diameter of 1 ft.

The diameter of the tube is therefore 0.184 ft.

Table 22.1   Calculated results for Illustrative Example 22.12

D, ft	Re	k/D	f	D_{new}, ft
1	582,250	0.00015	0.0037	0.126
0.126	73,270	0.00119	0.0052	0.177
0.177	103,000	0.00085	0.0054	0.184
0.184	106,900	0.00082	0.0054	0.184

Determine the cross-sectional area from the diameter obtained above

$$S = \frac{\pi D^2}{4} = \frac{\pi (0.184)^2}{4} = 0.0266 \, \text{ft}^2$$

The volumetric flow rate of the fluid is then:

$$q = vS = 12.66(0.0266) = 0.337 \, \text{ft}^3/\text{s} = 151{,}26 \, \text{gal}/\text{min}$$

The mass flow rate of the fluid is

$$\dot{m} = \rho v S = (0.872)(62.4)(12.66)(0.0266) = 18.33 \, \text{lb}/\text{s}$$

The brake horse power (bhp) is

$$\text{bhp} = \frac{\dot{m} W_s}{\eta} = (18.33)(401.9)\left(\frac{1}{550}\right)\left(\frac{1}{0.74}\right)$$

$$= 18.1 \, \text{hp}$$

Illustrative Example 22.13   Hydrogen at 1 atm and 20°C flows at 400 cc/s through an 80 mm diameter horizontal pipe. Assume the z-axis to be along the pipe axis. Calculate the average velocity and mass flow rate. Determine if the flow is laminar. What is the pressure drop per unit length of pipe (pressure gradient)? What are the velocities at $r = 0$, $r = 20$ mm, and $r = 40$ mm? What is the ratio of the average velocity/maximum velocity? Calculate the Darcy and Fanning friction factors. Indicate the units. What is the friction loss, in m, and the friction power loss per unit length of pipe? If a pipe with twice the diameter is used instead, and the flow rate and pipe length remain the same, will the pressure drop increase, decrease, or remain the same? Why?

Solution   Obtain the properties of hydrogen at 20°C from Table A.3 in the Appendix.

$$\rho = 0.0838 \, \text{kg}/\text{m}^3$$

$$\mu = 9.05 \times 10^{-6} \, \text{kg}/\text{m} \cdot \text{s}$$

$$k = 1.41$$

Convert all data into SI units (for convenience).

$$D = 80\,\text{mm} = 0.08\,\text{m}$$
$$q = 400\,\text{cc/s} = 400\,\text{mL/s} = 0.400\,\text{L/s}$$
$$= 0.0004\,\text{m}^3/\text{s}$$
$$S = \pi D^2/4 = \pi(0.08)^2/4$$
$$= 0.000503\,\text{m}^2$$

Calculate the average velocity and mass flow rate.

$$v = \frac{q}{S} = \frac{0.0004}{0.000503} = 0.8\,\text{m/s}$$
$$\dot{m} = \rho q = (0.0838)(0.0004) = 33.52 \times 10^{-6}\,\text{kg/s}$$

Check the flow type using the Reynolds number.

$$\text{Re} = \frac{Dv\rho}{\mu} = \frac{(0.08)(0.8)(0.0838)}{9.05 \times 10^{-6}}$$
$$= 593 < 2100; \quad \text{laminar}$$

Calculate the pressure gradient. Since the tube is horizontal, $z_1 = z_2$, and from Equation (14.3) with $v = \pi D^2/4$

$$\frac{\Delta P}{L} = \frac{128\,\mu q}{\pi D^4} = \frac{128(9.05 \times 10^{-6})(0.0004)}{\pi(0.08)^4}$$
$$= 3.60 \times 10^{-3}\,\text{Pa/m}$$

Note that $\Delta P \propto D^{-4}$.

Calculate the velocity at $r = 0, 0.02,$ and 0.04 m using the parabolic (laminar flow) velocity equation.

$$v_{\text{max}} = 2v = 1.6\,\text{m/s}$$
$$v = 1.6[1 - (r/0.04)^2]$$
$$\text{At } r = 0, \qquad v = v_{\text{max}} = 1.6\,\text{m/s}$$
$$\text{At } r = 0.02, \quad v = 1.2\,\text{m/s}$$
$$\text{At } r = 0.04, \quad v = 0\,\text{m/s}$$

Calculate the Fanning friction factor. Since the flow is laminar, the Fanning friction factor is

$$f = \frac{16}{Re} = \frac{16}{593}$$

$$= 0.0269$$

The Darcy friction factor is

$$f_D = 4f = 4(0.0269) = 0.108$$

Calculate the friction loss using Equation (14.3) employing the above Fanning friction factor

$$h_f' = 4f \frac{L}{D} \frac{v^2}{2g} = \frac{4(0.0269)(1/0.08)(0.8)^2}{2(9.807)} = 4.39 \times 10^{-2} \, \text{m of hydrogen}$$

Calculate the friction power loss

$$\dot{W}_f = \dot{m}gh_f' = (33.52 \times 10^{-6})(9.807)(4.39 \times 10^{-2})$$

$$= 1.4 \times 10^{-5} \, \text{W}$$

Examine the effect of doubling pipe diameter. For constant q, μ, and L:

$$\Delta P \propto D^{-4}$$

Thus, by doubling D, the pressure drop will decrease to $1/16$ of its original value.

Illustrative Example 22.14   Gasoline at 20°C is pumped at 0.3 m³/s through 30 m of 20-cm diameter horizontal cast-iron pipe. Calculate the average velocity of gasoline. Compute the head loss and brake power required to pump the gasoline if the pump is 80% efficient. By what percentage are the head loss and power requirements increased due to the roughness of the tube? For gasoline at 20°C, $\rho = 680 \, \text{kg/m}^3$, $\mu = 2.92 \times 10^{-4} \, \text{kg/m·s}$ (Table A.2 in the Appendix).

Solution   Calculate the average velocity.

$$q = q_1 = q_2 = vS$$

$$S = \frac{\pi D^2}{4} = \frac{\pi (0.2)^2}{4} = 0.03142 \, \text{m}^2$$

$$v = \frac{q}{S} = \frac{0.3}{0.03142} = 9.5 \, \text{m/s}$$

Check the flow regime

$$Re = \frac{Dv\rho}{\mu} = \frac{(0.2)(9.5)(680)}{(2.92 \times 10^{-4})} = 4.42 \times 10^6 > 4000; \quad \text{turbulent}$$

Obtain the roughness, k, of cast iron pipe (see Table 14.1).

$$k = 0.26\,\text{mm} = 0.00026\,\text{m}$$

Calculate the relative roughness

$$k/D = \frac{0.00026}{0.2} = 0.0013$$

Obtain the Fanning friction factor, f, from Fig. 14.2:

$$f = 0.00525$$

Note that the flow corresponds to complete turbulence in the rough pipe. Calculate the head loss

$$h_f' = 4f\frac{L}{D}\frac{v^2}{2g} = 4(0.00525)\frac{30}{0.2}\frac{(9.5)^2}{2(9.807)} = 14.50\,\text{m of gasoline}$$

Apply Bernoulli's equation to the fluid in the pipe. In the present case, the pipe is horizontal ($z_1 = z_2$) with constant diameter ($v_1 = v_2$) and no shaft head ($h_s = 0$). First convert the friction head to a pressure difference

$$\Delta P = \rho g h_f' = (680)(9.807)(14.647) = 97.68 \times 10^3\,\text{Pa} = 0.9640\,\text{atm}$$

Calculate the ideal shaft work.

$$\dot{W}_{s,id} = q\Delta P = 0.3(97.68 \times 10^3) = 29{,}304\,\text{W} = 39.297\,\text{hp}$$

Calculate the actual shaft work rate.

$$\dot{W}_s = \frac{\dot{W}_{s,id}}{\eta} = \frac{29{,}304}{0.8} = 36{,}630\,\text{W} = 49.121\,\text{hp}$$

Calculate the increase in power requirement due to pipe roughness

$$f_{\text{smooth}} = 0.009$$
$$f_{\text{rough}}/f_{\text{smooth}} = 0.021/0.009 = 2.333$$

The % increase in f due to pipe roughness is:

$$100(2.333 - 1) = 133.3\%$$

Illustrative Example 22.15 Liquid benzene flows steady at 4000 gal/min (gpm) through a 480 ft long horizontal smooth iron pipe that has an inside diameter of 2.3 m. The density of benzene is 899 kg/m³ (56.1 lb/ft³) and the viscosity is 0.0008 kg/m · s (0.000538 lb/ft · s). What is the average velocity of the benzene? What is the Reynolds number? Is the flow laminar or turbulent? What is the Fanning friction factor for benzene. What is the pressure drop? What are the friction power losses?

Solution   Calculate the cross-sectional area of the pipe

$$S = \frac{\pi D^2}{4} = \frac{\pi (2.3)^2}{4}$$

$$= 4.155 \, \text{m}^2$$

Calculate the average velocity

$$v = \frac{q}{S} = \frac{4000}{(4.155)(264.17)(60)}$$

$$= 6.074 \times 10^{-2} \, \text{m/s}$$

Calculate the Reynolds number

$$\text{Re} = \frac{Dv\rho}{\mu} = \frac{(2.3)(6.074 \times 10^{-2})(899)}{(0.0008)}$$

$$= 156{,}990$$

Since the Reynolds number falls in the turbulent regime, determine the Fanning friction factor from Figure 14.2.

$$f \approx 0.0032$$

Calculate the pressure drop with the assumption of no height and velocity change, and no pump work. Since only frictional losses are to be considered, apply Equation (14.3).

$$\frac{\Delta P}{\rho} = 4f \frac{L}{D} \frac{v^2}{2g_c}$$

$$\Delta P = 4f \frac{L}{D} \frac{v^2}{2g_c} \rho = 4(0.0032) \frac{(480)(0.3048)}{(2.3)} \frac{(6.074 \times 10^{-2})^2}{2(1)} (899)$$

$$\Delta P = 1.35 \, \text{Pa}$$

Write a friction power loss equation, employing both the volumetric flow rate and the pressure drop

$$\dot{W}_f = q \Delta P$$

$$= (4000) \left(\frac{1}{264.17} \right) \left(\frac{1}{60} \right) (1.35)$$

$$= 0.34 \, \text{W}$$

Illustrative Examples 22.16   A power plant employs steam to generate power and operates with a steam flowrate of 450,000 lb/h. For the adiabatic conditions listed below, determine the power produced in horsepower, kilowatts, Btu/h, and Btu/lb of steam. Data are given in Table 22.2.

Table 22.2 Power plant data

	Inlet	Outlet
Pressure, psia	100	1
Temperature, °F	1500	350
Steam velocity, ft/s	120	330
Steam vertical position, ft	0	−20

Solution   Apply the following energy equation:

$$z_1 \left(\frac{g}{g_c} \right) + \frac{v_1^2}{2g_c} + H_1 + Q = z_2 \left(\frac{g}{g_c} \right) + \frac{v_2^2}{2g_c} + H_2 + W_s$$

where z_1, z_2 = vertical position at inlet/outlet, respectively
v_1, v_2 = steam velocity at inlet/outlet, respectively
H_1, H_2 = steam enthalpy at inlet/outlet, respectively
W_s = work extracted from system (a negative quantity as written)

For adiabatic conditions, $Q = 0$. Substituting data from the problem statement yields

$$0 + \frac{(120)^2}{2(32.17)(778)} + 1505.4 + 0 = \frac{-20}{778} + \frac{(330)^2}{2(32.17)(778)} + 940.0 + W_s$$

$$0 + 0.288 + 1505.4 + 0 = -0.026 + 2.176 + 940.0 + W_s$$

$$W_s = 563.54 \, \text{Btu/lb}$$

The total power generated by the turbine is equal to

$$\dot{W}_s = (450{,}000 \, \text{lb/h})(563.54 \, \text{Btu/lb})$$

$$= 2.54 \times 10^8 \, \text{Btu/h}$$

Converting to horsepower gives

$$(2.54 \times 10^8 \, \text{Btu/h})(3.927 \times 10^{-4} \, \text{hp} \cdot \text{h/Btu})$$

$$= 9.98 \times 10^4 \, \text{hp}$$

REFERENCES

1. C. E. Lapple, "Fluid and Particle Dynamics," University of Delaware, Newark, Delaware, 1951.

2. J. Reynolds, J. Jeris, and L. Theodore, "Handbook of Chemical and Environmental Engineering Calculations," John Wiley & Sons, Hoboken, NJ, 2004.

NOTE: Additional problems are available for all readers at www.wiley.com. Follow links for this title.

V

FLUID-PARTICLE APPLICATIONS

Part V is concerned with fluid-particle dynamics and the applications that are associated with this subject. In addition to fluid-particle dynamics, the topics reviewed include primary fluid transport devices, sedimentation, centrifugation, flotation, porous media and packed beds, fluidization, and filtration. Although a detailed analysis of these topics is beyond the scope of this text, each subject area receives sufficient treatment that will allow the reader to tackle real-world applications that appear in several of the Illustrative Examples.

Fluid Flow for the Practicing Chemical Engineer. By J. Patrick Abulencia and Louis Theodore
Copyright © 2009 John Wiley & Sons, Inc.

23

PARTICLE DYNAMICS

23.1 INTRODUCTION

It should be noted that this chapter will primarily address gas-particle rather than liquid-particle behavior. The treatment of liquid–solid behavior/separation will appear in Chapter 24—Sedimentation, Centrifugation, Flotation. The general subject of particle classification and measurement receives treatment at the end of the chapter.

23.2 PARTICLE CLASSIFICATION AND MEASUREMENT

Particle size is uniquely defined by particle diameter only for the case of spherical particles. Unfortunately, except for liquid droplets, certain metallurgical fumes, and combustion emissions, particles are usually not spherical. This may also be the case with nanoparticles. To deal with nonspherical particles, it becomes necessary to define an *equivalent diameter* term that depends upon the various geometrical and/or physical properties of the particles.

Some of the methods used to express the size of a nonspherical particle measured by microscopy are illustrated in Fig. 23.1. With reference to this figure, Ferret's diameter is the mean length between two tangents on opposite sides of the particle perpendicular to the fixed direction of the microscopic scan. Martin's diameter measures the diameter of the particle parallel to the microscope scan that divides

Fluid Flow for the Practicing Chemical Engineer. By J. Patrick Abulencia and Louis Theodore
Copyright © 2009 John Wiley & Sons, Inc.

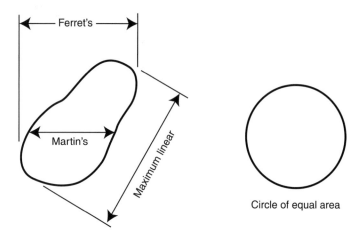

Figure 23.1 Diameters of nonspherical particles.

the particle into two equal areas. The diameter of a circle of equal area is obtained by estimating the projected area of the particle and comparing it with a sphere that approximates its size.

The most popular choice is that sphere diameter (of the same density) that will settle with the same velocity as the particle in question under the influence of gravity. Other diameters that are occasionally/rarely employed are listed in Table 23.1.

The *aerodynamic* diameter of a particle is defined as the diameter of a sphere of unit density (specific gravity $= 1.0$) having the same falling speed in air as the particle. It is most useful in evaluating particle motion in a fluid. The aerodynamic diameter is a function of the physical size, shape, and density of the particle. The aerodynamic diameter is useful when designing certain recovery/control devices and is usually measured by a device called an impactor.

The aerodynamic diameter ($d_{p,a}$) is defined by:

$$d_{p,a} = d_p \sqrt{\rho_p C} \tag{23.1}$$

Table 23.1 Equivalent diameters of particles

Name	Definition
Surface diameter	The diameter of a sphere having the same surface area as the particle
Volume diameter	The diameter of a sphere having the same volume as the particle
Drag diameter	The diameter of a sphere having the same resistance to motion as the particle in a fluid of the same viscosity and at the same velocity
Specific surface diameter	The diameter of a sphere having the same ratio of surface area to volume as the particle

where $d_{p,a}$ = aerodynamic diameter, consistent units; d_p = actual (equivalent) diameter, consistent units; ρ_p = particle specific gravity, dimensionless; and C = Cunningham correction factor (CCF), dimensionless (to be discussed shortly).

Illustrative Example 23.1 Calculate the aerodynamic diameter (μm) for the following two particles:

1. Solid sphere, equivalent diameter = 1.4 μm, specific gravity = 2.0.
2. Hollow sphere, equivalent diameter = 2.8 μm, specific gravity = 0.5.

Solution   Employ Equation (23.1).

1. For the solid sphere,

$$d_{p,a} = 1.4(2.0)^{0.5}$$
$$= 1.98 \, \mu m$$

2. For the hollow sphere,

$$d_{p,a} = 2.80(0.51)^{0.5}$$
$$= 2.0 \, \mu m$$

Illustrative Example 23.2 Calculate the aerodynamic diameter (μm) of an irregular-shaped "sphere" with an equivalent diameter = 1.3 μm and specific gravity = 2.35.

Solution   Once again employ Equation (23.1). For the irregular shape,

$$d_{p,a} = 1.3(2.35)^{0.5}$$
$$= 1.99 \, \mu m$$

Based on the results of this and the previous illustrative example, one concludes that particles with different specific gravity, but the same equivalent size, can have different aerodynamic diameters. For example, if $d_p = 2 \, \mu$m, the reader is left the exercise of showing that the aerodynamic diameter for particles with specific gravity 1.0, 2.0, 4.0, and 8.0 is 2.00 μm, 2.83 μm, 4.00 μm, and 5.66 μm, respectively. Thus, particles of different size and shape can have the same aerodynamic diameter while particles of the same size can have different aerodynamic diameters.

A common method of specifying large particle sizes is to designate the screen mesh that has an aperture corresponding to the particle diameter. Since various screen scales are in use, confusion may result unless the screen scale involved is specified. The screen mesh generally refers to the number of screen openings per unit of length or area. The aperture for a given mesh will depend on the wire size

Table 23.2   Tyler and U.S. standard screen scales

Tyler Mesh	Aperture, Microns	U.S. Mesh	Aperture, Microns
400	37	400	37
325	43	325	44
270	53	270	53
250	61	230	62
200	74	200	74
170	88	170	88
150	104	140	105
100	147	100	149
65	208	70	210
48	295	50	297
35	417	40	420
28	589	30	590
20	833	20	840
14	1168	16	1190
10	1651	12	1680
8	2362		
6	3327		
4	4699		
3	6680		

employed. The Tyler and the U.S. Standard Screen Scales in SI units (Table 23.2) are the most widely used in the United States. The screens are generally constructed of wire mesh cloth, with the diameters of the wire and the spacing of the wires being closely specified. These screens form the bottoms of metal pans about 8 in. in diameter and 2 in. high, whose sides are so fashioned that the bottom of one sieve nests snugly on the top of the next. Additional information is provided in Chapter 25—refer to Table 25.2.

The clear space between the individual wires of the screen is termed the screen aperture. As indicated above, the term *mesh* is applied to the number of apertures per linear inch; for example, a 10-mesh screen will have 10 openings per inch, and the aperture will be 0.1 in. minus the diameter of the wire.

A typical particulate size distribution analysis method of representation employed in the past is provided below in Table 23.3. The numbers in Table 23.3 mean that 40% of the particles by weight are greater than 5 μm (microns or micrometers) in size, 27% are less than 5 μm but greater than 2.5 μm, 20% are less than 2.5 μm but greater than 1.5 μm, and the remainder (13%) are less than 1.5 μm.

Table 23.3   Particle size distribution

	>5.0	μm	40%
<5	>2.5	μm	27%
<2.5	>1.5	μm	20%
<1.5		μm	13%
			100%

Most industrial techniques used for the separation of particles from gases involve the relative motion of the two phases under the action of various external forces. The collection methods for particulates are based on the movement of solid particles (or liquid droplets) through a gas. The final objective is their removal and/or recovery for economic reasons. In order to accomplish this, the particle is subjected to external forces—forces large enough to separate the particle from the gas stream during its residence time in the unit. Perhaps the most important of these forces is the drag force.

23.3 DRAG FORCE

Whenever a difference in velocity exists between a particle and its surrounding fluid, the fluid will exert a resistive force upon the particle. Either the fluid (gas) may be at rest with the particle moving through it or the particle may be at rest with the gas flowing past it. It is generally immaterial which phase (solid or gas) is assumed to be at rest; it is the *relative* velocity between the two that is important. The resistive force exerted on the particle by the gas is called the *drag*.

In treating fluid flow through pipes, a friction factor term is used in many engineering calculations. An analogous factor, called the drag coefficient, is employed in drag force calculations for flow past particles. Consider a fluid flowing past a stationary solid sphere. If F_D is the drag force and ρ is the density of the gas, the drag coefficient, C_D, is defined as

$$C_D = \frac{F_D}{A_p} \frac{2g_c}{\rho v^2} \tag{23.2}$$

From dimensional analysis, one can then show that the drag coefficient is solely a function of the particle Reynolds number, Re, that is,

$$C_D = C_D(\text{Re}) \tag{23.3}$$

where

$$\text{Re} = \frac{d_p v \rho}{\mu}$$

The quantitative use of the equation of particle motion presented in the next section requires numerical and/or graphical values of the drag coefficient as a function of the Reynolds number. These are presented in Fig. 23.2 and Table 24.3 respectively.

In the following analysis, it is assumed that:

1. The particle is a rigid sphere (with a diameter d_p) surrounded by gas in an infinite medium (no wall or multiparticle effects).
2. The particle or fluid is not accelerating.

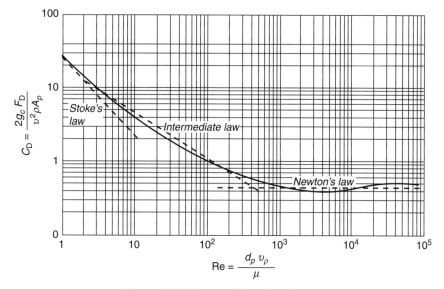

Figure 23.2 Drag coefficient for spheres.

A brief discussion of fundamentals is appropriate here because of the importance of air flow around particulates. No attempt will be made to develop the expressions for the distribution of momentum flux, pressure, and velocity. However, these expressions will be applied to develop some of the more important relationships.

The drag force, F_D, exerted on a particle by a gas at low Reynolds numbers is given by

$$F_D = \frac{6\pi\mu va}{g_c} = \frac{3\pi\mu v d_p}{g_c} \tag{23.4}$$

Equation 23.4 is known as *Stokes' law* and can be derived theoretically. However, keep in mind that Stokes' equation is valid only for very low Reynolds numbers—up to Re ≈ 0.1; at Re $= 1$, it predicts a value for the drag force that is nearly 10% too low. In practical applications, Stokes' law is generally assumed applicable up to a Reynolds number of 2.0. By rearranging Stokes' law in the form of Equation 23.2, the drag coefficient becomes

$$C_D = \frac{6\pi\mu va/\pi a^2}{\rho v^2/2}; \quad d_p = 2a \tag{23.5}$$

where a equals the particle radius. Hence, for *creeping flow* around a particle, Equation (23.5) reduces to

$$C_D = \frac{24}{\text{Re}} \tag{23.6}$$

This is the straight-line portion of the log-log plot of C_D vs. Re (Fig. 23.1). For higher values of the Reynolds number, it is almost impossible to perform purely theoretical calculations. However, several investigators have managed to estimate, with a considerable amount of effort, the drag and/or drag coefficient at higher Reynolds numbers.

In addition to the analytical equation (Eq. 23.6), one may use

$$C_D = 18.5/Re^{0.6}; \quad 2 < Re < 500 \tag{23.7}$$

for the *intermediate* range. This indicates a lesser dependence than Stokes' law on Re; it is less accurate than Stokes' law for Re < 2. At higher Re, the drag coefficient is approximately constant. This is the *Newton's* law range, for which

$$C_D \approx 0.44; \quad 500 < Re < 200,000 \tag{23.8}$$

In this region the drag force on the sphere is proportional to the square of the gas velocity. (Note that Newton's law for the drag force is not to be confused with Newton's law of viscosity or Newton's laws of motion.) A simple two-coefficient model of the form

$$C_D = \alpha \, Re^{-\beta} \tag{23.9}$$

can therefore be used over the three Reynolds-number ranges given in Equations (23.6)–(23.8). The numerical values of α and β are given below:

	α	β
Stokes range	24.0	1.0
Intermediate range	18.5	0.6
Newton range	0.44	0.0

Using the above model in Equation (23.2), the drag force becomes

$$F_D = \frac{\alpha \pi (d_p v)^{2-\beta} \mu^\beta \rho^{1-\beta}}{8 g_c} \tag{23.10}$$

The above equation reduces to

$$F_D = \frac{3 \pi \mu v d_p}{g_c} \tag{23.11}$$

for the Stokes' law range (Re < 2),

$$F_D = \frac{2.31 \pi (d_p v)^{1.4} \mu^{0.6} \rho^{0.4}}{g_c} \tag{23.12}$$

for the intermediate range (2 < Re < 500), and

$$F_D = \frac{0.055 \pi (d_p v)^2 \rho}{g_c} \tag{23.13}$$

Table 23.4 Calculated vs. experimental values for the drag coefficient as a function of the Reynolds number

Range	Re	C_D					
		Equation (23.6)	Equation (23.7)	Equation (23.8)	Equation (23.14)	Equation (23.15)	Experiment
Stokes' law	0.01	2400				2364	2100
	0.02	1200				1195	1050
	0.03	820				803	700
	0.05	490				488	420
	0.07	350				352	300
	0.10	290				249	240
	0.20	120				129	120
	0.30	82				88.3	80
	0.50	49				55.0	49.5
	0.70	35				40.5	36.5
	1.00	24			22.4	29.5	26.5
Intermediate	2		12.0			16.2	14.4
	3		9.5		10.5	11.6	10.4
	4		7.8			9.2	8.2
	5		6.9			7.7	6.9
	6				6.23		5.9
	7		5.4			5.97	5.4
	10		4.5		4.26	4.61	4.1
	20		3.0			2.90	2.55
	30		2.4			2.26	2.0
	40		2.0		1.71	1.93	1.6
	50		1.8			1.71	1.5
	70		1.5			1.45	1.27

90	1.14	1.23	1.11		
100	1.07	0.93	0.785		1.2
200	0.77	0.82			0.80
300	0.65	0.76			0.68
400	0.57	0.71			0.60
500	0.55	0.66	0.568		0.56
700	0.50	—	—	0.44	
900	0.46	0.61	0.482	0.44	
1000	0.45	0.54	—	0.44	
2000	0.42	0.51	—	0.44	
3000	0.40	0.49	0.393	0.44	
4000	0.39	0.48	—	0.44	
5000	0.385	0.47	—	0.44	
7000	0.39	—	0.386	0.44	
9000	0.40	0.46	0.398	0.44	
10,000	0.405	0.44	—	0.44	
20,000	0.45	0.43	0.451	0.44	
30,000	0.47	0.42	—	0.44	
40,000	0.48	0.42	—	0.44	
50,000	0.49	—	—	0.44	
60,000	0.49	0.42	0.520	0.44	
70,000	0.50	0.42	—	0.44	
100,000	0.48	0.41	0.466	0.44	
200,000	0.42	—	—	0.44	
300,000	0.20	—	—	—	
400,000	0.084	—	—	—	
600,000	0.10	—	—	—	
1,000,000	0.13	—	—	—	
3,000,000	0.20	—	—	—	

Newton's law

for the Newton's law range $(500 < \mathrm{Re} < 200{,}000)$. This two-coefficient, three-Reynolds-number range model will be used for drag force calculations in this and subsequent chapters. Numerical and experimental values for the drag coefficient from the model, Equations (23.6)–(23.8) are presented in Table 23.4. A comparison between the two indicates that these three equations are fairly consistent with the experimental values found in the literature.

Another empirical drag coefficient model[1] is given by Equation (23.14):

$$\log C_D = 1.35237 - 0.60810(\log \mathrm{Re}) - 0.22961(\log \mathrm{Re})^2$$
$$+ 0.098938(\log \mathrm{Re})^3 + 0.041528(\log \mathrm{Re})^4$$
$$- 0.032717(\log \mathrm{Re})^5 + 0.007329(\log \mathrm{Re})^6$$
$$- 0.0005568(\log \mathrm{Re})^7 \tag{23.14}$$

This is an empirical equation which has been obtained by the use of a statistical fitting technique. As is evident from Table 23.4, this correlation gives excellent results over the entire range of Reynolds numbers. An advantage of using this correlation is that it is not partitioned for application only to a specific Reynolds number range. However, the lengthy calculation warrants its use only as a subroutine in a computer program.

Still another empirical equation[2] is

$$C_D = \left[0.63 + (4.80/\sqrt{\mathrm{Re}})\right]^2 \tag{23.15}$$

This correlation is also valid over the entire spectrum of Reynolds numbers. Its agreement with literature values, as seen from Table 23.4, is generally good. However, in the range of $30 < \mathrm{Re} < 10{,}000$, there is considerable deviation. For $\mathrm{Re} < 30$ or $\mathrm{Re} > 10{,}000$, the agreement is excellent. This correlation lends itself easily to manual calculations.

23.4 PARTICLE FORCE BALANCE

Consider now a solid spherical particle located in a gas stream and moving in one direction with a velocity, v, relative to the gas. The net or resultant force experienced by the particle is given by the summation of all the forces acting on the particle. These forces include drag, buoyancy, and one or more external forces (such as gravity, centrifugal, and electrostatic). In order to simplify the presentation, the direction of particle movement relative to the gas is always assumed to be positive. Newton's law of motion is then

$$F_R = F - F_B - F_D \tag{23.16}$$

where F_R is the resultant or net force; F is the external force; F_B is the buoyant force; and F_D is the drag force. The net force results in acceleration of the particle, given by

$$F_R = \frac{m}{g_c}\left(\frac{dv}{dt}\right) \tag{23.17}$$

where m is the mass of the particle $(\pi d_p^3 \rho_p/6)$; and ρ_p is the particle density. The external force per unit mass is denoted as f. The external force, F, on the particle is then

$$F = mf \tag{23.18}$$

Unless the particle is in a vacuum, it will experience a buoyant force in conjunction with the external force(s); this is given by

$$F_B = m_f f \tag{23.19}$$

where m_f is the mass of gas (fluid) displaced by the particle. The equation of motion now becomes

$$\begin{aligned}
\left(\frac{dv}{dt}\right)\Big/g_c &= f - \left(\frac{m_f}{m}\right)f - \left(\frac{F_D}{m}\right) \\
&= f\left(1 - \frac{m_f}{m}\right) - \left(\frac{F_D}{m}\right) \\
&= f\left(\frac{m - m_f}{m}\right) - \left(\frac{F_D}{m}\right)
\end{aligned} \tag{23.20}$$

This equation may also be written as

$$\left(\frac{dv}{dt}\right)\Big/g_c = \left(\frac{fm_{eq}}{m}\right) - \left(\frac{F_D}{m}\right) \tag{23.21}$$

where $m_{eq} = (m - m_f)$, or

$$\begin{aligned}
\left(\frac{dv}{dt}\right)\Big/g_c &= f\left(1 - \frac{\rho}{\rho_p}\right) - \left(\frac{F_D}{m}\right) \\
&= f\left(\frac{\rho_p - \rho}{\rho_p}\right) - \left(\frac{F_D}{m}\right)
\end{aligned} \tag{23.22}$$

For gases, $\rho_p >>> \rho$, so that the bracketed terms in Equations (23.20) and (23.22) reduce to unity.

The particle may also be acted upon by one or more external forces. If the external force is gravity

$$f_g = \frac{g}{g_c}$$

with

$$F_g = m\left(\frac{g}{g_c}\right)$$

The describing equation for particle motion then becomes

$$\left(\frac{dv}{dt}\right)\Big/g_c = \left(\frac{g}{g_c}\right) - \left(\frac{F_D}{m}\right) \tag{23.23}$$

If the particle experiences another type of force, for example, an electrostatic force, F_E, then

$$F_E = mf_E$$

so that

$$\left(\frac{dv}{dt}\right)\Big/g_c = f_E - \left(\frac{F_D}{m}\right) \tag{23.24}$$

where f_E is the electrostatic force per unit mass of particle. If, for example, the external force is from a centrifugal field

$$f_C = \frac{r\omega^2}{g_c} = \frac{v_\phi^2}{g_c r}$$

where r is the radius of the path of the particle, f_C is the centrifugal force per unit mass of particle, ω is the angular velocity, and v_ϕ is the tangential velocity at that point. The centrifugal force, F_C, is then

$$F_C = mf_C$$

The describing equation becomes

$$\left(\frac{dv}{dt}\right)\Big/g_c = \left(\frac{r\omega^2}{g_c}\right) - \left(\frac{F_D}{m}\right) \tag{23.25}$$

or

$$\left(\frac{dv}{dt}\right)\Big/g_c = \left(\frac{v_\phi^2}{g_c r}\right) - \left(\frac{F_D}{m}\right) \tag{23.26}$$

The reader is reminded on the use of g_c. Any term or group of terms in the above equations may be indiscriminately multiplied or divided by this conversion constant.

If a particle is initially at rest in a stationary gas and is then set in motion by the application of a constant external force or forces, the resulting motion occurs in two stages. The first period involves acceleration, during which time the particle velocity increases from zero to some maximum velocity. The second stage occurs when the particle achieves this maximum velocity and remains constant. During the second stage, the particle is not accelerating. The left-hand side of Equations (23.20) and (23.26) are, therefore, zero. The final, constant, and maximum velocity attained is defined as the *terminal settling velocity* of the particle. Most particles reach their terminal settling velocity almost instantaneously.

Consider the equations examined above under terminal settling conditions. Since

$$\frac{dv}{dt} = 0$$

the general equation for particle motion becomes

$$0 = f - \left(\frac{F_D}{m}\right)$$

or

$$f = \frac{F_D}{m} \tag{23.27}$$

The units of f in this equation are those of acceleration, that is, length/(time)^2. The general equation for the terminal settling velocity is obtained by direct substitution of Equation (23.10) into Equation (23.27) and solving for v. Thus,

$$f = \frac{3\alpha v^2 \mu^\beta \rho}{4d_p(d_p v \rho)^\beta}$$

so that

$$v = \left[\frac{4 f d_p^{1+\beta} \rho_p}{3\alpha \mu^\beta \rho^{1-\beta}}\right]^{1/(2-\beta)} \tag{23.28}$$

For the Stokes' law range, Equation (23.28) becomes

$$v = \frac{f d_p^2 \rho_p}{18\mu} \tag{23.29}$$

For the intermediate range,

$$v = \frac{0.153 f^{0.71} d_p^{1.14} \rho_p^{0.71}}{\mu^{0.43} \rho^{0.29}} \tag{23.30}$$

Finally, for Newton's law range

$$v = 1.74(fd_p\, \rho_p\rho)^{0.5} \tag{23.31}$$

Keep in mind that f denotes the external force per unit mass of particle. One consistent set of units (English) for the above equations is ft/s^2 for f, ft for d_p, lb/ft^3 for ρ, $lb/ft \cdot s$ for μ, and ft/s for v.

Ordinarily, determining the settling velocity of a particle of known diameter would require a trial-and-error calculation since the particle's Reynolds number is unknown. Thus, one cannot select the proper describing drag force equation. This iterative calculation can be circumvented by rearrangement of the drag force equations and solving for the settling velocity directly. Both sides of Equations (23.29) and (23.31) are multiplied by

$$\frac{d_p\rho}{\mu}$$

A dimensionless constant, K, is defined as

$$K = d_p\left(\frac{f\rho_p\rho}{\mu^2}\right)^{1/3} \tag{23.32}$$

Equations (24.29) and (24.31) can now be rewritten, respectively, as

$$Re = \frac{K^3}{18} \tag{23.33}$$

and

$$Re = 1.74K^{1.5} \tag{23.34}$$

Since K is not a function of the settling velocity, the choice of drag force equations may now be based on calculated K values. These new K range limits are given as follows:

$$K < 3.3 \qquad \text{Stokes}$$
$$43.6 > K > 3.3 \qquad \text{Intermediate range}$$
$$2360 > K > 43.6 \qquad \text{Newton}$$

If K is greater than 2360, the drag coefficient may change abruptly with small changes in fluid velocity.

Larocca[3] and Theodore[4], using the same approach employed above, defined a dimensionless term W that would enable one to calculate the diameter of a particle if the terminal velocity is known. This particular approach has found application in catalytic reactor particle size calculations. The term W—which does not depend on the particle diameter—is given by

$$W = \frac{v^3 \rho^2}{g \mu \rho_p} \tag{23.35}$$

The two key values of W that are employed in a manner similar to that for K are 0.2222 and 1514, that is, for $W < 0.2222$, the Stokes' law region applies, for $W > 1514$, the Newton's law region applies and in between, the intermediate law region applies.

23.5 CUNNINGHAM CORRECTION FACTOR

When particles approach sizes comparable to the mean free path of other fluid molecules, the medium can no longer be regarded as continuous since particles can fall between the molecules at a faster rate than predicted by aerodynamic theory. To allow for this "slip," Cunningham's correction factor[5] is introduced to Stokes' law

$$v = \frac{g d_p^2 \rho_p}{18 \mu} C \tag{23.36}$$

where C is the Cunningham correction factor (CCF), and

$$C = 1 + \frac{2A\lambda}{d_p} \tag{23.37}$$

The term A is $1.257 \times 10^{0.40} \exp(-1.10 d_p / 2\lambda)$ and λ is the mean free path of the fluid molecules (6.53×10^{-6} cm for ambient air). The CCF is usually applied to particles equal to or smaller than 1 micron. Applications include particulate air pollution and nanotechnology[6] studies.

Illustrative Example 23.3   Calculate the CCF for particle size variation from 1.0 nm to 10^4 nm at temperatures of 70°F, 212°F, and 500°F. Include a sample calculation for a particle diameter of 400 nm (0.4 μm) at 70°F, 1 atm.

Solution   Employ the equations presented above. The calculated results are provided in Table 23.5 along with a sample calculation.

Table 23.5   Cunningham correction factors

d_p (nm)	d_p (μm)	C (70°F)	C (212°F)	C (500°F)
1	0.001	216.966	274.0	405.32
10	0.01	22.218	27.92	39.90
100	0.1	2.867	3.61	5.14
250	0.25	1.682	1.952	2.528
500	0.5	1.330	1.446	1.711
1000	1	1.164	1.217	1.338
2500	2.5	1.066	1.087	1.133
5000	5	1.033	1.043	1.067
10,000	10	1.016	1.022	1.033

For a d_p of 0.4 μm, the CCF should be included. Employing the equation given above

$$A = 1.257 + 0.40e^{-1.10d_p/2\lambda}$$

$$= 1.257 + 0.40e^{-\left[\frac{(1.10)(0.4)}{(2)(6.53 \times 10^{-2})}\right]}$$

$$= 1.2708$$

Therefore

$$C = 1 + \frac{2A\lambda}{d_p}$$

$$= 1 + \frac{(2)(1.2708)(6.53 \times 10^{-2})}{0.4}$$

$$= 1.415$$

The results clearly demonstrate that the CCFs become more pronounced for nano-sized particles in the 10–1000 nm range. In addition, an increase in temperature also leads to an increase in this effect.

The reader should also note that a comparable effect does not exist for particles settling in liquids until the diameter become less than 10 nm (0.01 μm).

Illustrative Example 23.4   Three different diameter sized fly-ash particles—0.4, 40, and 400 microns—settle through air. You are asked to calculate the particle terminal velocity and determine how far each will fall in 30 seconds. Assume the particles are spherical. The air temperature and pressure are 238°F and 1 atm, respectively. The specific gravity of fly-ash is 2.31.

Solution   Calculate the particle density using the specific gravity given

$$\rho_p = SG(62.4) = 2.31(62.4) = 144.14 \, \text{lb/ft}^3$$

Determine the properties of the air

$$\rho = \frac{PM}{RT} = \frac{(1)(29)}{(0.7302)(238 + 460)} = 0.0569 \, \text{lb/ft}^3$$

For the viscosity of air (see Table A.9 in the Appendix)

$$\mu = 0.021 \, \text{cP} = 1.41 \times 10^{-5} \, \text{lb/ft-s}$$

Determine the value for K for each fly-ash particle size settling in air.
For d_p of 0.4 microns

$$K = d_p \left(\frac{g \rho_p \rho}{\mu^2} \right)^{1/3} = \frac{0.4}{25,400(12)} \left(\frac{32.174(144.14)(0.0569)}{(1.41 \times 10^{-5})^2} \right)^{1/3} = 0.0144$$

For d_p of 40 microns

$$K = \frac{40}{25,400(12)} \left(\frac{32.174(144.14)(0.0569)}{(1.41 \times 10^{-5})^2} \right)^{1/3} = 1.44$$

For d_p of 400 microns

$$K = \frac{400}{25,400(12)} \left(\frac{32.174(144.14)(0.0569)}{(1.41 \times 10^{-5})^2} \right)^{1/3} = 14.4$$

Determine which fluid-particle dynamic law applies for the above values of K.
For a d_p of 0.4 microns, Stokes' law applies; for a d_p of 40 microns, Stokes' law applies; for a d_p of 400 microns, the Intermediate law applies.
Calculate the terminal settling velocity for each particle size in ft/s using the appropriate velocity equation.
For a d_p of 0.4 microns

$$v = \frac{g d_p^2 \rho_p}{18 \mu} = \frac{32.2(0.4)^2 144.14}{(25,400(12))^2 (18)(1.41 \times 10^{-5})} = 3.15 \times 10^{-5} \, \text{ft/s}$$

For a d_p of 40 microns

$$v = \frac{g d_p^2 \rho_p}{18 \mu} = \frac{32.2(40)^2 144.14}{(25,400(12))^2 (18)(1.41 \times 10^{-5})} = 0.315 \, \text{ft/s}$$

For a d_p of 400 microns

$$v = \frac{0.153g^{0.71}d_p^{1.14}p_p^{0.71}}{\mu^{0.43}\rho^{0.29}} = \frac{0.153(32.2)^{0.71}[400/25,400(12)]^{1.14}(144.14)^{0.71}}{(1.41 \times 10^{-5})^{0.43}(0.0569)^{0.29}}$$

$$= 8.76\,\text{ft/s}$$

Calculate how far, x, the fly-ash particles will fall in 30 seconds.
For a d_p of 40 microns

$$x = vt = 0.315(30) = 9.45\,\text{ft}$$

For a d_p of 400 microns

$$x = vt = 8.76(30) = 262.8\,\text{ft}$$

For a d_p of 0.4 microns (0.4×10^{-6} m), $K = 0.0144$ and $v = 3.15 \times 10^{-5}$ ft/s, without the CCF. With the correction factor ($\lambda = 6.53 \times 10^{-8}$), one obtains

$$A = 1.257 + 0.40 \; \exp\!\left(\frac{-1.10d_p}{2\lambda}\right)$$

$$= 1.257 + 0.40 \; \exp\!\left(\frac{-1.10(0.4 \times 10^{-6})}{2(6.53 \times 10^{-8})}\right) = 1.2708$$

$$C = 1 + \frac{2A\lambda}{d_p} = 1 + \frac{2(1.2708)(6.53 \times 10^{-8})}{0.4 \times 10^{-6}} = 1.415$$

Equation (23.36) may now be employed.

$$v_{\text{corrected}} = vC = 3.15 \times 10^{-5}(1.415) = 4.45 \times 10^{-5}\,\text{ft/s}$$

$$x = v_{\text{corrected}}t = 4.45 \times 10^{-5}(30) = 1.335 \times 10^{-3}\,\text{ft}$$

Illustrative Example 23.5   Refer to Illustrative Example 23.4. Calculate the size of the a fly-ash particle that will settle with a velocity of 1.384 ft/s.

Solution   First calculate the dimensionless number, W, using Equation (24.35):

$$W = \frac{(1.384)^3(0.0569)^2}{32.2(144.14)(1.41 \times 10^{-5})} = 0.1312$$

Since $W < 0.2222$, Stokes' law applies

$$d_p = \sqrt{\frac{18\mu v}{g\rho_p}} = \sqrt{\frac{18(1.41 \times 10^{-5})(1.384)}{(32.2)(144.14)}} = 2.751 \times 10^{-4}\,\text{ft}$$

Illustrative Example 23.6   Appropriate Drag Force Equation   In order to calculate the terminal settling velocity of a particle in a gravity field, one must decide which of the three approximate drag force equations (Stokes, Intermediate, or Newton) is applicable. Explain why, when all three equations are used to calculate values of the terminal velocities for a given Reynolds number, the correct value is always the smallest of the three.

Solution   Refer to Fig. 23.3, which is a slight modification of Fig. 23.2. Irrespective of whether one is in region I, II, or III, the calculated drag coefficient from any of these describing equations produces the highest drag for the correct (and applicable) drag force equation. The higher drag provides greater resistance to flow, which in turn corresponds to a smaller (or lower) velocity.

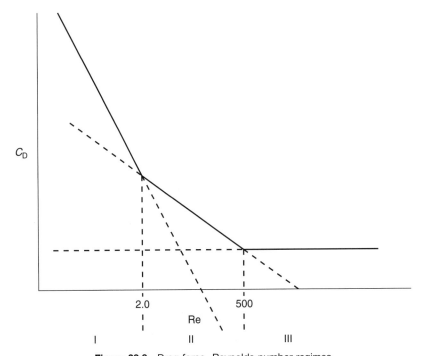

Figure 23.3   Drag force–Reynolds number regimes.

Illustrative Example 23.7   A plant manufacturing Ivory Soap detergent explodes one windy day. It disperses 100 tons of soap particles (SG = 0.8) into the atmosphere (70°F, $\rho = 0.0752$ lb/ft^3). If the wind is blowing 20 miles/h from the west and the particles range in diameter from 2.1 to 1000 μm, calculate the distance from the plant where the soap particles will start to deposit and where they will cease to deposit. Assume the particles are blown vertically 400 ft in the air before they start to settle. Also, assuming even ground-level distribution through an average 100 ft wide path of settling, calculate the average height of the soap particles on the ground in the settling area. Assume the bulk density of the settled particles equals half the actual density.

Solution   The smallest particle will travel the greatest distance while the largest will travel the least distance. For the minimum distance, use the largest particle:

$$d_p = 1000\,\mu\text{m} = 3.28 \times 10^{-3}\,\text{ft}$$

$$K = d_p\left(\frac{g(\rho_p - \rho)\rho}{\mu^2}\right)^{1/3} = 3.28 \times 10^{-3}\left(\frac{32.174(0.8(62.4)) - (0.0752)0.0752}{(1.18 \times 10^{-5})^2}\right)^{1/3} = 31.3$$

The value of K indicates the intermediate range applies. The settling velocity is given by

$$v = \frac{0.153g^{0.71}d_p^{1.14}\rho_p^{0.71}}{\mu^{0.43}\rho^{0.29}} = \frac{0.153(32.2)^{0.71}[3.28 \times 10^{-3}]^{1.14}(0.8(62.4))^{0.71}}{(1.18 \times 10^{-5})^{0.43}(0.0752)^{0.29}}$$

$$= 11.9\,\text{ft/s}$$

The descent time is

$$t = \frac{H}{v} = \frac{400}{11.9} = 33.6\,\text{s}$$

The horizontal distance traveled is

$$L = tv = 33.6\left(\frac{20}{3600}\right)5280 = 986\,\text{ft}$$

For the maximum distance, use the smallest particle

$$d_p = 2.1\,\mu\text{m} = 6.89 \times 10^{-6}\,\text{ft}$$

$$K = d_p\left(\frac{g(\rho_p - \rho)\rho}{\mu^2}\right)^{1/3} = 6.89 \times 10^{-6}\left(\frac{32.174(0.8(62.4)) - (0.0752)0.0752}{(1.18 \times 10^{-5})^2}\right)^{1/3}$$

$$= 0.066$$

The velocity is in the Stokes regime and is given by

$$v = \frac{g d_p^2 \rho_p}{18\mu} = \frac{(32.2)(6.89 \times 10^{-6})^2 (0.8(62.4))}{(1.18 \times 10^{-5})} = 3.59 \times 10^{-4} \, \text{ft/s}$$

The descent time is

$$t = \frac{H}{v} = \frac{400}{3.59 \times 10^{-4}} = 1.11 \times 10^6 \, \text{s}$$

The horizontal distance traveled is

$$L = tv = 1.11 \times 10^6 \left(\frac{20}{3600} \right) 5280 = 3.26 \times 10^7 \, \text{ft}$$

To calculate the depth D, the volume of particles (actual), V_{act}, is first determined.

$$V_{\text{act}} = \frac{m}{\rho_p} = \frac{100(2000)}{0.8(62.4)} = 4006 \, \text{ft}^3$$

The bulk volume is (with 50% voids)

$$V_b = \frac{V_{\text{act}}}{\varepsilon} = \frac{4006}{0.5} = 8012 \, \text{ft}^3$$

The length of the drop area, L_d, is

$$L_d = 3.2 \times 10^7 - 994 = 3.2 \times 10^7 \, \text{ft}$$

Since the width is 100 ft, the deposition area A is

$$A = L_d W = (3.2 \times 10^7)(100) = 3.2 \times 10^9 \, \text{ft}^2$$

The deposition height H is then

$$H = \frac{V_b}{A} = \frac{8012}{3.2 \times 10^9} = 2.5 \times 10^{-6} \, \text{ft}$$

The deposition height can be, at best, described as a "sprinkling."

23.6 LIQUID-PARTICLE SYSTEMS

As indicated in the introduction to this chapter, the general treatment of liquid-particle dynamics, as it applies to liquid–solid separation, appears later in this Part (Chapter 24). However, the reader should note that the equations developed earlier in this chapter may be applied directly with only one minor change. Equation (23.38) contains the density ratio, DR,

$$DR = \frac{\rho_p - \rho}{\rho} \tag{23.38}$$

For gases, $\rho_p >>> \rho$ so that the above term reduces to

$$DR = \frac{\rho_p}{\rho} \tag{23.39}$$

since the ρ term can be neglected in comparison to ρ_p. However, if the fluid medium is a liquid rather than a gas, the ρ term must be retained in all the equations. This change is demonstrated in the next Illustrative Example.

Illustrative Example 23.8   A small sphere (6 mm diameter) is observed to fall through castor oil at a terminal speed of 42 mm/s. At the operating temperature of 20°C, the densities of castor oil and water are 970 kg/m^3 and 1000 kg/m^3, respectively. The viscosities of castor oil and water are 900 cP and 1 cP, respectively. Determine the density of the spherical particle, compute the drag coefficient for the spherical particle, calculate the drag and buoyant forces and, if the same sphere is dropped in water, would the sphere fall slower or faster and why? Also, calculate the Reynolds number and the terminal settling velocity.

Solution   Calculate the particle density, ρ_p, assuming Stokes' law to apply

$$v_t = \frac{g d_p^2 (\rho_p - \rho_f)}{18 \mu_f}$$

Solving for ρ_p

$$\rho_p = \frac{18 \mu_f v_t}{g d_p^2} + \rho_f = \frac{18(0.9)(0.042)}{(9.807)(0.006)^2} + 970 = 2897 \, \text{kg/m}^3$$

Check on Stokes' law validity with $\mu/\rho = 9.28 \times 10^{-4}$.

$$\text{Re} = \frac{d_p v_t}{v_f} = \frac{0.006(0.042)}{9.28 \times 10^{-4}} = 0.272 < 0.3$$

Alternatively, calculate the settling criterion factor, K.

$$K = d_p \left(\frac{g \rho_p (\rho_s - \rho_f)}{\mu_f^2} \right)^{1/3} = 0.006 \left(\frac{9.807(970)(2897 - 970)}{(0.9)^2} \right)^{1/3} = 1.7$$

Since $1.7 < 3.3$, Stokes' law applies. The drag coefficient, C_D, for the Stokes' law regime is

$$C_D = \frac{24}{Re} = \frac{24}{0.272}$$
$$= 88.2$$

Calculate the drag force, F_D, for the Stokes' law regime using Equation (23.11).

$$F_D = 3\pi\mu_f d_p v_t = 3\pi(0.9)(0.006)(0.042)$$
$$= 0.00213 \, N$$

Calculate the buoyancy force

$$F_b = V_p \rho_f g = \frac{\pi d_p^{\,3}}{6}\rho_f g = \frac{\pi(0.006)^3}{6}970(9.807)$$
$$= 0.001076 \, N$$

Consider the case when the same spherical particle is dropped in water. For water, ρ_f is $1000 \, kg/m^3$, and μ is $0.001 \, kg/m \cdot s$. The particle will move faster because of the lower viscosity of water. Stokes' law will almost definitely not apply.

Calculate the settling criterion factor once again.

$$K = d_p\left(\frac{g\rho_p(\rho_s - \rho_f)}{\mu_f^{\,2}}\right)^{1/3} = 0.006\left(\frac{9.807(970)(2897 - 970)}{(0.001)^2}\right)^{1/3}$$
$$= 158$$

Since $158 > 43.6$, the flow is in the Newton's law regime. Employ Equation (23.31) but include the (buoyant) density ratio factor. Therefore,

$$v_t = 1.75\sqrt{\left(\frac{\rho_s - \rho_f}{\rho_f}\right)gd_p} = 1.75\sqrt{\left(\frac{2897 - 1000}{1000}\right)(9.807)(0.006)} = 0.58 \, m/s$$

23.7 DRAG ON A FLAT PLATE

The previous sections treated fluid particle dynamics where the particle was assumed to be a sphere. The drag force and the development that followed keyed solely on spheres. However, there are other applications involving drag that address other bodies. One such body is a flat plate.

The drag on a body submerged in a moving fluid depends on the body shape and size, the speed of the flow, and the properties of the fluid (its viscosity and density). One of the simpler bodies to study is a flat plate aligned parallel to the flow. Several semi-empirical relationships for the drag coefficient for this geometry have been proposed. Two such equations are presented in Equations (23.40) and (23.41)

$$C_D = \frac{1.33}{Re^{0.5}}; \quad 10^4 < Re < 5 \times 10^5 \tag{23.40}$$

$$C_D = \frac{0.031}{Re^{1/7}}; \quad 10^6 < Re < 10^9 \tag{23.41}$$

The first equation applies to a laminar flow, and the second (note the 1/7th power) when the flow is turbulent. The Reynolds number for a flat plate is given by:

$$Re = \frac{\rho v L}{\mu} \tag{23.42}$$

where L is the plate length parallel to the flow.

The drag coefficient can be used to find the drag on the plate using Equation (23.43)

$$F_D = \frac{1}{2} C_D \rho v^2 L W \tag{23.43}$$

where W is the plate width perpendicular to the flow.

Note that these formulas compute the force on one side of a plate. The drag based on these equations should be doubled to compute the drag on a plate in which both sides are exposed to the fluid. There are also other similar formulas available in the literature.

Illustrative Example 23.9 The bottom of a ship, moving at $12\,\text{m/s}$, can be modeled as a flat plate of length 20 m and width 5 m. The water density is 1000 kg/m^3, and the viscosity is $10^{-3}\,\text{N} \cdot \text{s/m}^2$. Calculate the drag on the bottom of the ship.

Solution Compute the flow Reynolds number

$$Re = \frac{\rho v L}{\mu} = \frac{(1000)(12)(20)}{(10^{-3})}$$

$$= 2.4 \times 10^8; \quad \text{turbulent flow}$$

Compute the drag coefficient employing the appropriate equation.

$$C_D = \frac{0.031}{Re^{1/7}} = \frac{0.031}{(2.4 \times 10^8)^{1/7}}$$

$$= 0.002$$

Calculate the drag on area LW is

$$F_D = \frac{1}{2} C_D \rho v^2 LW$$

$$= \frac{1}{2}(0.002)(1000)(12)^2(20)(5) = 14{,}180\,N$$

$$= 14.2\,kN$$

REFERENCES

1. L. Theodore and A. J. Buonicore, personal notes, 1982.
2. S. Barnea and I. Mizraki, Ph.D. thesis, Haifa University, Haifa, Israel, 1972.
3. M. Larocca, *Chem. Eng*, April 2, 1987.
4. L. Theodore, personal notes, 1984.
5. E. Cunningham, Proc. R. Soc. London Ser., 17, 83, 357, 1910.
6. L. Theodore, "Nanotechnology: Basic Calculations for Engineers and Scientists," John Wiley & Sons, Hoboken, NJ, 2007.

NOTE: Additional problems for each chapter are available for all readers at www.wiley.com. Follow links for this title.

24

SEDIMENTATION, CENTRIFUGATION, FLOTATION

This chapter examines three industrial separation techniques that exploit the density difference between a liquid and a solid. The driving force in these processes is usually the result of gravity, centrifugal action, and/or buoyant effects. Unlike the previous chapter that primarily treated gas-particle dynamics, this presentation will address the three above titled topics, with the bulk of the material keying on liquid–solid separation. The above captioned headings will be addressed in this chapter. Filtration is treated separately in Chapter 27.

24.1 SEDIMENTATION

Gravity sedimentation is a process of liquid–solid separation that separates, under the effect of gravity, a feed slurry into an underflow slurry of higher solids concentration and an overflow of substantially clearer liquid. A difference in density between the solids and the suspended liquid is, as indicated, a necessary prerequisite.

Nearly all commercial equipment for continuous sedimentation is built with relatively simple settling tanks. Distinction is commonly made depending on the purpose of the separation. If the clarity of the overflow is of primary importance, the process is called clarification and the feed slurry is usually dilute. If a thick underflow is the primary aim, the process is called thickening and the feed slurry is usually more concentrated.[1]

Fluid Flow for the Practicing Chemical Engineer. By J. Patrick Abulencia and Louis Theodore
Copyright © 2009 John Wiley & Sons, Inc.

The most commonly used thickener is the circular-basin type (shown in Fig. 24.1). The flocculant-treated feedstream enters the central feedwell, which dissipates the stream's kinetic energy and disperses it gently into the thickener. The feed finds its height in the basin, where its density matches the density of the suspension inside (the concentration increases downward in an operating thickener, giving stability to the process) and spreads out at that level. The settling solids move downward as does some liquid, the liquid amount being determined by the underflow withdrawal rate. Most of the liquid goes upward and into the overflow. Typically, a thickener has three operating zones: clarification, zone settling and compression (see Fig. 24.1).[1]

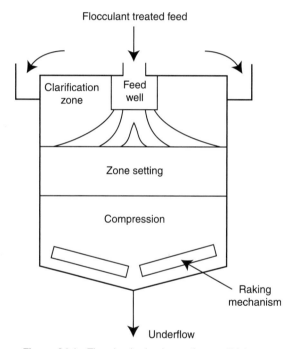

Figure 24.1 The circular-basin continuous thickener.

Gravity clarifiers sometimes resemble circular thickeners but more often are rectangular basins, with feed at one end and overflow at the other. Settled solids are pushed to a discharge trench by paddles or blades on a chain mechanism. Flocculent may be added prior to the clarifier. Conventional thickeners are also used for clarification, but the typically low feed concentrations hinder the benefits of zone settling, so the basin area is based on clarification-zone demands.[1] Thus, in a general sense, sedimentation is the process of removing solid particles heavier than water by gravity settling. It is the oldest and most widely used unit operation in water and wastewater treatment. The terms sedimentation, settling, and clarification are often used interchangeably. The sedimentation basin unit may also be referred to as a sedimentation tank, clarifier, settling basin, or settling tank.

In wastewater treatment (to be discussed in Part VI, Chapter 28), sedimentation is used to remove both inorganic and organic materials which are settleable in continuous flow conditions. It removes grit, particulate matter in the primary settling tank, and chemical flocculants from a chemical precipitation unit. Sedimentation is also used for solids concentration in sludge thickeners.[2]

Details on discrete, or individual, particle settling is available in Chapter 23. Here, Stokes' law assumes that particles are present in relatively dilute concentrations and in a fluid medium of relatively large cross-section. When there are a large number of particles, the particles in close proximity will retard other particles. This is termed hindered settling. The effect is not significant at volumetric concentrations below 0.1%. However, when the particle diameter becomes appreciable with respect to the diameter of the container in which it is settling, the container walls will exert an additional retarding effect known as the wall effect.

Farag[3] has reviewed and developed equations for the above situation. For hindered flow, the settling velocity is less than would be calculated from Stokes' law. The density of the fluid phase becomes the bulk density of the slurry, ρ_M, which is defined as follows:

$$\rho_M = \varepsilon_f \rho_f + (1 - \varepsilon_f)\rho_s \qquad (24.1)$$

where ρ_M is the density of slurry in kg (solid plus liquid)/m^3 and ε_f is the volume fraction of the liquid in the slurry mixture. (See Chapter 25 for more details on void volume.) The density difference is then:

$$\rho_s - \rho_M = \rho_s - [(\varepsilon_f \rho_f + (1 - \varepsilon_f)\rho_s] = \varepsilon_f(\rho_s - \rho_f) \qquad (24.2)$$

The effective viscosity of the mixture, μ_m, is defined as:

$$\mu_m = \mu_f b \qquad (24.3)$$

where μ_f is the pure fluid viscosity and b is a dimensionless correction factor which is a function of ε_f. The term b can be evaluated from Equation (24.4)

$$b = 10^{1.82(1-\varepsilon_f)}$$
$$\log_{10} b = 1.82(1 - \varepsilon_f) \qquad (24.4)$$

The settling velocity, v, with respect to the unit is ε_f times the velocity calculated by Stokes' law. Substituting Equations (24.2) and (24.3) into Stokes' law terminal velocity equation and multiplying by ε_f for the relative velocity effect leads to:

$$v = \frac{gd_p^2(\rho_s - \rho_f)\varepsilon_f^2}{18\mu_f}\frac{1}{b} \qquad (24.5)$$

The Reynolds number, based on the settling velocity relative to the fluid is then:

$$Re = \frac{\rho_m v d_p}{\mu_m \varepsilon_f} = \frac{g d_p^2 (\rho_s - \rho) \rho_m}{18 \mu_f^2} \frac{\varepsilon_f}{b^2} \tag{24.6}$$

When the Reynolds number is less than 1.0, the settling is in the Stokes' law regime.

When the diameter, d_p, of the particle becomes appreciable with respect to the diameter of the container, D_w, the terminal velocity is reduced. This is termed the aforementioned wall effect. In the case of settling in the Stokes' law regime, the calculated terminal velocity is multiplied by a correction factor, k_w. This k_w is approximated as follows:

$$k_w = \frac{1}{1 + 2.1(d_p/D_w)^{1.5}} \quad \text{for } d_p/D_w < 0.05 \tag{24.7}$$

For $0.5 < d_p/D_w < 0.8$, employ the values provided in Table 24.1. For the turbulent flow regime, Equation (24.8) should be used

$$k_w = 1 - (d_p/D_w)^{1.5} \tag{24.8}$$

Table 24.1   Wall correction factors

d_p/d_w	k_w
0.1	0.792
0.2	0.596
0.3	0.422
0.4	0.279
0.5	0.170
0.6	0.0945
0.7	0.0468
0.8	0.0205

Illustrative Example 24.1   Glass spheres are settling in water at 20°C. The slurry contains 60 wt% solids and the particle diameter is 0.1554 mm. The glass density is 2467 kg/m³. Find the volume fraction, ε_f, of the liquid, the bulk density of the slurry, ρ_m, the terminal velocity, and the effective mixture viscosity, μ_m. For water at 20°C, $\rho_f = 998$ kg/m³ and $\mu_f = 0.001$ kg/m · s.

Solution   Start by assuming a basis of 100 kg of slurry. To determine the volume fraction of the liquid, divide the mass of liquid by its density. Since 100 kg of slurry is the basis and the slurry contains 60 wt% solid,

$$m_f = 40 \, \text{kg}$$

The volume of the fluid (water) is

$$V_f = \frac{m_f}{\rho_f} = \frac{40}{998} = 0.040 \, \text{m}^3$$

Similarly,

$$m_s = 60 \, \text{kg}$$

$$V_s = \frac{m_s}{\rho_s} = \frac{60}{2467}$$

$$= 0.0243 \, \text{m}^3$$

Therefore,

$$V = V_f + V_s = 0.040 + 0.0243$$

$$= 0.0643 \, \text{m}^3$$

and

$$\varepsilon_f = \frac{V_f}{V} = \frac{0.040}{0.643}$$

$$= 0.622$$

For the glass particles,

$$\varepsilon_p = 1 - \varepsilon_f$$

$$= 0.378$$

Calculate ρ_m from Equation (24.1)

$$\rho_m = \varepsilon_f \rho_f + \varepsilon_p \rho_p = 0.622(998) + 0.378(2467)$$

$$= 1553 \, \text{kg/m}^3$$

Calculate b and the terminal velocity. See Equations (24.4) and (24.5).

$$b = 10^{1.82(1-\varepsilon_f)} = 10^{1.82(0.378)}$$

$$= 4.875$$

$$v = \frac{g d_p^2 (\rho_p - \rho_f)\varepsilon_f^2}{18\mu_f b} = \frac{9.807(0.0001554)^2(2467 - 998)(0.622)^2}{18(0.001)(4.875)}$$

$$= 0.00153 \, \text{m/s}$$

Compute μ_m by employing Equations (24.3) and (24.6).

$$\mu_m = \mu_f b \qquad (24.3)$$
$$= 0.001(4.875)$$
$$= 0.0049 \, \text{kg/m} \cdot \text{s}$$

Illustrative Example 24.2   Refer to Illustration Example (24.1). Calculate the Reynolds number.

Solution   Employ Equation (24.6):

$$\text{Re} = \frac{\rho_m v d_p}{\mu_m \varepsilon_f} \qquad (24.6)$$

Substituting,

$$\text{Re} = \frac{1553(0.00153)(1.554 \times 10^{-4})}{(0.0049)(0.622)}$$

$$= 0.121$$

Illustrative Example 24.3   Classify small spherical particles of charcoal with a specific gravity of 2.2. The particles are falling in a vertical tower against a rising current of air at 25°C and atmospheric pressure. Calculate the minimum size of charcoal that will settle to the bottom of the tower if the air is rising through the tower at the rate of 15 ft/s.

Solution   The particles that have the settling velocity differentially greater than the rising air velocity will be the smallest diameter particles to settle. First note that

$$\rho = 0.075 \, \text{lb/ft}^3$$

and

$$\mu = 1.23 \times 10^{-5} \, \text{lb/ft} \cdot \text{s}$$

Assume Stokes' Law to apply

$$d_p = \left(\frac{18 \mu v}{g \rho_p} \right)^{0.5}$$

Substituting the data (in consistent units),

$$d_p = \frac{(18)(1.23 \times 10^{-5})(15)}{(32.2)2.2 \times 62.4}$$

$$d_p = 8.67 \times 10^{-4} \, \text{ft} = 264.19 \, \mu\text{m}$$

Check K,

$$K = d_p \left[\frac{g\rho\rho_p}{\mu^2} \right]^{1/3}$$

Substituting the data gives

$$K = 8.67 \times 10^{-4} \text{ ft} \left[\frac{(32.2 \text{ ft/sec}^2)(0.074 \text{ lb/ft}^3)(2.2 \times 62.4 \text{ lb/ft}^3)}{(1.23 \times 10^{-5} \text{ lb/ft-s})^2} \right]^{1/3}$$

$$= 11.2$$

Stokes' law does not apply. Therefore, assume the Intermediate range law applies:

$$v_t = \frac{0.153 \, g^{0.71} d_p^{1.14} \rho_p^{0.71}}{\rho^{0.29} \mu^{0.43}}$$

Rearranging gives

$$d_p^{1.14} = \frac{v_t \rho^{0.29} \mu^{0.43}}{0.153 (g\rho_p)^{0.71}}$$

Substituting the data,

$$d_p^{1.14} = \frac{(1.5 \text{ ft/s})(0.074 \text{ lb/ft}^3)^{0.29} (1.23 \times 10^{-5} \text{ lb/ft-s})^{0.43}}{0.153 (32.2 \text{ ft/s}^2 \times 2.2 \times 62.4 \text{ lb/ft}^3)^{0.71}}$$

$$= 9.20 \times 10^{-4} \text{ (ft)}^{1.14}$$

or

$$d_p = 2.17 \times 10^{-3} \text{ ft} = 662 \, \mu\text{m}$$

Checking K,

$$K = \frac{2.17 \times 10^{-3} \text{ ft}}{8.6677 \times 10^{-4} \text{ ft}} (11.2081) = 28.077$$

Then, $3.3 < K < 43.6$, and the result is correct for the intermediate range

$$d_p = 2.17 \times 10^{-3} \text{ ft} = 662 \, \mu\text{m}$$

24.2 CENTRIFUGATION

Centrifugal force is widely used when a force greater than that of gravity as in settling, is desired for the separation of solids and fluids of different densities. Two terms need to be defined. A centrifugal force is created by moving a mass in a curved path and is exerted in the direction away from the center of curvature of the path. The centripetal force is the force applied to the moving mass in the direction toward the center of curvature that causes the mass to travel in a curved path. If these forces are equal, the particle continues to rotate in a circular path around the center.[4]

Centrifugation is therefore another process that uses density differences to separate solids from liquids (or an immiscible liquid from other liquids). The feed is subjected to centrifugal forces that make the solids move radially through the liquid (outward if heavier, inward if lighter). In a sense, centrifugation is an extension of gravity sedimentation to particle sizes and to emulsions that are normally stable in a gravity field. The describing equations developed earlier in the previous chapter again apply. The gravity force is replaced by a centrifugal force, F_c (force/mass) where

$$F_c = \frac{r\omega^2}{g_c} = \frac{v_t^2}{g_c r} \tag{24.9}$$

where r is the radius of curvature of the particle or heavier phase, ω the angular velocity and v_t the tangential velocity at the point in question.

As indicated above, centrifugation attempts to increase the particle "settling" velocity many times higher than that due to gravity by applying a centrifugal force. The ratio of these two forces has been defined by some as the number of "Gs", where

$$G = \frac{r\omega^2}{g} \tag{24.10}$$

Note that the g term in the denominator in Equation (24.10) is the acceleration due to gravity and not the term g_c (the conversion constant) that appears in Equation (24.9).

There are two main classes of centrifugation equipment: cyclones and centrifuges. Cyclones are generally classified as air recovery/control equipment. The unit is primarily used for separating gas–solid systems. Centrifuges are primarily employed for liquid–solid separation; units in this category include basket, tubular, scroll-type, dish and multiple chamber. The units may operate in the batch or continuous mode. Details on this equipment are available in the literature.[4–6]

Illustrative Example 24.4   A particle is spinning in a 3-inch ID centrifuge with an angular velocity of 30 rad/s. Calculate the number of Gs for the particle.

Solution   Employ Equation (24.10).

$$G = \frac{r\omega^2}{g}$$

Substituting yields:

$$G = \frac{r\omega^2}{g} = \frac{(3/12)(30)^2}{32.2} = 7.0$$

24.3 HYDROSTATIC EQUILIBRIUM IN CENTRIFUGATION

Another extended topic of interest in centrifugation is hydrostatic centrifugation equilibrium. When a contained liquid of constant ρ and μ is rotated around the vertical z-axis at a constant angular speed, ω, it is thrown outward from the axis by centrifugal force. The free surface of the liquid develops a paraboloid of revolution, the cross-section of which has a parabolic shape. When the fluid container has been rotated long enough, the whole volume of fluid will be rotating at the same angular velocity. Then, there is no sliding of one layer of liquid over the other. This condition is termed rigid-body rotation. The pressure distribution can be calculated from the principles of fluid statics. Further, one may determine the shape of the surface and the pressure distribution. Figure 24.2 shows a cross-section of this system.

The pressure at any point in the liquid has to be calculated in two directions: the axial (z) and radial (r) directions. In the vertical z-direction, the pressure gradient is given by

$$\frac{dP}{dz} = -\rho \frac{g}{g_c} \tag{24.11}$$

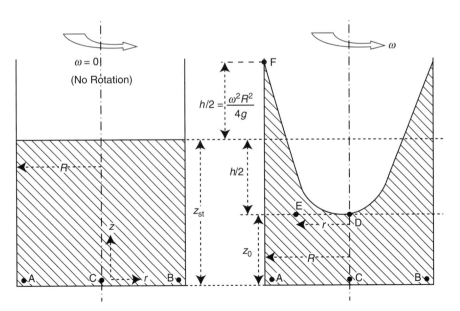

Figure 24.2 Rigid body rotation.

Therefore, the gauge pressure at any point in the liquid is

$$P_g = \rho \frac{g}{g_c} \Delta z \qquad (24.12)$$

where Δz is the height of liquid above the point. Note that at point D (or at any point on the paraboloid surface), the gauge pressure is zero since it is exposed to the atmosphere. At point E, the gauge pressure, P_g, is

$$P_g = \rho \frac{g}{g_c} (z - z_0) \qquad (24.13)$$

In the radial direction, the pressure gradient is given by

$$\frac{dP}{dr} = \frac{-\rho a_r}{g_c} \qquad (24.14)$$

where

$$a_r = \text{(radial centripetal acceleration)} = -\omega^2 r / g_c \qquad (24.15)$$

Substitution of Equation (24.15) into Equation (24.14) yields

$$\frac{dP}{dr} = \frac{\rho \omega^2 r}{g_c} \qquad (24.16)$$

Integrating from point D (where $r = 0$ and $P_g = 0$) to any radius r (e.g., point E) gives

$$P_g = \frac{\rho \omega^2 r^2}{2 g_c} \qquad (24.17)$$

Equating Equations (24.13) and (24.17) and simplifying results in

$$\frac{\rho \omega^2 r^2}{2} = \rho g (z - z_0) \qquad (24.18)$$

or

$$z - z_0 = \frac{\omega^2 r^2}{2g} \qquad (24.19)$$

Equation (24.19) indicates that the free surface of the rotating liquid is a parabola. At point F on the wall surface, $r = R$, $z - z_0 = h$. Thus, Equation (24.19) becomes

$$h = \frac{\omega^2 R^2}{2g} \qquad (24.20)$$

Since the volume of the liquid is conserved (remains the same), the volume of liquid in the cylinder of height $z_{st} - z_0$ is equal to the volume of liquid in the paraboloid of radius R and height h which reduces to

$$V_{cyl} = V_{para}$$

$$\pi R^2(z_{st} - z_0) = \frac{1}{2} \pi R^2 h \tag{24.21}$$

Solving of Equation (24.21) and replacing h by Equation (24.20) yields

$$z_{st} - z_0 = \frac{1}{2}h = \frac{\omega^2 R^2}{4g} \tag{24.22}$$

Equation (24.22) indicates that the increase in elevation at the container wall equals the decrease in elevation at the centerline, with both being equal to $\omega^2 R^2/4g$. Finally, the gauge pressure distribution at any point in the liquid can be shown to be

$$P = \rho \frac{g}{g_c}(z_0 - z) + \frac{\rho \omega^2 r^2}{2g_c} \tag{24.23}$$

Illustrative Example 24.5 Consider the case of a circular cylinder (diameter = 6 in., height = 1 ft), that is filled with water (density 62.4 lb/ft³). If it is rotated at a uniform, steady angular speed about its central axis in rigid-body motion, calculate the rate in rpm for which the water will start spilling out. Also calculate the rpm for which one third of the water will spill out.

Solution Since the cylinder is full, the water will spill the moment the cylinder starts to spin. Therefore, spilling occurs when $\omega > 0$ rpm. To determine the angular speed for $\frac{1}{3}$ of the water to spill, consider the cylinder at rest when $\frac{1}{3}$ of the water has already been spilled. The stationary height, z_{st}, is obviously $\frac{2}{3}$ ft. The increase in height, $h/2$, due to rigid-body motion is

$$h/2 = 1 - 0.667 = 0.333 \text{ ft}$$

Employ Equation (24.22) to calculated the angular velocity:

$$h/2 = \frac{\omega^2 R^2}{4g}$$

Substitution yields:

$$\omega = \sqrt{\frac{4g(h/2)}{R^2}} = \sqrt{\frac{4(32.174)(0.333)}{(0.25)^2}} = 26.2 \text{ rad/s} = 250 \text{ rpm}$$

Illustrative Example 24.6 Refer to Illustrative Example 24.5. Determine the equation describing the pressure distribution within the system.

Solution   The (gauge) pressure distribution equation is obtained from Equation (24.23):

$$P = \rho \frac{g}{g_c}(z_0 - z) + \frac{\rho \omega^2 r^2}{2g_c}$$

since

$$z_0 = z_{st} - h/2 = 0.333 \, \text{ft}$$

and substituting

$$P = \rho \frac{g}{g_c}(z_0 - z) + \frac{\rho \omega^2 r^2}{2g_c} = 62.4(0.333 - z) + \frac{62.4(26.2)^2 r^2}{2(32.174)}$$

$$P = 62.4(0.333 - z) + 665.3 r^2; \, \text{psfg}$$

Illustrative Example 24.7   A cylindrical cup (diameter $= 6 \, \text{cm}$, height $= 10 \, \text{cm}$) open to the atmosphere is filled with a liquid (density $= 1010 \, \text{kg/m}^3$) to a height of 7 cm. It is placed on a turn-table and rotated around its axis. Find:

1. The angular speed that causes the liquid to start spilling.
2. The gauge pressures at points A, B, and C along the bottom of the cup. Point C is at the center, and points A and B are at the wall (see Fig. 24.2).
3. The thickness of liquid film at the original height of the liquid during rotation.

Solution   Calculate an angular velocity that will cause the liquid to start spilling

$$h = 10 - 7 = 3 \, \text{cm} = 0.03 \, \text{m}$$
$$R = \text{radius} = 0.03 \, \text{m}$$

Applying Equation (24.22)

$$\omega^2 = \frac{(2)(0.03)(9.807)}{(0.03)^2}$$

$$\omega = 36.2 \, \text{rad/s} = 345 \, \text{rpm}$$

Calculate the pressure at A and B, that is, P_A and P_B, respectively

$$P_A = P_B \, \text{(because of symmetry)}$$
$$P_A = P_B = \rho g \, \Delta z = (1010)(9.807)(0.1) = 990 \, \text{Pa gauge (Pag)}$$

This may be converted to psig noting that 14.696 psig $= 101{,}325$ Pa. The liquid height above point C is

$$z_0 = z_{st} - h/2 = 7 - 3 = 4\,\text{cm} = 0.04\,\text{m}$$

The gauge pressure at point C, P_C is therefore

$$P_C = (1010)(9.807)(0.04) = 396\,\text{Pag}$$

To obtain the film thickness, determine the original height.

$$z_{st} = 0.07\,\text{m} = 7\,\text{cm}$$
$$z_0 = z_{st} - h/2 = 7 - 3 = 4\,\text{cm} = 0.04\,\text{m}$$

Substitute in Equation (24.18) to obtain r.

$$0.07 - 0.04 = \frac{36.2^2 r^2}{2(9.807)}$$
$$r = 0.0212\,\text{m} = 2.12\,\text{cm}$$

Therefore

$$\text{film thickness} = R - r = 3 - 2.12 = 0.88\,\text{cm} = 8.8\,\text{mm}$$

24.4 FLOTATION

Flotation processes are useful for the separation of a variety of species, ranging from molecular and ionic to microorganisms and mineral fines, from one another for the purpose of extraction of valuable products as well as cleaning of waste waters. These processes are particularly attractive for separation problems involving very dilute solutions where most other processes usually fail. The success of flotation processes is dependent primarily on the tendency of surface-active species to concentrate at the water–fluid interface and on their capability to make selected non-surface-active materials hydrophobic by means of adsorption on them or association with them. Under practical conditions, the amount of interfacial area available for such concentration is increased by generating air bubbles or oil droplets in the aqueous solution.[6]

Flotation is a gravity separation process based on the attachment of air or gas bubbles to solid (or liquid) particles that are then carried to the liquid surface where they accumulate as float material and can be skimmed off. The process consists of two stages: the production of suitably small bubbles, and their attachment to the particles. Depending on the method of bubble production, flotation is classified as dissolve-air, electrolytic, or dispersed-air, with dissolve-air primarily employed by industry.[1]

The separation of solid particles into several fractions based upon their terminal velocities is called hydraulic classification. By placing the particles of different densities in an upward-flowing stream of fluid (often the fluid is water) particles of materials of different densities but of the same size may also be separated by the method of hydraulic separation or classification. If the velocity of the water is adjusted so that it lies between the terminal falling velocities (or settling velocities) of the two particles, the slower particles will be carried upward and the particles of higher terminal velocity than the water velocity will move downward, and a separation is thereby attained.

Illustrative Example 24.8   It is desired to separate quartz (q) particles (SG $= 2.65$, with a size range of 40–90 μm) from galena (g) particles (SG $= 7.5$, with a similar size range of 40–90 μm). The mixture will be placed in a rising water (w) flow (density, $\rho = 1000$ kg/m^3; viscosity, $\mu = 0.001$ kg/m $\cdot$ s). See Fig. 24.3. Calculate the water velocity to obtain pure galena. Will this pure galena be a top or bottom product?

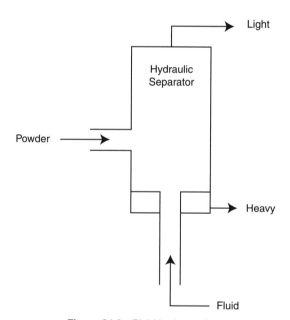

Figure 24.3   Rigid body rotation.

Solution   Calculate the settling velocity of the largest quartz particle, with a diameter, $d_p = 90$ μm. Employ Equation (23.32) from Chapter 23.

$$K = d_p \left(\frac{g(\rho_s - \rho_f)\rho_f}{\mu_f{}^2} \right)^{1/3} = 9 \times 10^{-5} \left(\frac{9.807(2650 - 1000)1000}{0.001^2} \right)^{1/3} = 2.27 < 3.3$$

Stokes' flow regime applies. Therefore, from Equation (23.36),

$$v_q = \frac{g d_p{}^2 (\rho_s - \rho_f)}{18 \mu_f} = \frac{9.807 (9 \times 10^{-5})^2 (2650 - 1000)}{18(0.001)}$$

$$= 0.0073 \text{ m/s}$$

Calculating the settling velocity of the smallest galena particle with a diameter of $d_p = 4 \times 10^{-5}$ m.

$$K = d_p \left(\frac{g(\rho_s - \rho_f) \rho_f}{\mu_f{}^2} \right)^{1/3} = 4 \times 10^{-5} \left(\frac{9.807(7500 - 1000)1000}{0.001^2} \right)^{1/3} = 1.6 < 3.3$$

Stokes' flow regime again applies. Therefore,

$$v_g = \frac{g d_p{}^2 (\rho_s - \rho_f)}{18 \mu_f} = \frac{9.807 (4 \times 10^{-5})^2 (7500 - 1000)}{18(0.001)} = 0.00567 \text{ m/s}$$

To obtain pure galena, the upward velocity of the water must be equal to or greater than the settling velocity of the largest quartz particle. Therefore,

$$v_w = 0.0073 \text{ m/s} = 7.3 \text{ mm/s}$$

Since the water velocity of 7.3 mm/s is greater than the settling velocity of the smallest galena particle, some galena will be washed up with the quartz. One may conclude that pure galena will be the bottom product; the top product will be the quartz plus some galena.

Illustrative Example 24.9   Refer to Illustrative Example 24.8. Determine the size range of the galena in the top product.

Solution   To determine the size range of the galena product, calculate the galena particle size that has a settling velocity of 7.33 mm/s. Assume Stokes' law applies

$$v = \frac{g d_p{}^2 (\rho_s - \rho_f)}{18 \mu}$$

$$d_p = \sqrt{\frac{18 \mu_f v}{g(\rho_s - \rho_f)}} = \sqrt{\frac{18(0.001)(0.0073)}{9.807(7500 - 1000)}} = 4.54 \times 10^{-5} \text{ m} = 45.4 \text{ } \mu\text{m}$$

Check on the validity of Stokes' flow by calculating the K factor.

$$K = d_p \left(\frac{g(\rho_s - \rho_f)\rho_f}{\mu_f{}^2} \right)^{1/3} = 4.54 \times 10^{-5} \left(\frac{9.807(7500 - 1000)1000}{0.001^2} \right)^{1/3}$$

$$= 1.82 < 3.3$$

Therefore, the flow is in the Stokes' law range.

The size ranges for the galena are 40–45.4 μm for the top washed product.

Illustrative Example 24.10   Air is being dried by bubbling (in very small bubbles) through concentrated NaOH (specific gravity of 1.34 and viscosity equal to 4.3 cP). The base fills a 4.5-ft tall, 2.5-in. ID tube to a depth of 1.0 ft. The air above the base is at ambient conditions. If the air rate is 4.0 ft^3/min, what is the maximum diameter of a base spray droplet that would be carried out of the apparatus by entrainment in the air stream?

Solution   The velocity of air is

$$v = \frac{q}{S} = \frac{4 \dfrac{\text{ft}^3}{\text{min}} \times \dfrac{\text{min}}{60\,\text{s}}}{\dfrac{\pi}{4}(2.5/12\,\text{ft})^2} = 1.956\,\text{ft/s}$$

Assuming that the Intermediate range applies, calculated $d_p(\text{max})$ by

$$d_p(\text{max})^{1.14} = \frac{v\rho^{0.29}\mu^{0.43}}{0.153(g\rho_p)^{0.71}}$$

Substituting the data (in consistent units) gives

$$d_p(\text{max})^{1.14} = \frac{(1.956)(0.0775)^{0.29}(1.23 \times 10^{-5})^{0.43}}{(0.153)(32.2 \times 1.34 \times 62.4)^{0.71}}$$

$$= 1.7291 \times 10^{-4}(\text{ft})^{1.14}$$

or

$$d_p(\text{max}) = 5.01 \times 10^{-4}\,\text{ft}$$

Check the original assumption by calculating K.

$$K = d_p \left[\frac{g\rho\rho_p}{\mu^2} \right]^{1/3}$$

Substituting the data,

$$K = 5.01 \times 10^{-4} \left[\frac{(32.2)(0.0775)(1.34)(62.4)}{(1.23 \times 10^{-5})^2} \right]^{1/3}$$

$$= 5.58$$

Thus, the result for $d_p(\max)$ is correct.

REFERENCES

1. L. Svarousky, "Sedimentation, Centrifugation and Flotation", *Chem. Eng.*, New York, July 16, 1979.
2. S. Lin, "Water and Wastewater Evaluations Manual", McGraw-Hill, New York, 2001.
3. I. Farag, "Fluid Flow," A Theodore Tutorial, Theodore Tutorials, East Williston, NY, 1996.
4. D. Green and R. Perry (editors), "Perry's Chemical Engineers' Handbook," 8th edition, McGraw-Hill, New York, 2008.
5. J. Reynolds, J. Jeris, and L. Theodore, "Handbook of Chemical and Environmental Engineering Calculations", John Wiley & Sons, Hoboken, NJ, 2002.
6. R. Rousseau (editor), "Handbook of Separation Process Technology", John Wiley & Sons, Hoboken, NJ, 1987.

NOTE: Additional problems are available for all readers at www.wiley.com. Follow links for this title.

25

POROUS MEDIA AND PACKED BEDS

25.1 INTRODUCTION

The flow of a fluid through porous media and/or a packed bed occurs frequently in chemical process applications. Some examples include flow through a fixed-bed catalytic reactor, flow through an adsorption tower, and flow through a filtration unit. An understanding of this type of flow is also important in the study of fluidization and some particle dynamics applications.[1]

This introductory section provides key definitions in this area and information on porous media flow regimes. The chapter concludes with six Illustrative Examples.

A porous medium consists of a solid phase with many void spaces that some refer to as pores. Examples include sponges, paper, sand, and concrete. As indicated above, *packed beds* (see Fig. 25.1) or porous materials are used in a number of engineering operations. These porous media are divided into two categories.

1. Impermeable media: solid media in which the pores are not interconnected, e.g., foamed polystyrene.
2. Permeable media: solid media in which the pores are interconnected, e.g., packed columns and catalytic reactors.

Before proceeding to the study of fluid flow through a porous medium (e.g., packed bed) the following key variables are defined.

Fluid Flow for the Practicing Chemical Engineer. By J. Patrick Abulencia and Louis Theodore
Copyright © 2009 John Wiley & Sons, Inc.

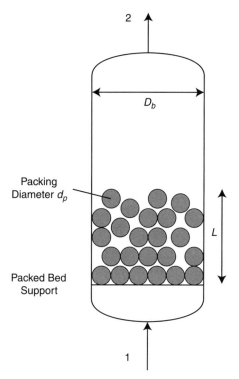

Figure 25.1   Packed bed.

25.2   DEFINITIONS

By definition, the *porosity* or *void fraction*, ε, is

$$\varepsilon = \frac{V_f}{V} = \frac{S_f}{S} \tag{25.1}$$

where V_f is the void volume (occupied by the fluid), V is the total system volume, S_f is the cross-sectional area of the void space, and S is the total cross-sectional area. Bed porosities are fractional numbers that range from approximately 0.4 to 0.6. *Solid fraction* is the fraction of the total bed volume occupied by solids and given by $1 - \varepsilon$. The *empty bed cross-section* (S) is

$$S = \frac{\pi D_b{}^2}{4} \tag{25.2}$$

where D_b is the bed diameter (see Fig. 25.2). The *actual bed cross-section* for flow (*open or void bed cross-section*), S_f, is

$$S_f = \frac{\pi D_b{}^2 \varepsilon}{4} \qquad (25.3)$$

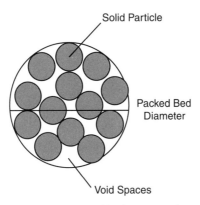

Figure 25.2 Packed bed cross-section.

The *ultimate* or *true density*, ρ_s, is the density of the solid material. The *bulk density*, ρ_B,

$$\rho_B = (1 - \varepsilon)\rho_s + \varepsilon \rho_f \qquad (25.4)$$

where ρ_f is the fluid density. The *superficial velocity* (empty-tower velocity), v_s, is (with this subscript is referring to superficial, not solid)

$$v_s = \frac{q}{S} = \frac{4q}{\pi D_b{}^2} \qquad (25.5)$$

The superficial velocity may be thought of as the average fluid velocity based on an empty cross-section. The *interstitial velocity*, v_I, is the actual velocity of fluid through pores

$$v_I = \frac{q}{S_v} = \frac{q}{\pi D_b{}^2 \varepsilon} = \frac{v_s}{\varepsilon} \qquad (25.6)$$

The particle size, d_p, is expressed in different units depending on the size range involved. These units used are summarized in Table 25.1 with relation to particle size. *Standard screen mesh size* is often used to measure particles in the range of 3 inches to about 0.0015 inches. The testing sieves are standardized and have square openings.

Table 25.1 Measurement units of particle sizes

Particle Size	Measurement Units
Coarse particles	Inches or centimeters
Fine particles	Standard screen mesh size
Very fine particles	Micrometers or nanometers
Ultrafine particles	Specific surface area, or surface area per unit mass

A common group of screens is the Tyler standard screen scales, listed in Table 25.2. As noted in Chapter 23, a common method of specifying large particle sizes is to designate the screen mesh that has an aperture corresponding to the particle diameter. Since various screen scales are in use, confusion may result unless the screen scale involved is specified. The screen mesh generally refers to the number of screen openings per unit of length or area. The aperture for a given mesh will depend on the wire size employed. An expanded version, the Tyler Standard Screen Scales in Table 23.2, including English units, is provided in Table 25.2.

The clear space between the individual wires of the screen is termed the screen aperture. As indicated above, the term *mesh* is applied to the number of apertures per linear inch; for example, a 10-mesh screen will have 10 openings per inch, and the aperture will be 0.1 in. minus the diameter of the wire.

The *particle specific surface area*, a_p, is the surface area per unit volume of the packed bed particles. For a spherical particle of diameter d_p,

$$a_p = \frac{S_p}{V_p} = \frac{\pi d_p^2}{(\pi/6)d_p^3} = \frac{6}{d_p} \tag{25.7}$$

The *bed specific surface area* (a_b) is

$$a_b = a_p(1 - \varepsilon) = \frac{6(1 - \varepsilon)}{d_p} \tag{25.8}$$

The *effective diameter* $(d_{p,e})$ of non-spherical particles is the diameter of a sphere that has the same specific surface area, a_p, as that of a non-spherical particle. In effect, it is the surface area of a sphere having a volume equal to that of the particle, divided by the surface area of the particle. If all the particles are not the same size, one should employ the *mean effective diameter*, $\overline{d_{p,e}}$, where

$$\overline{d_{p,e}} = \frac{1.0}{\sum (w_i)(d_{p,e})_i} \tag{25.9}$$

and w_i is the mass fraction in size of particle i. The *hydraulic diameter* provides a measure of the space in which the fluid flows, either through the particle pores or the spaces between particles. By definition, the hydraulic diameter, D_h, is given by

$$D_h = 4r_h = \frac{\text{volume open to flow}}{\text{total wetted surface}} = \frac{2}{3}\left(\frac{\varepsilon}{1-\varepsilon}\right)d_p \qquad (25.10)$$

where r_h is the hydraulic radius.

Table 25.2 Tyler standard screen scale

Mesh	Inches	Clear Opening, mm	Wire Diameter, in
2.5	0.312	7.925	0.088
3	0.263	6.680	0.070
3.5	0.221	5.613	0.065
4	0.185	4.699	0.065
5	0.156	3.962	0.065
6	0.131	3.327	0.036
7	0.110	2.794	0.0328
8	0.093	2.362	0.032
9	0.078	1.981	0.033
10	0.065	1.651	0.035
12	0.055	1.397	0.028
14	0.046	1.168	0.025
16	0.039	0.991	0.0235
20	0.0328	0.833	0.0172
24	0.0276	0.701	0.0141
28	0.0232	0.589	0.0125
32	0.0195	0.495	0.0118
35	0.0164	0.417	0.0122
42	0.0138	0.351	0.0100
48	0.0116	0.295	0.0092
60	0.0097	0.246	0.0070
65	0.0082	0.208	0.0072
80	0.0069	0.174	0.0056
100	0.0058	0.147	0.0042
115	0.0059	0.124	0.0038
150	0.0041	0.104	0.0026
170	0.0035	0.088	0.0024
200	0.0029	0.074	0.0021
230	0.0024	0.061	0.0016
270	0.0024	0.061	0.0016
325	0.0017	0.043	0.0014
400	0.0015	0.038	0.0010

25.3 FLOW REGIMES

The Reynolds number (Re) is based on the actual (interstitial) velocity and the hydraulic diameter:

$$Re = \frac{d_h v_I \rho_f}{\mu_f} \qquad (25.11)$$

Substituting for v_I and d_h leads to

$$Re = \frac{2}{3} \frac{d_p v_s \rho_f}{(1 - \varepsilon)\mu_f} \qquad (25.12)$$

Ergun[2] defined a porous medium Reynolds number as $Re_p = 1.5\,Re$, so that

$$Re_p = \frac{d_p v_s \rho_f}{(1 - \varepsilon)\mu_f} \qquad (25.13)$$

Flow regimes in porous media may be laminar, transition, or turbulent, based on the porous medium Reynolds number. For laminar flow, $Re_p < 10$. For transition flow, $10 < Re_p < 1000$ and for turbulent flow, $Re_p > 1000$. The effect of media *roughness* is less significant than the other variables but may become more important in the highly turbulent region. Generally, for flow in the laminar and early turbulent region, roughness has little effect on pressure drop and should not be included in most correlations for porous media.[3]

Orientation is an important variable in special cases. However, variations in orientation do not occur with random packing as encountered in most industrial practices. Oriented beds have been occasionally used in some absorbers and for other specialized applications where the packing is stacked by hand rather than dumped into the vessel; however, real world applications of this case are rare.[3]

Illustrative Example 25.1   Calculate the effective particle diameter for a set of packing with the following given characteristics (obtained during a packed column experiment at Manhattan College):

Weight of packing:	5.86 g
Total volume:	21 mL
Packing volume:	0.2 mL
Number of packing particles:	99
Average surface area:	2.18 mm² (from 20 randomly drawn packing)

Solution   The volume of a single particle is calculated by the weight of the packing by the number of particles, n (assumed to be 100 for the purposes of calculation)

$$V_p = \frac{V}{n}$$

$$V_p = 0.2\,\text{mL}/100 = 0.2\,\text{cm}^3/100$$

$$V_p = 0.002\,\text{cm}^2 = 2.0\,\text{mm}^2$$

The specific surface area, a_p, is calculated by dividing the average surface area of the particle by the volume of a single particle. See Equation (25.7).

$$a_p = \frac{S_p}{V_p}$$

$$a_p = \frac{2.18\,\text{mm}^2}{2.0\,\text{mm}^3}$$

$$a_p = 1.09\,(\text{mm})^{-1}$$

The effective particle diameter, d_p, is also calculated by employing the same equation.

$$d_p = \frac{6}{a_p}$$

$$d_p = \frac{6}{1.09\,\text{mm}}$$

$$d_p = 5.50\,\text{mm}$$

Illustrative Example 25.2   For the packing in Illustrative Example 25.1, calculate the Reynolds number if a fluid is flowing through a column containing the same packing with an interstitial velocity of 10 cm/s. The fluid density and viscosity are 0.235 g/cm^3 and 0.02 cP.

Solution   Employ the Reynolds number defined in Equation (25.11)

$$\text{Re} = \frac{D_h \rho v_I}{\mu}$$

while noting that

$$0.02\,\text{cP} = 2.0 \times 10^{-4}\,\text{g/cm} \cdot \text{s}$$

Substituting gives

$$\text{Re} = (5.50/10)(0.235)(10)/2.0 \times 10^{-4}$$
$$= 6463$$

This illustrates that at a velocity of 10 m/s, the Reynolds number is 646,300 and the flow of the fluid would be in the turbulent region.

Illustrative Example 25.3 Air ($\rho = 1.2\,\text{kg/m}^3$, $\mu = 1.25 \times 10^{-5}\,\text{kg/m} \cdot \text{s}$) flows across a packed bed. The bed porosity is 0.4 and the superficial velocity of the air is 0.1 m/s. The packing consists of solid cylindrical particles of diameter equal to 1.5 cm and height of 2.5 cm (see Fig. 25.3). Calculate the particle specific surface and the effective diameter of the particle.

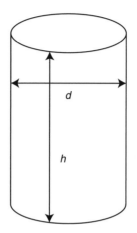

Figure 25.3 Cylindrical packing.

Solution Calculate the volume, V_p, of the cylindrical particle

$$V_p = \frac{\pi d_p^2 h}{4} = \frac{\pi (1.5)^2 (2.5)}{4}$$
$$= 4.418\,\text{cm}^3$$

Calculate the cylindrical particle surface area, S_p

$$S_p = \pi d_p h + 2\left(\frac{\pi d_p^2}{4}\right) = \pi (1.5)(2.5) + \frac{\pi (1.5)^2}{2}$$
$$= 15.32\,\text{cm}^2$$

Calculate the particle specific surface, a_p

$$a_p = \frac{S_p}{V_p} = \frac{15.32}{4.418}$$
$$= 3.467\,\text{cm}^{-1}$$

Calculate the effective particle diameter, $d_{p,e}$

$$d_{p,e} = \frac{6}{3.467} = 1.73\,\text{cm}$$

Illustrative Example 25.4 An absorber bed packing consists of cube-shaped particles of edge length equal to $\frac{3}{4}$ inch. Calculate the particle specific surface and the effective particle diameter.

Solution If the particle side length is L, then

$$V_p = L^3$$

and the surface area is

$$S_p = 6L^2$$

The specific packing (particle) surface area is defined as [see Eq. (25.7)]:

$$a_p = \frac{S}{V}$$

Substituting,

$$a_p = \frac{6L^2}{L^3} = \frac{6}{L} = \frac{6}{3/4} = 8\,\text{in}^2/\text{in}^3$$

The effective particle diameter, $d_{p,e}$, is calculated by equating a_p and $a_{p,e}$. Since

$$a_p = \frac{6}{L}$$

and

$$a_{p,e} = \frac{6}{d_{p,e}}$$

it is clear that the effective particle diameter is equal to L and

$$d_{p,e} = \frac{3}{4}\,\text{in} = 0.75\,\text{in}$$

Illustrative Example 25.5 A catalyst tower, 50 ft high and 20 ft in diameter, is packed with one-inch diameter spheres (ultimate density $= 77.28\,\text{lb/ft}^3$). Gas enters at the top of the vertical bed at a temperature of 500°F, flows downward, and leaves at the same temperature. The absolute pressure at the bottom of the catalyst bed is 4320 psf. The bed porosity is 0.4. The gas has average properties similar to propane and the time of contact between the gas and the catalyst is 10 s. Assuming incompressible flow, calculate the interstitial velocity, the superficial velocity, the gas flow rate, the bulk density of the bed, and the particle and bed specific surface area.

Solution   Calculate the density of propane assuming the ideal gas law to apply.

$$M = 44.1 \quad T = 500 \deg F = 960 \deg R$$

$$\rho = \frac{PM}{RT} = \frac{4320(44.1)}{(10.73)(144)(960)}$$

$$= 0.0128 \, \text{lb/ft}^3$$

Also, estimate the viscosity of propane from the gas viscosity monogram (see Fig. B.2 in the Appendix) at 500°F

$$\mu \cong 1.33 \times 10^{-4} \, P = 8.54 \times 10^{-7} \, \text{lb/ft} \cdot \text{s}$$

$$= 0.0133 \, \text{cP}$$

Calculate the bed volume.

$$V = \frac{\pi D^2}{4} L = \frac{\pi (20)^2 (50)}{4}$$

$$= 15,708 \, \text{ft}^3$$

Use the contact time, θ, to calculate the volumetric flow rate.

$$q = \frac{V_f}{\theta} = \frac{V \varepsilon}{\theta} = \frac{(15,708)(0.4)}{10}$$

$$= 628.3 \, \text{ft}^3/\text{s}$$

Calculate the superficial velocity, or empty tower velocity.

$$v_s = \frac{4q}{\pi D^2} = \frac{4(628.3)}{\pi (20)^2}$$

$$= 2.0 \, \text{ft/s}$$

The interstitial velocity is therefore [see Eq. (25.6)]:

$$v_I = \frac{v_s}{\varepsilon} = \frac{2.0}{0.4}$$

$$= 5.0 \, \text{ft/s}$$

Calculate the bulk density from Equation (25.4).

$$\rho_B = (1 - \varepsilon)\rho_s + \varepsilon\rho_f$$

Neglecting the latter term,

$$\rho_B = (0.6)(77.28)$$
$$= 46.4 \, \text{lb/ft}^3$$

Finally, calculate the particle specific surface area, a_p, and bed specific surface a_b from Equations (25.7) and (25.8).

$$a_p = \frac{6}{d_p} = \frac{6}{0.0833}$$
$$= 72 \, \text{ft}^{-1}$$
$$a_b = a_p(1 - \varepsilon) = (72)(0.6) = 43.2 \, \text{ft}^{-1}$$

Illustrative Example 25.6   Refer to Illustrative Example 25.5. Calculate the hydraulic radius and hydraulic diameter.

Solution   The hydraulic radius and diameter may now be calculated from the results of Illustrative Example 25.5. See Equation (25.10)

$$D_h = \frac{2}{3}\left(\frac{\varepsilon}{1 - \varepsilon}\right)d_p = \frac{2}{3}(0.0833)\frac{0.4}{0.6}$$
$$= 0.037 \, \text{ft}$$

By definition

$$r_h = \frac{D_h}{4} = \frac{0.037}{4}$$
$$= 0.00926 \, \text{ft}$$

REFERENCES

1. C. Bennett and J. Myers, "Momentum, Heat, and Mass Transfer," McGraw-Hill, New York, 1962.
2. S. Ergun, "Fluid Flow Through Packed Columns," CEP, New York, 48:89, 1952.
3. G. Brown & Associates, "Unit Operations," John Wiley & Sons, Hoboken, NJ, 1950.

NOTE: Additional problems are available for all readers at www.wiley.com. Follow links for this title.

26

FLUIDIZATION

26.1 INTRODUCTION

Fluidization is the process in which fine solid particles are transformed into a fluid-like state through contact with either a gas or liquid, or both. Fluidization is normally carried out in a vessel filled with solids. The fluid is introduced through the bottom of the vessel and forced up through the bed. At a low flow rate, the fluid (liquid or gas) moves through the void spaces between the stationary and solid particles and the bed is referred to as *fixed*. (This topic is treated first in the development that follows.) As the flow rate increases, the particles begin to vibrate and move about slightly, resulting in the onset of an *expanded bed*. When the flow of fluid reaches a certain velocity, the solid particles become suspended because the upward frictional force between the particle and the fluid balances the gravity force associated with the weight of the particle. This point is termed *minimum fluidization* or *incipient fluidization* and the velocity at this point is defined as the *minimum* or *incipient fluidization velocity*. Beyond this stage, the bed enters the fluidization state where bubbles of fluid rise through the solid particles, thereby producing a circulatory and/or mixing pattern.[1]

From a force balance perspective, as the flow rate upward through a packed bed is increased, a point is reached at which the frictional drag and buoyant force is enough to overcome the downward force exerted on the bed by gravity. Although the bed is supported at the bottom by a screen, it is free to expand upward, as it will if the velocity is increased above the aforementioned minimum fluidization velocity. At this point, the particles are no longer supported by the screen, but rather are suspended

---

Fluid Flow for the Practicing Chemical Engineer. By J. Patrick Abulencia and Louis Theodore
Copyright © 2009 John Wiley & Sons, Inc.

in the fluid in equilibrium and act and behave as the fluid. The bed is then said to be *fluidized*. From a momentum or force balance perspective, the sum of the drag, buoyancy, and gravity forces must be equal to zero.

The terminal settling velocity can be evaluated for the case of flow past one bed particle. By superimposition, this case is equivalent to that of the terminal velocity that a particle would attain flowing through a fluid. Once again, a force balance can be applied and empirical data used to evaluate a friction coefficient (see Chapter 23 for more details).

At intermediate velocities between the minimum fluidization velocity and the terminal velocity, the bed is expanded above the volume that it would occupy at the minimum value. Note also that above the minimum fluidization velocity, the pressure drop stays essentially constant.

One of the novel characteristics of fluidized beds is the uniformity of temperature found throughout the system. Essentially constant conditions are known to exist in both the horizontal and vertical directions in both short and long beds. This homogeneity is due to the turbulent motion and rapid circulation rate of the solid particles within the fluid stream described above. In effect, excellent fluid-particle contact results. Temperature variations can occur in some beds in regions where quantities of relatively hot or cold particles are present but these effects can generally be neglected. Consequently, fluidized beds find wide application in industry, e.g., oil cracking, zinc coating, coal combustion, gas desulfurization, heat exchangers, plastics cooling and fine powder granulation.

26.2 FIXED BEDS[2]

The friction factor f for a "fixed" packed bed is defined as:

$$\frac{\Delta P}{\frac{1}{2}\rho v_s^2} = \left(\frac{L}{d_p}\right)4f \tag{26.1}$$

in which d_p is the particle diameter (defined presently) and v_s is the superficial velocity defined in the previous chapter as the average linear velocity that the fluid would have in the column if no packing were present. The term L is the length of the packed column. The friction factor for laminar flow and that for turbulent flow can now be estimated separately.

For laminar flow in circular tubes of radius R, it was shown that

$$v = \frac{\Delta P R^2}{8\mu L} \tag{26.2}$$

Now imagine that a packed bed is just a tube with a very complicated cross-sectional area with hydraulic radius r_h. The average flow velocity in the cross section available for flow is then

$$v = \frac{\Delta P r_h^2}{2\mu L}; \quad r_h = \frac{\text{cross-section available for flow}}{\text{wetted perimeter}} \tag{26.3}$$

The hydraulic radius may be expressed in terms of the void fraction ε and the wetted surface "a" per unit volume of bed in the following way:

$$
\begin{aligned}
r_h &= \frac{\text{cross-section available for flow}}{\text{wetted perimeter}} \\[4pt]
&= \frac{\text{volume available for flow}}{\text{total wetted surface}} \\[4pt]
&= \frac{\left(\dfrac{\text{volume of voids}}{\text{volume of bed}}\right)}{\left(\dfrac{\text{wetted surface}}{\text{volume of bed}}\right)} = \frac{\varepsilon}{a}
\end{aligned}
\tag{26.4}
$$

The quantity "a" is related to the "specific surface" a_v (total particle surface/volume of the particles) by

$$a = a_v(1 - \varepsilon) \tag{26.5}$$

The quantity a_v is in turn used to define the mean particle diameter d_p:

$$d_p = \varepsilon/a_v \tag{26.6}$$

This definition is chosen because, for spheres, Equation (26.6) reduces to just d_p as the diameter of sphere. Finally, note that the average value of the velocity in the interstices, v_I, is not of general interest to the engineer but rather the aforementioned superficial velocity v_s; these two velocities are related by

$$v_s = v_I \varepsilon \tag{26.7}$$

If the above definitions are combined with the modified Hagen–Poiseuille equation, the superficial velocity can be expressed as

$$
\begin{aligned}
v_s &= \frac{\Delta P r_h^2}{2\mu L} \\[4pt]
&= \frac{\Delta P \varepsilon^2}{2\mu L a^2} \\[4pt]
&= \frac{\Delta P \varepsilon^2}{2\mu L a_v^2 (1 - \varepsilon)^2} \\[4pt]
&= \frac{\Delta P d_p^2}{2L(36\mu)} \frac{\varepsilon^2}{(1 - \varepsilon)^2}
\end{aligned}
$$

or finally

$$v_s = \frac{\Delta P}{L} \frac{d_p^2}{2(36\mu)} \frac{\varepsilon^3}{(1-\varepsilon)^2} \tag{26.8}$$

In laminar flow, the assumption of mean hydraulic radius frequently gives a throughput velocity too large for a given pressure gradient. Because of this assumption, one would expect that the right side of Equation (26.8) should be somewhat smaller. A second assumption implicitly made in the foregoing development is that the path of the fluid flowing through the bed is of length L, i.e., it is the same as the length of the packed column. Actually, of course, the fluid traverses a very tortuous path, the length of which may be approximately twice as long as the length L. Here, again, one would expect that the right side of Equation (26.8) should be somewhat diminished.

Experimental measurements indicate that the above theoretical formula can be improved if the 2 in the denominator on the right-hand side is changed to a value somewhere between 4 and 5. Analysis of a great deal of data has led to the value $25/6$, which is accepted here. Insertion of that value into Equation (26.8) then gives

$$v_s = \frac{\Delta P}{L} \frac{d_p^2}{150\mu} \frac{\varepsilon^3}{(1-\varepsilon)^2} \tag{26.9}$$

which some have defined as the Blake–Kozeny equation. This result is generally good for void fractions less than 0.5 and is valid only in the laminar region where the particle Reynolds number is given by $d_p G_s/\mu(1-\varepsilon) < 10$; $G_s = \rho v_s$.[3] Note that the Blake–Kozeny equation corresponds to a bed friction factor of

$$f = \left[\frac{(1-\varepsilon)^2}{\varepsilon^3}\right] \frac{75}{d_p G_s/\mu} \tag{26.10}$$

Exactly the same treatment can be repeated for highly turbulent flow in packed columns. One begins with the expression for the friction-factor definition for flow in a circular tube. This time, however, note that for highly turbulent flow in tubes with any appreciable roughness, the friction factor becomes a function of the roughness only. Assuming that all packed beds have similar roughness characteristics, a unique friction factor f_0 may be used for turbulent flow. This leads to the following results if the some procedure as before is applied:

$$\frac{\Delta P}{L} = \frac{1}{D} \frac{\rho v_s^2}{2} 4f_0 = 6f_0 \frac{1}{d_p} \frac{\rho v_s^2}{2} \left(\frac{1-\varepsilon}{\varepsilon^3}\right) \tag{26.11}$$

Experimental data indicate that $6f_0 = 3.50$. Hence Equation (26.11) becomes

$$\frac{\Delta P}{L} = 3.50 \frac{1}{d_p} \frac{\rho v_s^2}{2} \frac{1-\varepsilon}{\varepsilon^3}$$

$$= 1.75 \frac{\rho v_s^2}{d_p} \left(\frac{1-\varepsilon}{\varepsilon^3}\right) \tag{26.12}$$

which some have defined as the Burke–Plummer equation and is valid for $(d_p G_s/\mu)(1 - \varepsilon) > 1000$. This result corresponds to a friction factor given by

$$f_0 = 0.875 \frac{1 - \varepsilon}{\varepsilon^3} \tag{26.13}$$

Note that this dependence on ε is different from that given for laminar flow.

When the Blake–Kozeny equation for laminar flow and the Burke–Plummer equation for turbulent flow are simply added together, the result is

$$\frac{\Delta P}{L} = \frac{150 \mu v_s}{d_p{}^2} \frac{(1 - \varepsilon)^2}{\varepsilon^3} + \frac{1.75 \rho v_s{}^2 (1 - \varepsilon)}{d_p} \frac{}{\varepsilon^3} \tag{26.14}$$

This may be rewritten in terms of dimensionless groups (numbers):

$$\left(\frac{\Delta P \rho}{G_0{}^2}\right) \left(\frac{d_p}{L}\right) \left(\frac{\varepsilon^3}{1 - \varepsilon}\right) = 150 \frac{1 - \varepsilon}{d_p G_0/\mu} + 1.75 \tag{26.15}$$

This is the Ergun equation. It has been applied with success to gases by using the density of the gas at the arithmetic average of the end pressures. For large pressure drops, however, it seems more reasonable to use Equation (26.14) with the pressure gradient in differential form. Note that G_s is a constant through the bed whereas v_s changes through the bed for a compressible fluid. The d_p used in this equation is that defined in Equation (26.6).

Equation (26.14) may be written in the following form

$$\Delta P = 150 \frac{v_s \mu (1 - \varepsilon)^2}{g} \frac{}{\varepsilon^3} \left(\frac{L}{d_p{}^2}\right) + 1.75 \frac{v_s{}^2 (1 - \varepsilon)}{g} \frac{}{\varepsilon^3} \left(\frac{L}{d_p}\right) \rho \tag{26.16}$$

Equation (26.15) may also be written in a similar form. Other terms have been used to represent ΔP, including

$$\Delta P = \Delta P_f = h_f{}' \tag{26.17}$$

where the subscript f is a reminder that the pressure drop term represents friction due to the flowing fluid. Thus,

$$h_f{}' = 150 \frac{v_s \mu (1 - \varepsilon)^2}{g \rho} \frac{}{\varepsilon^3} \left(\frac{L}{d_p{}^2}\right) + 1.75 \frac{v_s{}^2 (1 - \varepsilon)}{g} \frac{}{\varepsilon^3} \left(\frac{L}{d_p}\right) \tag{26.16}$$

The reader should note that the pressure drop term in Equation (26.16) has units of height of flowing fluid, e.g., in H_2O. This may be converted into units of force per unit area (e.g., psf), by applying the hydrostatic pressure equation

$$\Delta P = \frac{\rho g h}{g_c} \tag{26.18}$$

This equation can then be employed to rewrite Equation (26.16) in the following form:

$$\Delta P = \frac{v_s \mu (1 - \varepsilon)^2}{g_c} \frac{L}{\varepsilon^3} \frac{L}{d_p^2} + 1.75 \frac{v_s^2 \rho (1 - \varepsilon)}{g_c} \frac{L}{\varepsilon^3} \left(\frac{L}{d_p}\right) \tag{26.19}$$

The units of ΔP are then those of pressure (i.e., force per unit area).

Illustrative Example 26.1 Comment on the relationship between the Ergun equation and the Burke–Plummer and Blake–Kozeny equations.

Solution Note that for high rates of flow, the first term on the right-hand side drops out and the equation reduces to the Burke–Plummer equation. At low rates of flow, the second term on the right-hand side drops out and the Blake–Kozeny equation is obtained. It should be emphasized that the Ergun equation is but one of many that have been proposed for describing the pressure drop across packed columns.

26.3 PERMEABILITY

In porous medium applications involving laminar flow, the Carmen–Kozeny equation is rewritten as

$$h_f' = 150 \frac{v_s \mu (1 - \varepsilon)^2}{g \rho} \frac{L}{\varepsilon^3} \frac{L}{d_p^2} = \frac{1}{k} \frac{v_s \mu L}{g \rho} \tag{26.20}$$

where k is the permeability of the medium. The permeability may then be written as

$$k = \frac{1}{150} \frac{\varepsilon^3}{(1 - \varepsilon)^2} d_p^2 \tag{26.21}$$

The permeability may be expressed in units of darcies, where 1 darcy $= 0.99 \times 10^{-12}$ m$^2 = 1.06 \times 10^{-11}$ ft^2.

Illustrative Example 26.2 A water softener unit consists of a large diameter tank of height $h = 0.25$ ft. The bottom of the tank is connected to a vertical ion-exchange pipe of length $L = 1$ ft and a diameter D of 2 inches. The ion exchange resin particle diameter is 0.05 in. $= 0.00417$ ft, and the bed porosity is 0.25. The water has an absolute viscosity of 6.76×10^{-4} lb/ft · s and a density of 62.4 lb/ft^3. Calculate the water flow rate and the superficial velocity (see Fig. 26.1).

Solution First determine the total fluid height, $h_L = h_f'$

$$h_L = h_f' = z_1 - z_2 = 1.25 \text{ ft of water}$$

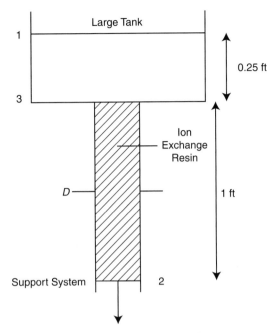

Figure 26.1 Ion-exchange softener.

Assume turbulent flow to calculate the superficial velocity, v_s. Employ a revised version of the Burke–Plummer equation [Equation (26.11)].

$$v_s = \sqrt{\frac{gh_f'}{1.75}\frac{\varepsilon^3}{1-\varepsilon}\frac{d_p}{L}} = \sqrt{\frac{(32.174)(1.25)}{(1.75)}\frac{(0.25)^3}{(1-0.25)}\frac{(0.00417)}{(1.0)}}$$

$$= 0.0446 \, \text{ft/s}$$

Check the turbulent flow assumption

$$\text{Re}_p = \frac{d_p v_s \rho}{(1-\varepsilon)\mu} = \frac{(0.00417)(0.0446)(62.4)}{(1-0.25)6.76 \times 10^{-4}}$$

$$= 22.9 < 1000$$

Since the Reynolds number is low, the calculation is not valid. Assume laminar flow and use a modified form of the Blake–Kozeny equation [see Equation (26.9)].

$$v_s = \frac{\rho g h_L}{150\mu}\frac{\varepsilon^3}{(1-\varepsilon)^2}\frac{d_p^2}{L} = \frac{(62.4)(32.174)(1.25)}{150(6.76\times 10^{-4})}\frac{(0.25)^3}{(1-0.25)^2}\frac{(0.00417)^2}{1}$$

$$= 0.0119 \, \text{ft/s}$$

Once again, check the porous medium Reynolds number

$$\text{Re}_p = \frac{d_p v_s \rho}{(1 - \varepsilon)\mu} = \frac{(0.00417)(0.0119)(62.4)}{(1 - 0.25)6.76 \times 10^{-4}} = 6.11 < 10$$

The flow is therefore laminar.

Calculate the empty cross-sectional area, S

$$S = \frac{\pi D^2}{4} = \frac{\pi (0.167)^2}{4} = 0.0218 \, \text{ft}^2$$

The volumetric flow rate of water, q, is then

$$q = v_s S = (0.0119)(0.0218) = 0.000252 \, \text{ft}^3/\text{s}$$

Illustrative Example 26.3   Refer to Illustrative Example 26.2. Calculate the pressure drop due to friction and the pressure drop across the resin bed.

Solution   Calculate the packed bed permeability, k, using Equation (26.21).

$$k = \frac{1}{150} \frac{\varepsilon^3}{(1 - \varepsilon)^2} d_p{}^2 = \frac{1}{150} \frac{(0.25)^3}{(1 - 0.25)^2} (0.00417)^2$$

$$= 3253 \, \text{Darcies}$$

The friction pressure drop across the resin bed, ΔP_{fr}, may also be calculated noting that $h_L = h_f'$.

$$\Delta P_{fr} = \frac{\rho_f g h_f'}{g_c} = 62.4(1.25) = 78.0 \, \text{psf}$$

Finally, calculate the pressure drop across the resin bed by applying Bernoulli's equation across the resin bed (between points 2 and 3):

$$\frac{P_3}{\rho} + \frac{v_3{}^2}{2g_c} + \frac{g}{g_c} z_3 = \frac{P_2}{\rho} + \frac{v_2{}^2}{2g_c} + \frac{g}{g_c} z_2 + h_f - h_s$$

Noting that $v_3 = v_2$ and $h_s = 0$,

$$\frac{\Delta P}{\rho} = \frac{P_3 - P_2}{\rho} = (z_2 - z_3)\frac{g}{g_c} + h_f' = -1 + 1.25 = 0.25 \, \text{ft of liquid}$$

$$\Delta P = \rho \left(\frac{\Delta P}{\rho}\right) = 62.4(0.25) = 15.6 \, \text{psf} = 0.108 \, \text{psi}$$

The total pressure drop represents both the friction drop and the height of the fluid (water).

26.4 MINIMUM FLUIDIZATION VELOCITY

Figure 26.2 is a photograph of the fluidization experimental unit at Manhattan College. Figure 26.3 shows the kinds of contact between solids and a fluid, starting from a packed bed and ending with "pneumatic" transport. At a low fluid velocity, one observes a fixed bed configuration of height L_m, a term that is referred to as the slumped bed height. As the velocity increases, fluidization starts, and this is termed the *onset of fluidization*. The superficial velocity (that velocity which would occur if the actual flow rate passed through an empty vessel) of the fluid at the onset of fluidization is noted again as the *minimum fluidization velocity*, v_{mf}, and the bed height is L_{mf}. As the fluid velocity increases beyond v_{mf}, the bed expands and the bed void volume increases. At low fluidization velocities (fluid velocity $> v_{mf}$), the operation is termed *dense phase fluidization*.

The onset of fluidization (or *minimum fluidization* condition) in a packed bed occurs when drag forces due to friction by the upward moving gas equal the

Figure 26.2 Fluidization experiment.

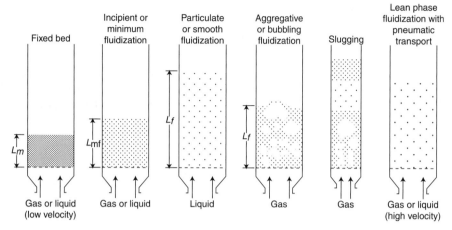

Figure 26.3   Types of particle-fluid contact in a bed.

gravity force of the particles minus the buoyancy force on the particles. This can be represented in equation form as

$$F_D = W_{net} = \left(W - F_{bouy}\right)_{particle} \tag{26.22}$$

where W is the weight of the particle. The drag force, F_D, exerted by the upward gas is a product of the friction pressure drop in gas flow across the bed and the bed cross-section area. From Bernoulli's equation, the total pressure drop ΔP is given by

$$\Delta P = \Delta P_{fr} + \rho_f \frac{g}{g_c} L_{mf}$$

or

$$\Delta P_{fr} = \Delta P - \rho_f \frac{g}{g_c} L_{mf} \tag{26.23}$$

with the latter term representing the fluid head. Therefore,

$$F_D = \Delta P_{fr} S_b = \left(\Delta P - \rho_f \frac{g}{g_c} L_{mf}\right) S_b \tag{26.24}$$

The net gravity force, W_{net}, is due to the gravity of the solid particles and to the fluid buoyancy:

$$W_{net} = S_b L_{mf}\left((1 - \varepsilon_{mf})(\rho_s - \rho_f)\frac{g}{g_c}\right) \tag{26.25}$$

Combining Equations (26.24) and (26.25), one obtains the condition for minimum fluidization:

$$\Delta P_{fr} = \rho_f \frac{g}{g_c} h_f' = (1 - \varepsilon_{mf})(\rho_s - \rho_f)\frac{g}{g_c} L_{mf} \tag{26.26}$$

or

$$h_f' = (1 - \varepsilon_{mf})\left(\frac{\rho_s - \rho_f}{\rho_f}\right)L_{mf} \tag{26.27}$$

The equations for minimum fluidization are similar to those presented for a fixed bed, i.e., Equations (26.16)–(26.19). For laminar flow conditions ($Re_p < 10$), the Blake–Kozeny equation is used to express h_f' in terms of the superficial gas velocity at minimum fluidization, v_{mf}, and other fluid and bed properties. This equation is obtained from Equations (26.9) and (26.27)

$$h_f' = (1 - \varepsilon_{mf})\left(\frac{\rho_s - \rho_f}{\rho_f}\right)L_{mf} = 150\frac{v_{mf}\mu_f}{\rho_f g}\frac{(1 - \varepsilon_{mf})^2}{\varepsilon_{mf}^3}\frac{L_{mf}}{d_p^2} \tag{26.28}$$

Rearranging, one obtains the minimum fluidization velocity:

$$v_{mf} = \frac{1}{150}\left(\frac{\varepsilon_{mf}^3}{1 - \varepsilon_{mf}}\right)\frac{g(\rho_s - \rho_f)d_p^2}{\mu_f}\,(Re_p < 10) \tag{26.29}$$

with (once again)

$$Re_p = \frac{d_p v_{mf}\rho}{\mu_f(1 - \varepsilon_{mf})} \tag{26.30}$$

The Burke–Plummer equation is used to express the head loss, h_f'. For turbulent flow conditions ($Re_p > 1000$). For this condition, the result is:

$$v_{mf} = \sqrt{\frac{1}{1.75}\varepsilon_{mf}^3\left(\frac{\rho_s - \rho_f}{\rho_f}\right)g d_p} \tag{26.31}$$

In the absence of ε_{mf} data, the above equations can be approximated as

$$v_{mf} = \frac{1}{1650}\frac{g(\rho_s - \rho_f)d_p^2}{\mu_f}\quad Re_p < 10 \tag{26.32}$$

$$v_{mf} = \sqrt{\frac{1}{24.5}\left(\frac{\rho_s - \rho_f}{\rho_f}\right)g d_p}\quad Re_p > 1000 \tag{26.33}$$

where Re_p is the particle Reynolds number at minimum fluidization and is equal to:

$$Re_p = \frac{d_p v_{mf}\rho}{\mu_f} \tag{26.34}$$

Illustrative Example 26.4 Air is used to fluidize a bed of spherical particles. The particles are 200 mesh uniform spheres; bed diameter, D_b, is 0.2 m; ultimate solids density, ρ_s, is 2200 kg/m^3; voidage at minimum fluidization, ε_{mf}, equals 0.45; bed length (height) at minimum fluidization, L_{mf}, is 0.3 m; and air properties are $\rho_f =$ 1.2 kg/m^3 and $\mu_f = 1.89 \times 10^{-5}$ kg/m-s.

Calculate the minimum fluidization mass flow rate of air and the pressure drop of air across the bed at minimum fluidization.

Solution   Obtain the diameter of a 200 mesh particle from Table 23.2.

$$d_p = 74\,\mu\text{m} = 7.4 \times 10^{-5}\,\text{m}$$

Assume turbulent flow to apply and calculate v_{mf} from Equation (26.31).

$$v_{mf} = \sqrt{\frac{(0.45)^3}{1.75}\left(\frac{2200-1.2}{1.2}\right)(9.807)(7.4 \times 10^{-5})} = 0.263\,\text{m/s}$$

Check the flow regime. Employ Equation (26.30):

$$\text{Re}_p = \frac{v_{mf}d_p}{v_f(1-\varepsilon_{mf})} = \frac{(0.263)(7.4 \times 10^{-5})}{(1.89 \times 10^{-5})(1-0.45)} = 1.87 < 1000$$

Therefore, turbulent flow conditions do not apply.

Assume laminar flow, with $\rho_s - \rho_f = \rho_s$, and employ Equation (26.29).

$$v_{mf} = \frac{1}{150}\frac{(1-0.45)}{(0.45)^3}\frac{9.807(2200)(7.4 \times 10^{-5})^2}{1.89 \times 10^{-5}} = 0.25\,\text{m/s}$$

Once again, check the flow regime

$$\text{Re}_p = \frac{v_{mf}d_p}{\mu_f(1-\varepsilon_{mf})} = \frac{(0.25)(7.4 \times 10^{-5})}{(1.89 \times 10^{-5})(1-0.45)}$$

$$= 1.79 < 10$$

The flow is indeed laminar.

The mass flow rate of air is

$$\dot{m} = \frac{\pi D^2}{4}v_{mf}\rho_f = \frac{\pi(0.2)^2}{4}(0.25)(1.2) = 9.40^{-3}\,\text{kg/s}$$

Calculate the gas pressure drop across the bed. Use Equation (26.28).

$$\Delta P_{fr} = (1-0.45)(2200)(9.807)(0.3) = 3560\,\text{Pa} = 0.0351\,\text{atm}$$

Both Equations (26.16) and (26.19) may be rewritten in a slightly different form and viewed as contributing terms to a more general equation. An equation covering the entire range of flow rates but for *various shaped* particles can be obtained by assuming that the laminar and turbulent effects are additive. This result is also referred to as the Ergun equation.[3] Thus,

$$\frac{\Delta P}{L} = \frac{150 v_0 \mu}{g_c \phi_s^2 d_p^2} \frac{(1-\varepsilon)^2}{\varepsilon^3} + \frac{1.75 \rho v_0^2}{g_c \phi_s d_p} \frac{(1-\varepsilon)}{\varepsilon^3} \qquad (26.35)$$

where ϕ_s is the sphericity or shape factor of the fluidized particles. Typical values for the sphericity of typical fluidized particles are in the 0.75–1.0 range. In lieu of any information on ϕ_s, one should employ a value of 1.0, typical for spheres, cubes and cylinders $(L = d)$.

Another approach that may be employed to calculate the minimum fluidization velocity is to employ Happel's equation.[4] Happel's equation is given by:

$$\frac{v_{mf}}{v_t} = \frac{3 - 4.5(1 - \varepsilon_{mf})^{1/3} + 4.5(1 - \varepsilon_{mf})^{5/3} - 3(1 - \varepsilon_{mf})^2}{3 + 2(1 - \varepsilon_{mf})^{5/3}} \qquad (26.36)$$

where v_{mf} is the minimum fluidization velocity, v_t the terminal velocity, v, and ε_{mf} the bed porosity at minimum fluidization.

Illustrative Example 26.5   Determine the pressure drop of 60°F air flowing through a 3-inch diameter 10-ft packed bed with 0.24-inch protruded packing made of 316 stainless steel. The superficial velocity is 4.65 ft/s. The protruded packing has a fraction void volume, effective particle diameter and surface area per unit packing of 0.89, 0.0078.5 ft, and 3305 ft^{-1}, respectively.

Solution   Use the Ergun equation

$$\Delta P = \left[\left(\frac{150 \overline{V}_0 \mu_g}{g_c(\Phi_s D_p)^2} \right) \left(\frac{(1-\varepsilon)^2}{\varepsilon^3} \right) + \left(\frac{1.75 \rho_g \overline{V}_0^2}{g_c(\Phi_s D_p)} \right) \left(\frac{(1-\varepsilon)^2}{\varepsilon^3} \right) \right] L \qquad (26.35)$$

For air at 60°F, Appendix A.9, indicates

$$\mu = 1.3 \times 10^{-5} \, \text{lb/ft} \cdot \text{s}$$

$$\rho = 0.067 \, \text{lb/ft}^3$$

Plugging in values from the problem statement, one obtains

$$\Delta P = \left[\left(\frac{(150)(4.65)(1.3E^{-5})}{(32.2)(0.007815)^2} \right) \left(\frac{(1-0.89)^2}{(0.89)^3} \right) \right.$$

$$\left. + \left(\frac{(1.75)(0.67)(4.65)^2}{32.2(0.007815)} \right) \left(\frac{(1-0.89)^2}{(0.89)^3} \right) \right] 10$$

$$= 10.25 \, \text{lb/ft}^2$$

26.5 BED HEIGHT, PRESSURE DROP AND POROSITY

The above development is now extended above and beyond the state of minimum fluidization. As described above, when a fluid moves upward in a packed bed of solid particles, it exerts an upward drag force. Minimum fluidization occurs at a point when the drag force equals the net gravity force. By increasing the fluid velocity above minimum fluidization, the bed expands, the porosity increases, and the pressure drop remains the same.

The fluidized bed height, L_f, at any voidage, ε, can be found from the minimum fluidization conditions (L_{mf} and ε_{mf}), or from the bed height at zero porosity, L_0, that is,

$$m = \rho_s S_b L_0 = \rho_s S_b L_{mf}(1 - \varepsilon_{mf})$$
$$= \rho_s S_b L_f(1 - \varepsilon)$$

so that

$$L_f = L_{mf}\frac{1 - \varepsilon_{mf}}{1 - \varepsilon} = \frac{L_0}{1 - \varepsilon} \qquad (26.37)$$

The pressure drop at minimum fluidization remains constant at any fluidization height, L_f, [see Equation (26.17)] so that

$$\Delta P_{fr} = \frac{g}{g_c}(\rho_s - \rho_f)L_{mf}(1 - \varepsilon_{mf}) = \frac{g}{g_c}(\rho_s - \rho_f)L_f(1 - \varepsilon) \qquad (26.38)$$

The effect of bed pressure drop on the superficial velocity is now briefly discussed. Initially, the bed pressure drop increases rapidly with a slight increase in velocity. Then, the pressure drop begins to level off. This point, as defined earlier, is called incipient fluidization. Beyond this point, the pressure drop remains fairly constant as the superficial velocity increases. This is the fluidized region.

The variation of porosity (and hence bed height) with the superficial velocity is calculated from the Blake–Kozeny equation, assuming laminar flow and $\rho_f \ll \rho_s$.

$$v_s = \frac{d_p^2 g}{150 \mu_f}\frac{\varepsilon^3}{1 - \varepsilon}(\rho_s - \rho_f) \quad \text{for } \varepsilon < 0.8 \qquad (26.39)$$

It should be noted that the bed density is a function of superficial velocity. As the velocity increases, the bed density decreases. This occurs because the volume of the expanding fluidized bed increases while the mass of the bed remains constant. Therefore, the bed density decreases. An increase in gas velocity causes a greater force on the particles and thus drives them further apart. The increased distance between the particles causes an increase in bed volume. After the point of incipient fluidization, the decrease in density is more dramatic as the bed volume increases rapidly during fluidization. The bed porosity is also a function of superficial velocity.

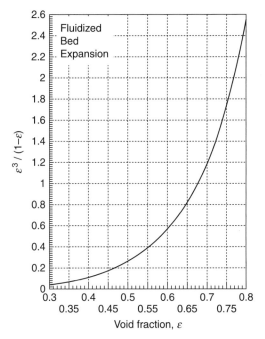

Figure 26.4 Expanded bed porosity.

The bed porosity also increases with an increase in superficial velocity. (Once again, the porosity is a measure of the empty space existing between the particles in the bed.) Initially, the porosity increases gradually; however, after incipient fluidization, the particles are rapidly forced further apart and the porosity increases at a greater rate.

The term,

$$\frac{\varepsilon^3}{1 - \varepsilon}$$

appears often in packed bed and fluidized bed equations. To simplify the calculations, it is plotted versus the void fraction, ε, in Fig. 26.4. This plot may be useful to some in estimating the expanded bed porosity and height without trial-and-error.

26.6 FLUIDIZATION MODES

There are two modes of fluidization. When the fluid and solid densities are not too different, or the particles are very small, and therefore, the velocity of the flow is low, the bed is fluidized evenly with each particle moving individually through the bed. This is called smooth fluidization, and is typical of liquid–solid systems. If the fluid and solid densities are significantly different, or the particles are large, the velocity of the flow must be relatively high. In this case, fluidization is uneven, and the fluid passes through the bed mainly in large bubbles. These bubbles burst at the surface, spraying solid particles above the bed. Here, the bed has many of the characteristics of a liquid with the fluid phase acting as a gas bubbling through it. This is called bubbling (or aggregative) fluidization; it is typical of gas–solid systems and is due to the large density difference between the solid and gas. The approximate criterion to estimate the transition from bubbling to smooth fluidization is expressed in terms of the dimensionless Froude number at minimum fluidization. This is expressed in terms of the minimum fluidization velocity, v_{mf}, the particle diameter, d_p, and the acceleration due to gravity, g, as:

$$Fr = \frac{v_{mf}^2}{g d_p} \tag{26.40}$$

If $Fr < 0.13$, the fluidization mode is smooth; if $Fr > 0.13$, then the fluidization is bubbling.

Illustrative Example 26.6   A bed of 35 mesh pulverized coal is to be fluidized with a liquid oil. The bed diameter is 4 ft. At minimum fluidization, the bed height is 8 ft and its void fraction in 0.38. The coal particle density is 84 lb/ft³. The liquid oil properties are: density = 55 lb/ft³ and dynamic viscosity = 15 cP. What is the pressure drop required for fluidization?

Solution   Obtain the particle diameter from Table 23.2.

$$35 \text{ mesh}; \quad d_p = 0.417 \text{ mm} = 0.0164 \text{ in} = 0.00137 \text{ ft}$$

Calculate the pressure drop from Equation (26.15):

$$\Delta P = \Delta P_{fr} - \rho_f \frac{g}{g_c} L_{mf}$$

Note that since the fluidizing fluid is a liquid, with a specific gravity comparable to the specific gravity of the solid particles, the fluid gravity term may not be neglected. In this case, the overall pressure drop, ΔP, is not the same as the friction pressure drop. ΔP_{fr} can be obtained from Equation (26.38). Therefore,

$$\Delta P = \frac{g}{g_c}(\rho_s - \rho_f)L_{mf}(1 - \varepsilon_{mf}) - \rho_f \frac{g}{g_c} L_{mf}$$
$$= (1 - 0.38)(84 - 55)(8) - (55)(8) = 583.7 \text{ psf}$$

Illustrative Example 26.7   Refer to Illustrative Example 26.6. If the bed is fluidized such that the bed height is 10 ft, calculate the volumetric flow rate of oil (in gpm).

Solution Calculate the bed voidage employing Equation (26.37).

$$L_f = L_{mf} \frac{1 - \varepsilon_{mf}}{1 - \varepsilon}$$

$$10 = 8 \frac{1 - 0.38}{1 - \varepsilon}$$

$$\varepsilon = 0.504$$

Calculate the superficial velocity of the oil, assuming laminar flow [see Equation (26.29)].

$$v_s = \frac{1}{150} \frac{d_p^2 g}{\mu_f} \frac{\varepsilon^3}{1 - \varepsilon} (\rho_s - \rho_f)$$

$$= \frac{1}{150} \frac{(0.00137)^2 (32.174)}{(3.13 \times 10^{-4})} \frac{0.504^3}{1 - 0.504} (84 - 55)$$

$$= 9.6 \times 10^{-3} \text{ ft/s}$$

Calculate the volumetric flow rate.

$$q = \frac{\pi D^2}{4} v_s = \frac{\pi (4)^2}{4} 9.6 \times 10^{-3} = 0.121 \text{ ft}^3/\text{s}$$

Check on the laminar flow assumption:

$$Re_p = \frac{d_p v_s \rho_f}{\mu_f (1 - \varepsilon)} = \frac{(0.00137)(9.6 \times 10^{-3})(55)}{(0.01)(1 - 0.504)} = 0.145 < 10$$

The flow is therefore laminar.

Illustrative Example 26.8 Refer to Illustrative Example 25.6. Calculate the following:

1. The porous medium friction factor.
2. The Reynolds number.
3. The absolute pressure of the inlet gas.
4. The permeability of the catalyst bed in darcies.

Solution Obtain the porous medium friction factor using the Burke–Plummer equation. Since the flow is turbulent, Equation (26.6) applies and

$$f_{PM} = 1.75$$

The head loss, h_f, is [see Equation (26.7)]

$$h_f' = 1.75 \frac{v_s^2 (1 - \varepsilon)}{g} \frac{L}{\varepsilon^3 d_p} = 1.75 \frac{2^2}{32.174} \frac{0.6}{(0.4)^3} \frac{50}{0.0833} = 1224.3 \text{ ft of propane}$$

Check on the assumption of neglecting the dynamic head (kinetic effects)

$$\frac{v_I^2}{2g} = \frac{5^2}{2(32.174)} = 9.71 \ll 1224.3 \text{ ft}$$

The assumption is justified.

Write Bernoulli's equation between the entrance and gas exit. Neglect the dynamic head.

$$P_1 - P_2 = \rho_f \frac{g}{g_c}[(z_2 - z_1) + h_f]$$

$$P_1 = P_2 + \rho_f \frac{g}{g_c}[(z_2 - z_1) + h_f]$$

$$= 4320 + 0.0128[(1)(-50) + 1224.3] = 4335 \text{ psf}$$

$$= 30.10 \text{ psi} = 2.048 \text{ atm}$$

The permeability of the medium, k, is defined only for laminar flow. Since the flow is turbulent, k cannot be calculated.

Illustrative Example 26.9   What is the minimum pressure drop in an activated carbon bed (0.5 m in depth, particle diameters of 0.001 m, bed porosity of 0.25) for turbulent flow of water through the bed?

Solution   At turbulent flow, Re is >1000. For minimum pressure drop, set

$$\text{Re} = 1000$$

Therefore

$$1000 = \frac{d_p v_s \rho}{\mu(1 - \varepsilon)}$$

$$v_s = \frac{(1000)(\mu)}{d_p \rho}(1 - \varepsilon)$$

Assume for water at room temperature (see Table A.4 in the Appendix): $\rho = 1000$ kg/m^3 and $\mu = 1 \times 10^{-3}$ kg/m · s. Therefore

$$v_s = \frac{(1000)(1 \times 10^{-3} \text{ kg/m · s})(1 - 0.25)}{(0.001 \text{ m})(1000 \text{ kg/m}^3)}$$

$$= 0.75 \text{ m/s}$$

Then for turbulent flow [Equation (26.12)],

$$\Delta P = \frac{1.75\rho L v_s{}^2(1-\varepsilon)}{g_c\phi_s d_p\varepsilon^3}$$

$$= \frac{(1.75)(1000\,\text{kg/m}^3)(0.5\,\text{m})(0.75\,\text{m/s})^2(0.75)}{(\text{kg}\cdot\text{m/N}\cdot\text{s})(1.0)(0.001\,\text{m})(0.25)^3}$$

$$= 2.36\times 10^7\,\text{Pa}$$

Illustrative Example 26.10   A bed of 200 mesh particles is fluidized with air at 20°C. The bed has a diameter $D = 0.2\,\text{m}$. The bed height and porosity at minimum fluidization are 0.3 and 0.45, respectively. The bed is operated with a super-ficial air velocity of 0.05 m/s. Determine the zero porosity bed height, the air pressure drop, the operating bed height and porosity, and the bed mass.

Solution   The particle diameter, d_p, is again obtained from Table 23.2.

$$d_p = 74\,\mu\text{m}$$

Calculate L_0 (the zero porosity bed height) from Equation (26.29).

$$L_0 = L_{\text{mf}}(1-\varepsilon_{\text{mf}}) = 0.3(1-0.45) = 0.165\,\text{m}$$

Calculate the minimum fluidization velocity, v_{mf}, assuming laminar flow. Use Equation (26.9):

$$v_{\text{mf}} = \frac{1}{150}\left(\frac{\varepsilon_{\text{mf}}{}^3}{1-\varepsilon_{\text{mf}}}\right)\frac{g(\rho_s-\rho_f)d_p{}^2}{\mu_f}$$

$$= \frac{1}{150}\left(\frac{(0.45)^3}{1-0.45}\right)\frac{9.807(2200-1.2)(7.4\times 10^{-5})^2}{1.89\times 10^{-5}} = 0.0069\,\text{m/s}$$

The terminal falling velocity of the particle was calculated as 0.35 m/s in Illustrative Example 26.3. Calculate the porosity of the expanded bed from Equation (26.9).

$$0.35 = \left(\frac{1}{150}\right)\left(\frac{\varepsilon^3}{1-\varepsilon}\right)\left(\frac{9.807(2200-1.2)(7.4\times 10^{-5})^2}{1.89\times 10^{-5}}\right)$$

$$\varepsilon = 0.91$$

Calculate the expanded bed height L and the bed inventory m.

$$L_f = \frac{L_0}{1-\varepsilon} = \frac{0.165}{1-0.91} = 1.833\,\text{m}$$

$$m = \rho_s\pi d_b{}^2 L_0 = (2200)(\pi)(0.2)^2(0.165) = 45.6\,\text{kg}$$

Illustrative Example 26.11   Refer to Illustrative Example 26.9. Specify whether the fluidization mode is smooth or bubbling.

Solution   Determine the fluidization mode [see Equation (26.40)]:

$$Fr_{mf} = \frac{v_{mf}^2}{gd_p} = \frac{(0.0069)^2}{9.807(7.4 \times 10^{-5})} = 0.066 < 0.13$$

The fluidization is smooth.

26.7 FLUIDIZATION EXPERIMENT DATA AND CALCULATIONS

One of the experiments conducted in the Chemical Engineering Laboratory at Manhattan College is concerned with fluidization. Students perform the experiment and later submit a report. In addition to theory, experimental procedure, discussion of results, etc., the report contains sample calculations. The following is an (edited) example of those calculations that cover a wide range of fluidization principles and applications.

The fluidization experiment consists of two parts. The first part, which is examined here, determines the particle characteristics of sand in the fluidized bed. There are four parts to determine the sand characteristics. First, the bulk density of the particles is calculated, then the particle density, particle size distribution, and finally the "hydraulic" particle diameter.

In order to calculate the bulk density, the mass and volume of the particles need to be measured. A sample of the glass particles was placed in a 1-L graduated cylinder. The weight and volume of the particles in the cylinder were determined as follows:

$$m_{cyl} = 623 \text{ g}$$
$$m_{cyl+sand} = 877 \text{ g}$$
$$m_{sand} = m_{cyl+sand} - m_{cyl} = 877 \text{ g} - 623 \text{ g}$$
$$= 254 \text{ g}$$

The volume occupied by the particles in the cylinder was 170 ml. The bulk density of any substance or particle is given by:

$$\rho_B = \frac{m_{glass}}{V_{glass}}$$

Substituting the above gives

$$\rho_B = \frac{254 \text{ g}}{170 \text{ mL}} = 1.494 \text{ g/mL} = 1494 \text{ kg/m}^3$$

The area S of the bed has a width of 24 in and a depth of 2 in. Therefore,

$$S = (W)(D) = (24)(2)$$
$$= 48 \text{ in}^2$$
$$= 0.0310 \text{ m}^2$$

The bulk density of the particles is used to determine the total mass, m, of the sand beads in the bed of static height L:

$$m = (S)(L)(\rho_B)$$

The static bed height is obtained by taking an average of three height measurements at different points in the bed

$$L_1 = 24.09 \text{ in}$$
$$L_2 = 23.62 \text{ in}$$
$$L_3 = 25.98 \text{ in}$$
$$L = \frac{L_1 + L_2 + L_3}{3} = \frac{24.09 + 23.62 + 25.98}{3}$$
$$= 24.56 \text{ in} = 0.624 \text{ m}$$

The mass is therefore

$$m = (0.0310 \text{ m}^2)(0.624 \text{ m})(1494 \text{ kg/m}^3)$$
$$= 28.9 \text{ kg}$$

The second necessary measurement in the experiment consists of finding the particle density. Approximately 75 g of sand particles were placed in a 100-ml volumetric flask. The mass of the glass is determined in the same way as it is found previously.

$$m_{\text{flask}} = 0.068 \text{ kg}$$
$$m_{\text{flask+sand}} = 0.143 \text{ kg}$$
$$m_{\text{sand}} = m_{\text{flask+sand}} - m_{\text{flask}} = 0.143 \text{ kg} - 0.068 \text{ kg}$$
$$= 0.075 \text{ kg}$$

Water is added to the flask—first up to $\frac{3}{4}$ full and later additional volume is added to the 100-ml mark. This reduces the void spaces between the particles. The volume of water added was 68.8 mL. Therefore, the volume of the glass particles is given by

$$V_{\text{sand}} = V_{\text{flask}} - V_{\text{water}} = 100 \text{ mL} - 68.8 \text{ mL}$$
$$= 31.2 \text{ mL} = 3.12 \times 10^{-5} \text{ m}^3$$

The particle density is now determined

$$\rho_p = \frac{m_{\text{sand}}}{V_{\text{sand}}} = \frac{0.075 \text{ kg}}{3.12 \times 10^{-5} \text{ m}^3}$$
$$= 2403.85 \text{ kg/m}^3$$

This particle density is used to calculate the height of the bed with no void spaces, L_0. This height is needed to determine the bed porosity. The following equation is used for this calculation

$$L_0 = \frac{m}{\rho_p S}$$

$$= \frac{(28.9\,\text{kg})}{(2403.85\,\text{kg/m}^3)(0.0310\,\text{m}^2)}$$

$$= 0.388\,\text{m}$$

The bed porosity is determined as follows:

$$\varepsilon = 1 - \frac{L_0}{L_{\text{static}}}$$

$$= 1 - \frac{0.388\,\text{m}}{0.624\,\text{m}}$$

$$= 0.378$$

The third part of the experiment consists of determining the particle size distribution. Here, approximately 1000 grams of sand were placed in the Tyler shaker (see Chapter 23 for more details) and the screen test was studied. Table 26.1 shows the screen numbers and sizes.

Two runs were performed in this part. After each run, the mass of the particles left in the trays were measured and an average was calculated as follows:

Run 1:

Tray No. 80

$$m_T = \sum m_{\text{tray+sand}} - m_{\text{tray}}$$

$$= (153\,\text{g} - 1.4989\,\text{g}) + (145\,\text{g} - 1.4981\,\text{g}) + (155\,\text{g} - 1.4753\,\text{g})$$
$$+ (145\,\text{g} - 1.5213\,\text{g}) + (13\,\text{g} - 1.4915\,\text{g})$$
$$= 603.5\,\text{g}$$

Table 26.1 Sieve trays screen number and sizes

No	Size, μm
50	297
60	250
80	177
100	149
120	125
140	105
170	88
200	74

The same calculation was made for the sand left on tray 80 on the second run:

Run 2:

Tray No. 80

$$m_T = 451.8 \, g$$

The average mass for tray 80 is therefore:

$$m_{ave} = [m_T \, (\text{Run 1}) + m_T \, (\text{Run 2})]/2$$
$$= 527.7 \, g$$

From the data, the cumulative mass percent of material smaller than each of the screen sizes was calculated. Using tray No. 80 again, the following percentage was obtained:

Tray No. 80

$$\text{mass} > 177 \, \mu m = \sum_{t=80}^{200} m_t; \quad t = \text{tray number}$$
$$= (373.02 \, g + 527.7 \, g) = 900.72 \, g$$

The 373.02 g represents the mass on trays 100 through 200. The total average mass of glass for the two runs was determined as follows:

Run 1:

Sand mass $= 986.3 \, g$

Run 2:

Sand mass $= 984.1 \, g$

Average mass $= (986.3 + 984.1)/2 = 985.2 \, g$

The mass of glass smaller than 177 μm is calculated:

Mass $< 177 \, \mu m =$ Average weight $-$ Weight $> 177 \, \mu m$
$$= 985.2 \, g - 900.72 \, g = 84.5 \, g$$

The cumulative percent smaller than 177 μm is therefore

$$\%W < 177 \, \mu m = \frac{\text{Weight} < 177 \, \mu m}{\text{Average weight}} \times 100 = \frac{84.5 \, g}{985.2} \times 100$$
$$= 8.85\%$$

This same procedure was used for each screen size. The final values obtained were plotted in a log-probability paper so as to show the curve for the particles size distribution.

The last part of the characteristics section is to determine the particles' hydraulic diameter. A sample of the glass captured on the screen nearest to the 50% mass position on the size distribution curve was taken in order to perform the terminal velocity experiment. The sample was taken from the particles left in tray No. 60. Six runs were made for the terminal velocity. The time the particles took to travel 24 in was measured for each run and the velocity was determined from this values. For the first run, the particles took 14 s to travel the given distance. The velocity was calculated as follows;

$$v_t = \frac{d}{t} = \frac{(24 \text{ in})(0.0254 \text{ m/in})}{(14 \text{ s})} = 0.044 \text{ m/s}$$

Table 26.2 shows the terminal velocities measured for the six runs.
The average terminal velocity in water is therefore:

$$v_t = \sum_{i=1}^{n} \frac{v_{ti}}{n}$$

$$= \frac{(0.044) + (0.043) + (0.044) + (0.038) + (0.038) + (0.039)}{6}$$

$$= 0.041 \text{ m/s}$$

The hydraulic diameter and the terminal velocity were also determined. The procedure set forth in Chapter 23 was employed. The calculated diameter was

$$d_p = 3.05 \times 10^{-4} \text{ m}$$

Using the calculated particle diameter, the terminal velocity of air can be obtained by substituting the properties of air instead of the ones for water in the appropriate Chapter 23 equations. The final result was:

$$v_{t,\text{air}} = 1.755 \text{ m/s}$$

Table 26.2   Terminal velocities for six runs

Run	Time (s)	v_t (m/s)
1	14	0.044
2	14.06	0.043
3	13.84	0.044
4	16.03	0.038
5	16.1	0.038
6	15.53	0.039

REFERENCES

1. D. Kunii and O. Levenspiel, "Fluidization Engineering," John Wiley & Sons, Hoboken, NJ, 1969.
2. L. Theodore, personal notes, 2008.
3. S. Ergun, "Fluid Flow Through Packed Column," CEP, 48:49, 1952.
4. J. Happel, personal communication, 2003.

NOTE: Additional problems for each chapter are available for all readers at www.wiley.com. Follow links for this title.

27

FILTRATION

27.1 INTRODUCTION

Filtration is one of the most common applications of the flow of fluids through packed beds. As carried out industrially, it is similar to the filtration carried out in the chemical laboratory using a filter paper in a funnel. The object is still the separation of a solid from the fluid in which it is carried and the separation is accomplished by forcing the fluid through a porous filter. The solids are trapped within the pores of the filter and (primarily) build up as a layer on the surface of this filter. The fluid, which may be either gas or liquid, passes through the bed of solids and through the retaining filter.

As noted above, solid particles are removed from a slurry in a filtration process by passing it through a filtering medium. (According to Webster, a slurry is defined as a "watery mixture of insoluble matter such as mud, lime or plaster of Paris.") The solids are deposited on the filtering medium, which is normally referred to as the filter. Filtration may therefore be viewed as an operation in which a heterogeneous mixture of a fluid and particles of a solid are separated by a filter medium that permits the flow of the fluid but retains the particles of the solid. Therefore it primarily involves the flow of fluids through porous media (see Chapter 25).

In all filtration processes, the mixture or slurry flows as a result of some driving force, that is, gravity, pressure (or vacuum), or centrifugal force. In each case, the filter medium supports the particles as a porous cake. This cake, supported by the filter medium, retains the solid particles in the slurry with successive layers added to the cake as the filtrate passes through the cake and medium.

Fluid Flow for the Practicing Chemical Engineer. By J. Patrick Abulencia and Louis Theodore
Copyright © 2009 John Wiley & Sons, Inc.

The several procedures for creating the driving force on the fluid, the different methods of cake deposition and removal, and the different means for removal of the filtrate from the cake subsequent to its formation, result in a great variety of filter equipment. In general, filters may be classified according to the nature of the driving force supporting filtration. The various equipment are described in the following section.

27.2   FILTRATION EQUIPMENT

There are various types of filtration equipment employed by industry. Included in this list are gravity filters, plate-and-frame filters, leaf filters, and rotary vacuum filters. Gravity filters are the oldest and simplest type. These filters consist of tanks with perforated bottoms filled with porous sand through which the slurry passes. The most common type, however, is the plate-and-frame filter where plates are held rigidly together in a frame. More details are provided in subsequent paragraphs. Leaf filters are similar to the plate-and-frame filters in that a cake is deposited on each side of the leaf and the filtrate flows to the outlet in the channels provided by the coarse drainage screen in the leaf between the cakes. The leaves are immersed in the slurry. Rotary vacuum filters are used where a continuous operation is desirable, particularly for large-scale operations. The filter drum is immersed in the slurry where a vacuum is applied to the filter medium that causes the cake to deposit on the outer surface of the drum as it passes through the slurry.

The plate-and-frame filter press is perhaps the most widely used type of filtering devices in the chemical industry. Plate-and-frame filter presses are used in a variety of industries. The chemical industry uses the filter press in order to separate the solid portion of slurry from the liquid. A chemical, for example zinc, builds up on the frames. The filter press is then opened and the wet cake, containing solid zinc, can be collected, removed and dried. The pharmaceutical industry also uses the filter press in similar applications. The solids collected on the inside of the frames can be dried and later sold as medication. The sugar industry also employs the filter press to separate solid sugar from a solution. A slurry is sent through a filter press and the solid cake is collected on the frames. This solid is later dried, crystallized, and sold as sugar. In the pottery industry, the filter press is used in order to make ceramic pieces. A slurry is sent though the filter press and the solid cake is collected. This cake is then used for the production of various pottery products. The filter press is also used in the wastewater treatment industry. A waste stream containing sludge is passed through a filter press. Once the solids are removed, a smaller volume of "liquid" will have to be disposed of. There are stringent rules that govern the disposal of liquid waste (see Chapter 28). However, it is easier to dispose of solid waste since it can be sent to a landfill. Therefore, the filter press reduces the amount of liquid waste that needs to be treated before it can be disposed of.[1,2]

Regarding the filter press, feed slurry is pumped to the unit under pressure and flows in the press and into the bottom-corner duct (see Fig. 27.1). This duct has outlets into each of the frames, so the slurry fills the frames in parallel. The plates

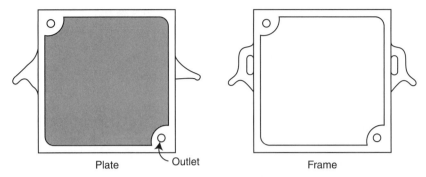

Figure 27.1 Plate and frame schematics.

and frames are assembled alternately with filter cloths over each side of each plate. The assembly is held together as a unit by mechanical force applied hydraulically or by a screw. The solvent, or filtrate, then flows through the filter media while the solids build up in a layer on the frame side of the media. The filtrate flows between the filter cloth and the face of the plate to an outlet duct. As filtration proceeds, the cakes build up on the filter cloths until the cake being formed on each face of the frame meet in the center. When this happens, the flow of filtrate, which has been decreasing continuously as the cakes build up, drops off abruptly to a trickle. Usually filtration is stopped well before this occurs.[3]

A photograph of a plate and frame filter located in the chemical engineering lab on the fourth floor of the Manhattan College Leo Engineering Building is provided in Figs. 27.2 and 27.3. It has a total of 18 plates and frames. The layout of these frames and plates can be seen in Fig. 27.4.

Figure 27.2 Filter press experiment: front view.

Figure 27.3 Filter press experiment: side view.

Figure 27.4 Frame and plate layout.

The process where the slurry flows through the press can be seen in Fig. 27.5. The slurry enters the lower right-hand side of frame 18. When frame 18 is filled with slurry, the excess slurry is forced through the filter media to the upper right-hand side of plate 17. When the slurry is in frame 18, cake builds up on the filter media. It then flows into plate 16. When in frame 16, the slurry flows down because of gravity. When the frame is filled with slurry, the cake forms and the filtrate is sent to the upper right-hand side of frame 16 and out through plate 15. Plate 15 leads to the filtrate collecting drum outside of the press.

A filter operation can be carried out using a centrifugal force rather than the pressure force used in the equipment described above. Filters using centrifugal

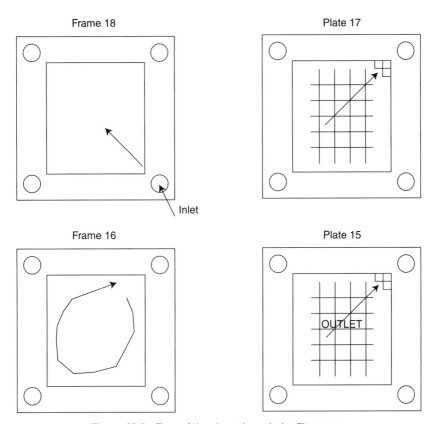

Figure 27.5   Flow of the slurry through the filter press.

force are usually used in the filtration of coarse granular or crystalline solids and are available primarily for batch operation.

Batch centrifugal filters most commonly consist of a basket with perforated sides rotated around a vertical axis. The slurry is fed into the center of the rotating basket and is forced against the basket sides by centrifugal force. There, the liquid passes through the filter medium, which is placed around the inside surface of the basket, and is caught in a "shielding" vessel, referred to as a curb, within which the basket rotates. The solid phase builds up a filter cake against the filter medium. When this cake is thick enough to retard the filtration to an uneconomical rate or to endanger the balance of the centrifuge, the operation is halted, and the cake is scraped into a bottom discharge or is scooped out of the centrifuge. In the *automatically discharging batch centrifugal filter*, unloading occurs automatically while the centrifuge is rotating, but the filtration cycle is still a batch one.[3]

Another process device that some have classified as a "hybrid" filtration unit is the mist eliminator. This class of separator allows gas (usually air) to pass through the unit while captured liquid droplets are returned to the emitting process equipment. These units are widely used to prevent the emission of undesirable amounts of liquid droplets from scrubbers and absorbers. An unfortunate consequence of intimate and vigorous liquid–gas contacting in the scrubber is that some of the scrubbing liquor is atomized, entrained by the gas that has been cleaned, and discharged from the unit. The droplets carried over by entrainment generally contain both suspended and dissolved solids. In many cases, excessive entrainment imposes a limitation upon the capacity of the scrubber. The describing equations to follow for the traditional filtration equipment bear no resemblence to those for mist eliminators.[4]

Irrespective of the type of equipment employed, washing and dewatering operations are usually accomplished with the filter cake in place. Since the cake thickness is now unchanged, the wash rate usually varies directly with the pressure drop. If the wash water follows the same path as the slurry and is fed at the same pressure, the wash rate will then be approximately equal to the final filtration rate. In the dewatering part of a filtration cycle, fluid is drawn through the filter cake, pulling the filtrate or wash water remaining in the pores of the cake out ahead of it.

As discussed earlier, filter media consisting of cloth, paper, or woven or porous metal may be used. The criteria upon which a filter medium is selected must include ability to remove the solid phase, high liquid throughput for a given pressure drop, mechanical strength, and chemical inertness to the slurry to be filtered and to any wash fluids.

Each of these considerations is tempered by the economics involved, so that the design engineer attempts to choose a medium that meets the required filtration requirement while contributing to the lowest possible overall filtration cost.

Filter-cake solids usually penetrate the filter medium and fill some of the pores. As filtration continues, these particles are thought to bridge across the pores as the cake begins to form on the face of the medium. As a result, the resistance to flow through the medium increases sharply. In some cases, the solids fill the filter medium to such an extent that the filtration rate is seriously reduced.

Of all the various filtration equipment, the plate-and-frame filter press is probably the cheapest per unit of filtering surface and requires the least floor space. However, the cost of labor for opening and dumping such presses is high, particularly in the large sizes. For this reason, they are not chosen when a large quantity of worthless solid is to be removed from the filtrate. If the solids have high value as in the pharmaceutical industry, the cost of labor per unit value of product is relatively low and the plate-and-frame filter press proves satisfactory. It has a high recovery of solids, and the solid in the form of a cake may be readily handled in a tray or shelf drier, which is frequently used for valuable products. The leaf filter offers the advantages of ease of handling, minimum labor with efficient washing, and discharge of cake without removing any leaves from the filter. The rotary continuous filter offers the additional advantages of continuous and automatic operation for feeding, filtering, washing, and cake discharge.

More recently, membranes have arrived on the scene. Although a detailed discussion on this subject is beyond the scope of this book, a brief introduction to membranes follows. A membrane can be described as a physical barrier between two fluids. In current industrial practice, membranes are typically made from polymers. Membrane separation is an integral technique among the various methods employed in industry for the separation of various materials and fits in well at the molecular level in food and drug processing applications. Ultrafiltration, reverse osmosis, and electrolyte dialysis are the membrane separation techniques that have been efficiently employed by industry to purify and concentrate desired products at low temperatures while maintaining basic qualities. For example, ultrafiltration is a membrane separation technique whereby a solution is introduced on one side of a membrane barrier, and water, salts, and low molecular weight materials pass through the unit under an applied pressure. Membrane separation processes can also be used to concentrate single solutes or mixtures of solutes. The most promising area for the expansion of ultrafiltration process applications is in the biochemical area (see Chapter 34). Some of its usage in this area includes purifying vaccines and blood fractions.

27.3 DESCRIBING EQUATIONS

Regarding the filtration process, two important considerations need to be addressed: pressure drop and filtration efficiency. Industry normally relies on certain simple guidelines and calculations to insure satisfactory separation efficiency. Although not discussed in this section, it will receive some treatment in the following chapter (Chapter 28—Environmental Concerns). Pressure drop, however, is another matter.

Perhaps the most convenient starting point to describe flow through porous media is with the classic Darcy equation:

$$\frac{dP}{dz} = \frac{-\mu v_s}{K} \tag{27.1}$$

where v_s is the superficial velocity in the filter and μ is the slurry viscosity. This may be integrated (for constant coefficients) to give

$$\frac{\Delta P}{L} = \frac{\mu v_s}{K} \tag{27.2}$$

where L is the filter thickness. The term K is defined as the permeability coefficient for the filtering medium in consistent units. See the permeability section (Section 26.3) in Chapter 26 for additional details.

Darcy's equation may be rearranged and written as

$$v_s = -\frac{K}{\mu}\frac{dP}{dz} \tag{27.3}$$

Multiplying both sides of this equation by the approach (face) area of the filter, S, gives

$$v_s S = q = -\frac{SK}{\mu}\frac{dP}{dz} \tag{27.4}$$

where q is the volumetric flow rate of the slurry passing S. Integrating yields,

$$\Delta P = \left(\frac{\mu \Delta z}{KS}\right) q \tag{27.5}$$

The term in parentheses represents the resistance to flow. It is a constant for the fabric since Δz is simply the filter thickness, L. The pressure drop across the fabric medium is then

$$\Delta P = \left(\frac{\mu L}{KS}\right) q \tag{27.6}$$

The development for the pressure drop across the filter cake is similar to that for the fabric. The bracketed term in Equation (27.5) is not a constant for the deposited particles since Δz is a variable. If the collection efficiency is close to 100%, which is usually the case in an industrial operation, then

$$\Delta z = \frac{V_s c}{\rho_p (1 - \varepsilon) S} = \frac{V_s c}{\rho_B S} \tag{27.7}$$

where V_s is the volume of slurry filtered for deposit thickness Δz, c is the inlet solids loading or concentration (mass/volume), ε is the void volume or void fraction, ρ_p is the true density of the solid and ρ_B is the bulk density of the cake of deposited solids. Substituting Equation (27.7) into Equation (27.5) yields an equation for the pressure drop across the cake

$$\Delta P = \left(\frac{\mu c}{\rho_B KS^2}\right) V_s q \tag{27.8}$$

It is assumed that Equations (27.6) and (27.8) are additive (i.e., the solids and fabric do not interact). Thus, the total or overall pressure drop across the filter system is obtained by combining these two equations:

$$\Delta P = \left(\frac{\mu c}{\rho_B KS^2}\right) V_s q + \left(\frac{\mu L}{KS}\right) q \tag{27.9}$$

Setting

$$B = \left(\frac{\mu c}{\rho_B KS^2}\right)$$

$$C = \left(\frac{\mu L}{KS}\right)$$

then

$$\Delta P = (BV_s + C)q \tag{27.10}$$

This represents the general relationship between ΔP, q, and V_s.

During a constant (overall) pressure (drop) operation, the flow rate is a function of time

$$q(t) = \frac{dV_s}{dt}$$

(27.11)

Substituting $q(t)$ into Equation (27.10) yields

$$\Delta P = (BV_s + C)\frac{dV_s}{dt}$$

or

$$(\Delta P)dt = (BV_s + C)dV_s$$

(27.12)

Integrating this from 0 to t and from 0 to V_s and solving for t leads to

$$t = \left(\frac{BV_s^2}{2\Delta P}\right) + \left(\frac{CV_s}{\Delta P}\right)$$

(27.13)

This equation can be rearranged so that V_s is an explicit function of t

$$V_s = \sqrt{\frac{2(\Delta P)t}{B} + \left(\frac{C}{B}\right)^2} - \left(\frac{C}{B}\right)$$

(27.14)

Also, V_s can be shown to be related to an instantaneous value of $q(t)$ by rearrangement of Equation (27.10)

$$V_s = \left(\frac{(\Delta P)t}{q(t)}\right) - \left(\frac{C}{B}\right)$$

(27.15)

Equating Equations (27.14) and (27.15) and solving for q yields

$$q = \frac{\Delta P}{B\sqrt{\frac{2(\Delta P)t}{B} + \left(\frac{C}{B}\right)^2}}$$

(27.16)

Numerical values for design coefficients B and C are usually obtained from experimental data. If Equation (27.13) is rewritten as

$$\frac{t}{V_s} = \left(\frac{BV_s}{2\Delta P}\right) + \left(\frac{C}{\Delta P}\right)$$

(27.17)

a plot of t/V_s vs. V_s will yield a straight line of slope $(B/2\Delta P)$ and an intercept $(C/\Delta P)$. Note that only two (V_s, t) data points are necessary to provide a first approximation of B and C.

Some industrial filter operations are conducted in a manner approaching constant flow rate. For this condition,

$$dq/dt = 0$$

and

$$q = dV_s/dt = \text{constant}$$

so that

$$V_s = qt$$

Equation (27.10) now becomes

$$\Delta P = Bq^2t + Cq \tag{27.18}$$

Thus, a plot of ΔP vs. t yields a straight line of slope Bq^2 and intercept Cq. At $t = 0$, the only resistance to flow is that of the filter medium; the pressure drop, however, is a linear function of time, and as time increases, the resistance due to the cake may predominate.

Some filter operations have both a constant pressure and constant rate period. At the beginning of a normal cycle, the pressure drop is held constant until the flow rate increases to a maximum value that is obtained by experiment. The flow rate is then maintained constant until the pressure drop increases above a predetermined limit that may be dictated by economics.

The coefficients B and C above have appeared in revised form in the literature in an attempt to relate them to physically measurable quantities. Equation (27.17) has been rewritten as:

$$\frac{t}{V_s} = \frac{K_c}{2}V_s + \frac{1}{q_0} \tag{27.19}$$

where K_c and q_0 are constants and (in English units), with

$$K_c = \frac{\mu c \alpha}{S^2 g_c \Delta P} \tag{27.20}$$

and

$$\frac{1}{q_0} = \frac{\mu R_m}{S g_c \Delta P} \tag{27.21}$$

Note that the S (once again) is the total surface of filtration cakes in the system (ft^2), α is the specific cake resistance (ft/lb) and R_m is the filter medium resistance (ft^{-1}). As before, a plot of t/V_s vs. V_s will give a straight line with a slope of $K_c/2$. This will allow the calculation of the specific cake resistance, α. Also, the y-intercept of the plot will be $1/q_0$, which means that filter medium resistance, R_m, can be calculated.

Theodore and Buonicore have developed equations to predict coefficients B and C (or α and R_m) from basic principles.[2] Despite the progress in developing pure filtration theory, and in view of the complexity of the phenomena, the most common methods of correlation are based on predicting a form of a final equation that can be verified by experiment.

Illustrative Example 27.1   Qualitatively, explain why the flow rate decreases during constant pressure filtration.

Solution   The flow rate is a function of both the pressure drop and the resistance to flow. During filtration, the solids build up within the filter and form a cake, which increases the resistance to flow. In addition, because of the buildup within the filter, the porosity decreases, further increasing the resistance to flow.

Illustrative Example 27.2   A plate-and-frame filter press is to be employed to filter a slurry containing 10% by mass of solids. If 1 ft^2 of filter cloth area is required to treat 5 lb/h of solids, what cloth area, in ft^2, is required for a slurry flowrate of 600 lb/min?

Solution   Convert the slurry flowrate, $\dot{m}$, to lb/h:

$$\dot{m} \text{ (slurry)} = (600 \text{ lb/min })(60 \text{ min/h}) = 36{,}000 \text{ lb/h}$$

Calculate the solids flowrate in the slurry:

$$\dot{m} \text{ (solids)} = (0.1)(36{,}000 \text{ lb/h}) = 3600 \text{ lb/h}$$

Calculate the filter cloth area, A, requirement:

$$A = (3600 \text{ lb/h})\left(\frac{1 \text{ h} \cdot \text{ft}^2}{5 \text{ lb}}\right) = 720 \text{ ft}^2$$

Illustrative Example 27.3   Engineers have designed a plate-and-frame filter press for a new plant. The design is based on pilot plant data, and the unit in the pilot plant was operated under conditions identical to those intended for the large plant. Engineers assumed that the filtration cycle would allow for cleaning and operating at constant pressure. Their design will provide an average filtration rate approximately equal to the required capacity. To add a factor of safety, Engineer A wants to use 25% more frames; B wants to make each frame 25% longer; C wants to make each frame with 25% more area; D wants to use 25% more filtration pressure. Rank these proposals in the order in which you consider they would increase the average hourly capacity of the filter. Explain and justify your rank by the use of appropriate theory or equations. For purposes of analysis, neglect the resistance of the filter.

Solution Refer to Equations (27.17)–(27.21) and note that $C = 0$. Engineer A's plan to use 25% more frames will increase the collection area by a factor of 1.25. The time of filtration will be decreased by a factor of $(1.25)^2$ since B is inversely proportional to the area squared. If the filtration capacity is

$$q = \frac{V_s}{t + t_c} \qquad t_c = \text{cleaning time}$$

The improved rate, QS, to insure a factor of safety, is then

$$QS = \frac{V_s}{[t/(1.25)^2] + t_c}$$

If cleaning time is neglected, one notes the rate is increased by a factor of 1.56, that is,

$$QS = 1.56\,q$$

The same improved rate is obtained using the plans of engineers B and C. For plan D, the time of filtration is decreased by 1.25. The improved rate is

$$QS = \frac{V_s}{(t/1.25) + t_c} = 1.25q$$

The proposals are therefore ranked (1) a, b, c (tied) and (2) d.

Illustrative Example 27.4 The following data were obtained during two constant pressure runs conducted on a plate-and-frame filter press. Calculated values for t/V are also included. The filtration area values for t/V are also included. The filtration area of the unit is 0.35 ft^2 and the slurry concentration (of solids) per volume of filtrate was previously calculated to be 4.142 lb/ft^3. See Table 27.1. Calculate the coefficients K_c and $(1/q_0)$.

Solution A plot of t/V vs. V (for the run at 20 psig) is shown in Fig. 27.6. The slope of this graph is $K_c/2$. Thus,

$$\frac{K_c}{2} = 2.4285 \text{ s/L}^2$$

$$= 1947 \text{ s/ft}^6$$

$$K_c = 2(1947)$$

$$= 3894 \text{ s/ft}^6$$

Also, the y-intercept is $1/q_0$. By extrapolation,

$$\frac{1}{q_0} = 7.6715 \text{ s/L}$$

$$= 217 \text{ s/ft}^3$$

Table 27.1   Illustrative Example 27.4

Filtrate		Run #1; 20 psig	Run #2; 15 psig
Weight, kg	Volume, L	Time, s	Time, s
0.0	0.0	0	0
0.5	0.5	7	9
1.0	1.0	13	14
1.5	1.5	19	20
2.0	2.0	27	27
2.5	2.5	37	36
3.0	3.0	47	46
3.5	3.5	59	57
4.0	4.0	72	69
4.5	4.5	86	83
5.0	5.0	101	98
5.5	5.5	118	115
6.0	6.0	135	133
6.5	6.5	154	152
7.0	7.0	175	172
7.5	7.5	202	194
8.0	8.0	229	220
8.5	8.5	259	
9.0	9.0	291	

[a]Calculate R_m and α for the 20 psig run.

Illustrative Example 27.5. Refer to the previous example. Calculate the filtration coefficients R_m and α. For water at 2°C, the viscosity (see Table A.2 in the Appendix) is converted to

$$\mu = 5.95 \times 10^{-4}\,\text{lb/ft}\cdot\text{s}$$

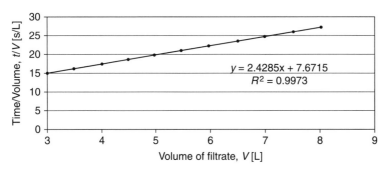

Figure 27.6 Relationship between t/V and V for a Pressure Drop of 20 psig; Illustrative Example 27.4.

Equation (27.21) is rearranged and solved for R_m

$$R_m = \frac{1}{q_0} \frac{S g_c \Delta P}{\mu}$$

Substitution yields

$$R_m = \frac{1}{q_0} \frac{S g_c \Delta P}{\mu} = (217) \frac{(0.35)(32.174)(20)(144)}{5.95 \times 10^{-4}} = 1.183 \times 10^{10} \, \text{ft}$$

Similarly, Equation (27.20) is solved for α

$$\alpha = \frac{K_c S^2 g_c \Delta P}{c\mu} = \frac{(3894)(0.35)^2(32.174)(20)(144)}{(4.142)(5.95 \times 10^{-4})} = 1.79 \times 10^{10} \, \text{ft/lb}$$

Illustrative Example 27.6 A filter unit is run at a constant rate for 30 min during which time 3000 ft^3 of slurry from a cement kiln operation is processed. The initial and final pressure in the unit is 0.5 and 5.0 in H_2O gauge, respectively. If the filter is further operated for 30 min at the final pressure, calculate the quantity of additional slurry treated.

Solution For constant rate filtration, Equation (27.18) is employed:

$$\Delta P = Bq^2 t + Cq$$

At time zero, the data indicates

$$0.5 \, \text{in} \, H_2O = 0 + C(100 \, \text{ft}^3/\text{min})$$

so that the value of C is 0.005 in $H_2O \cdot \text{min}/\text{ft}^3$.
  Substituting the data at t equal to 30 min gives:

$$5.0 = B(100)^2(30) + 0.005(100) \quad \text{or}$$

$$B = 1.5 \times 10^{-5} \, \text{in} \, H_2O \cdot \text{min}/\text{ft}^3$$

At constant pressure, the describing equation is given by Equation (27.13):

$$t = (B/2\Delta P)V_s^2 + (D/\Delta P)V_s$$

where D represents not only the resistance of the filter but also the resistance of the solids deposited during the constant rate period. The coefficient B remains the same since it is still proportional to the quantity of solids deposited during this constant pressure period. Coefficient D is calculated from Equation (27.10):

$$\Delta P = (BV_s + D)q$$

Since $\Delta P = 5$ in H_2O at $V_s = 0$, one obtains

$$D = 0.05 \text{ in } H_2O \cdot \text{min}/\text{ft}^3$$

After 30 min of constant pressure filtration

$$30 = (1.5 \times 10^{-5}/10)V_s^2 + (0.05/5)V_s$$

Solving for the positive root,

$$V_s = 2200 \text{ ft}^3$$

Therefore, the total quantity of slurry filtered is

$$(100)(30) + 2200 = 5200 \text{ ft}^3$$

27.3.1 Compressible Cakes

In actual practice, the deposited cake is usually assumed to be incompressible. However, all cakes are compressible to some degree. For large pressure drops, these effects can become important. These large changes in pressure tend to force the solids further into the interstices in the filtering medium, thereby increasing the resistance to flow and the value of α. If α is not constant, but is a function of ΔP, the cake is referred to as compressible. The cake resistance, α, is one of the more important variables in filtration applications; it is dependent on a host of factors including the filter area, pressure drop, viscosity, etc.

For the above situation, the following experimental relationship between α and ΔP is often assumed to apply:

$$\alpha = \alpha_0 \Delta P^b \qquad (27.22)$$

where a and b are empirically assumed constants that can be obtained from a best-fit straight line on a log-log plot of α vs. ΔP. The term α_0 is usually referred to as the specific cake resistance at zero pressure and b is the compressibility factor for the cake. Note that the term b is zero for incompressible (i.e., the cake resistance does not vary with pressure) sludges and is positive for compressible ones.

Constant pressure experiments are often used to determine the two coefficients of the cake. The first step in this process is to generate a logarithmic graph of α vs. ΔP. Note that the logarithmic form of Equation (27.22) indicates that if the data is regressed linearly, the slope of the regression equation will equal the value of b.

$$\log(\alpha) = \log(\alpha_0) + b\log(\Delta P) \qquad (27.23)$$

If a pressure drop and corresponding value of cake resistance (α) are obtained from the graph then a can then be solved mathematically since it is the only unknown.

Illustrative Example 27.7   The following results were obtained during the running of a filtration experiment in the Unit Operations Laboratory at Manhattan College. The equation of the linear regression line for the plot of cake resistance vs. pressure drop of the experimental filter cake was determined to be:

$$y = 0.210x + 10.99$$

where
$$y = \log \alpha$$
$$10.99 = \log \alpha_0$$
$$0.210x = b \log \Delta P$$

The units for α and ΔP are ft/lb and lb_f/ft^2, respectively. Based on the above results, calculate the specific cake resistance in ft/lb if the cake resistance is 4.57×10^{11} ft/lb at a pressure drop of 1554 lb_f/ft^2.

Solution   Refer to Equation 27.22 and solve for α_0. Note that the exponent b was obtained from the regression.

$$\alpha_0 = \alpha/(\Delta P)^{0.21}$$

Substituting

$$\alpha_0 = 4.57 \times 10^{11}/(1554)^{0.21}$$
$$= 9.73 \times 10^{10} \text{ ft/lb}$$

Illustrative Example 27.8.   Describe the factors that affect the choice of a filtration unit.

The choice of filter equipment depends largely on economics, but the economic advantages will vary depending on:

1. Fluid viscosity, density, and chemical reactivity, etc.?
2. Solid particle size, size distribution, and shape.
3. Flocculation tendencies.
4. Deformability.
5. Feed slurry concentration.
6. Slurry temperature.
7. Amount of material to be handled.
8. Absolute and relative values of liquid and solid products.
9. Completeness of separation required.
10. Relative costs of labor, capital, and power.

Illustrative Example 27.9   A filter press operates at a constant pressure of 50 psig. The initial rate of collection is 10 ft^3/min and 1 hr is needed to collect 100 ft^3 of filtrate. What is the hourly capacity of this filter if 15 ft³ of wash water is used (at the same operating pressure) and the time for dumping and cleaning is 30 min?

Solution Apply Equation (27.12):

$$q = \frac{\Delta P}{BV_s + C}$$

When $V_s = 0$,

$$q = \frac{P}{C}$$

so that

$$C = \frac{P}{q} = \frac{50}{10}$$
$$= 5$$

For the constant pressure process, apply Equation (27.13):

$$t = \frac{B}{2\Delta P} V_s^2 + \frac{C}{\Delta P} V_s$$

When $t = 60$ min,

$$V_s = 100; \ \Delta P = 50$$

Substituting

$$60 = \frac{B}{2(50)} (100)^2 + \frac{(5)(100)}{50}$$
$$B = 0.5$$

During the washing cycle,

$$t_w = \frac{V_w}{q_w}$$

and (with B and C remaining the same)

$$\Delta P_w = (BV_s + C)q_w$$

Therefore,

$$t_w = \frac{V_w(BV_s + C)}{\Delta P} = \frac{V_w}{q_w}$$
$$= \frac{(15)[(0.5)(100) + 5]}{50} = 16.5 \text{ min}$$

Thus,

$$t_c = t_f + t_w + t_d$$
$$= 60 + 16.5 + 30$$
$$= 106.5 \text{ min} = 1.775 \text{ hr}$$

Finally,

$$q = \frac{V_s}{t_c} = \frac{100}{1.775} = 56.3 \text{ gal/hr}$$

27.4 FILTRATION EXPERIMENTAL DATA AND CALCULATIONS

One of the experiments conducted in the Chemical Engineering Laboratory of Manhattan College is concerned with filtration. Students perform the experiment and later submit a report. In addition to theory, experimental procedure, discussion of results, and so on, the report contains sample calculations. The following is an (edited) example of these calculations that covers a wide range of filtration principles and applications.

During the experiment, two important measurements were determined: the mass increments of the filtrate, and the time for each increment. The scale measured the mass of the filtrate in grams and the time was recorded in seconds.

Four runs were carried out for this experiment. A sample of the cake from each frame was taken and placed in a previously measured dish after each run was stopped. Then, the new mass for the wet cake was obtained. The first run was performed at a constant pressure of 5 psig and is discussed below.

Two frames were used in this experiment: frame 16 and frame 18. The dishes corresponding to each frame were measured prior to addition of sample. Table 27.2 shows the mass of the dishes, wet cake and dry cake.

Table 27.2 Run 1 weights

Run–Frame	Dish (g)	Wet Cake + Dish (g)	Dry Cake + Dish (g)
1–16	1.0716	65.2296	37.4754
1–18	1.0679	66.0205	38.3801

The wet and dry cake masses were determined as follows:

Wet cake$_{1-16}$ = (Wet cake + Dish)$_{1-16}$ − Dish$_{1-16}$
Wet cake$_{1-16}$ = 65.2296 g − 1.0716 g = 64.158 g

Dry cake$_{1-16}$ = (Dry cake + Dish)$_{1-16}$ − Dish$_{1-16}$
Dry cake$_{1-16}$ = 37.4754 g − 1.0716 g = 36.4038 g

The same equations are used to determine the wet and dry cake masses for the subsequent runs. The masses are then converted to kilograms:

$$\text{Dry cake}_{1-16} = 36.4038\,\text{g}$$
$$= 0.0364\,\text{kg}$$

The moisture content of each filter cake sample is determined using the wet and dry cake masses calculated previously.

$$\text{Moisture}_{1-16} = \text{Wet cake}_{1-16} - \text{Dry cake}_{1-16}$$
$$\text{Moisture}_{1-16} = 0.0642\,\text{kg} - 0.0364\,\text{kg} = 0.0278\,\text{kg}$$

$$\%\text{Moisture}_{1-16} = \left(\frac{\text{Moisture}_{1-16}}{\text{Wet cake}_{1-16}}\right)100\% = \left(\frac{0.0278\,\text{kg}}{0.0642\,\text{kg}}\right)100\% = 43.3\%$$

One of the important results to be determined for this experiment is the mass of the particles deposited in the filter per unit volume of filtrate (c) for each run. The following equation is used to obtain this concentration:

$$c = \frac{c_s}{1 - [(m_F/m_c) - 1]c_s/\rho}$$

where c_s = mass fraction of solids in filter cake
m_F = mass of filter cake
m_C = mass of solids in filter cake

For run 1, c_s is calculated as follows:

$$c_s = \left(\frac{\text{kg}_{\text{solids}}}{\text{kg}_{\text{solids}} + \text{kg}_{\text{water}}}\right) = \frac{0.0364\,\text{kg}}{(0.0364 + 0.0278)\,\text{kg}} = 0.5669$$

The terms m_F and m_C are the mass of wet cake and mass of dry cake, respectively. Substituting gives:

$$c_{1-16} = \frac{c_s}{1 - [(m_F/m_C) - 1]c_s/\rho} = \frac{0.5669}{1 - [(0.0642/0.0364) - 1]0.5669/1000}$$
$$= 0.5671\,\text{kg/m}^3$$

An important value that needs to be calculated in this experiment is the filter medium resistance, R_m, α. A plot of time over volume vs. volume was constructed to obtain the necessary data to estimate R_m.

After plotting the data for the run, the slope and intercept of the line were obtained and used to calculate the filter medium resistance and the average specific cake resistance. Table 27.3 shows the values obtained from the plot.

Table 27.3 Slope and intercept for Run 1

ΔP (psig)	$K_C/2$ (s/L^2)	$1/q_0$ (s/L)
5	7.5823	13.974

The area for filtration is given by the two faces of the frames. This area S is calculated as follows:

$S = 2LW - 2\pi r^2$; r = hole opening

$L = 16\,\text{cm}$

$W = 16.2\,\text{cm}$

$r = 2.5\,\text{cm}$

$S = 2[(16\,\text{cm})(16.1\,\text{cm}) - (\pi(2.5\,\text{cm})^2)] = 475.9\,\text{cm}^2$

The values of R_m and α are obtained by first converting key terms to English units:

$$K_C = (2)(7.5823)(28.31)^2 = 1.215 \times 10^4\,\text{s/ft}^6$$

$$1/q_0 = (13.974)(28.31) = 395.7\,\text{s/ft}^3$$

$$\text{Area} = \frac{475.9}{929.03} = 0.5157\,\text{ft}^2$$

$$\Delta P = 5(144) = 720\,\text{lb}_\text{f}/\text{ft}^2$$

$$g_c = 32.2\,\text{lb} \cdot \text{ft}/\text{lb}_\text{f} \cdot \text{s}^2$$

$$\mu_{\text{H}_2\text{O}} = 5.94 \times 10^{-4}\,\text{lb/ft} \cdot \text{s}$$

Substituting into the following equation gives

$$R_m = \frac{(S)(\Delta P)(g_c)(1/q_0)}{\mu_{\text{H}_2\text{O}}}$$

$$= \frac{(0.5157)(720)(32.2)(395.7)}{5.94 \times 10^{-4}}$$

$$= 7.966 \times 10^9\,(\text{ft})^{-1}$$

REFERENCES

1. U.S. Filter, http://www.jwifilters.com/what_is_a_filter_press.htm.
2. ATR Corporation, http://atrcoporation.com/why_use_a_filter_press.htm.
3. A. Foust, L. Wenzel, C. Chung, L. Maus, and L. Andrews, "Principles of Unit Operations," John Wiley & Sons, Hoboken, NJ, 1950.

4. S. Calvert, "Guidelines for Selecting Mist Eliminators," Chemical Engineering, New York City, February 27, 1978.

5. L. Theodore and A. Buonicore, "Industrial Air Pollution Control Equipment for Particulates," CRC Press, Boca Raton, FL, 1976.

NOTE: Additional problems for each chapter are available for all readers at www.wiley.com. Follow links for this title.

VI

SPECIAL TOPICS

This last part of the book is concerned with seven topics that the Accreditation Board for Engineering and Technology (ABET) has recently indicated should be included in any engineering curriculum. The contents of these chapters are briefly described below.

Chapter 28 is concerned with Environmental Management; this chapter contains a broad discussion of environmental issues facing today's engineers and presents some of the more recent technology to deal with the issues at hand. Chapter 29 is concerned with Accident and Emergency Management; it deals with ways to ensure both employee and public safety, determination of the severity of accidents, and determining the causes and potential causes of accidents. Chapter 30 is concerned with Ethics; the case study approach is employed to make the reader think about the ethical questions involved, to reflect on their past decisions, and to project forward to their future decisions with a higher degree of thought and insight when faced with an ethical dilemma. Chapter 31 is concerned with Numerical Methods; a brief overview of numerical methods is given to provide the practicing engineer with some insight into what many of the currently used software packages (MathCad, Mathematica, MatLab) are actually doing. Chapter 32 is concerned with Economics and Finance; this subject area provides material that can ultimately dictate the decisions made by the practicing engineer and his/her company. Chapter 33 is concerned with the general subject area of Biomedical Engineering—a relatively new topic for many technical individuals. Finally, Chapter 34 introduces the reader to Open Ended Questions; although engineers at their very essence are problem solvers, most problems in real life do not come fully defined with a prescribed methodology to arrive at a solution. These open-ended questions are exercises in using brain power—and like any muscle, you use it or lose it. The authors believe that those who conquer or become adept with this topic will have taken the first step toward someday residing in an executive suite.

Fluid Flow for the Practicing Chemical Engineer. By J. Patrick Abulencia and Louis Theodore
Copyright © 2009 John Wiley & Sons, Inc.

28

ENVIRONMENTAL MANAGEMENT

28.1 INTRODUCTION

In the past four decades, there has been an increased awareness of a wide range of environmental issues covering all resources: air, land, and water. More and more people are becoming aware of these environmental concerns, and it is important that professional people, many of whom do not possess an understanding of environmental problems, have the proper information available when involved with environmental issues. All professionals should have a basic understanding of the technical and scientific terms related to these issues. In addition to serving the needs of the professional, this chapter examines how one can increase his or her awareness of and help solve the environmental problems facing both industry and society.

Chapter 28 is titled Environmental Management; it provides a broad discussion of environmental issues facing today's engineers and presents some of the more recent technology to deal with the issues at hand. Some of the topics covered in this chapter include air pollution, water pollution, solid waste, etc., and several Illustrative Examples dealing with these topics are presented.

This chapter is not intended to be all-encompassing. Rather, it is to be used as a starting point. Little is presented on environmental regulations because of the enormity of the subject matter; in a very real sense, it is a moving target that is beyond the scope of this text. Further, the material primarily keys on traditional environmental topics. Although much of the material is qualitative in nature, some quantitative material and calculations are presented in the Illustrative Examples that are presented in the last section.

Fluid Flow for the Practicing Chemical Engineer. By J. Patrick Abulencia and Louis Theodore
Copyright © 2009 John Wiley & Sons, Inc.

28.2   ENVIRONMENTAL MANAGEMENT HISTORY

BANG! The Big Bang. In 1948, physicist G. Gamow proposed the big bang theory of the origin of the universe. He believed that the universe was created in a gigantic explosion as all mass and energy were created in an instant of time. Estimates on the age of the universe at the present time range between 7 and 20 billion years, and with 13.5 billion years often mentioned as the age of the planet Earth.

The bang occurred in a split second and within a minute the universe was approximately a trillion miles wide and expanding at an unbelievable rate. Several minutes later all the matter known to humanity had been produced. The universe as it is known today was in place. Environmental problems, as they would later relate to living organisms and humans, were born.

Flash forward to the present. More than any other time in history, the 21st century will be a turning point for human civilization. Human beings may be facing ecological disasters that could affect their ability to survive. These crises could force them to reexamine the value system that has governed their lives for the past two million years (approximately) of existence.

28.2.1   Recent Environmental History

The year 1970 was a cornerstone year for modern environmental policy. The National Environmental Policy Act (NEPA), enacted on January 1, 1970, was considered a "political anomaly" by some. NEPA was not based on specific legislation; instead it referred in a general manner to environmental and quality-of-life concerns. The Council for Environmental Quality (CEQ), established by NEPA, was one of the councils mandated to implement legislation. April 22, 1970 brought Earth Day, where thousands of demonstrators gathered all around the nation. NEPA and Earth Day were the beginning of a long, seemingly never-ending debate over environmental issues.

The Nixon Administration became preoccupied with not only trying to pass more extensive environmental legislation, but also with implementing the laws. Nixon's White House Commission on Executive Reorganization proposed in the Reorganizational Plan #3 of 1970 that a single, independent agency be established, separate from the CEQ. The plan was sent to Congress by President Nixon on July 9, 1970, and this new U.S. Environmental Protection Agency (EPA) began operation on December 2, 1970. The EPA was officially born.

The aforementioned EPA works with the states and local governments to develop and implement comprehensive environmental programs. Federal laws such as the Clean Air Act, the Safe Drinking Water Act, the Resource Conservation and Recovery Act, the Comprehensive Environmental Response, Compensation, and Liability Act, etc., all mandate involvement by state and local government in the details of implementation. These laws, in a very real sense, have dictated the environmental management policies and procedures that are presently in place and serve as the subject matter for this chapter.

A waste management timetable that provides information on environmental approaches since World War II is provided in Table 28.1.

Table 28.1 Waste management timetable

Timeframe	Control
Prior to 1945	No control
1945–1960	Little control
1960–1970	Some control
1970–1975	Greater control (EPA founded)
1975–1980	More sophisticated control
1980–1985	Beginning of waste reduction management
1985–1990	Waste reduction management
1990–1995	Pollution Prevention Act
1995–2000	Sophisticated pollution prevention approaches
2000–2010	Green chemistry and engineering; Sustainability
2010–	?????

28.3 ENVIRONMENTAL MANAGEMENT TOPICS

There are two dozen topics that the authors consider to be integral parts of environmental management (there are, of course, more). Reviewing each subject area in any detail is beyond the scope of this text; the reader is referred to a key reference in the literature[1] for an extensive review of the entire field of environmental management. Additional and more specific references for each of the topics referred to above are provided below:

1. Air pollution control equipment, etc.[2–4]
2. Atmospheric dispersion modeling.[5]
3. Indoor air quality.[6]
4. Industrial wastewater management.[6,7]
5. Wastewater treatment technologies.[6,7]
6. Wastewater treatment processes.[5–7]
7. Solid waste management.[5–8]
8. Superfund.[5,6,8]
9. Municipal solid waste management.[5,6,8]
10. Hospital waste management.[5,6,8]
11. Nuclear waste management.[5,6]
12. Pollution prevention.[9–15]
13. Multimedia analysis and lifecycle cost analysis.[6,7]
14. Noise.[5,6,16,17]

15. ISO14000[5,6,18,19]
16. Environmental justice.[5,6,20-22]
17. Electromagnetic field.[5,6]
18. Acid rain.[5,6]
19. Greenhouse effect and global warning.[5,6]
20. Public perception of risk.[5,6,20-22]
21. Health risk assessment.[5-7,23]
22. Hazard risk assessment.[5-7,24]
23. Risk communication.[5,6,20-22]
24. Environmental implication of nanotechnology.[25,26]

28.4 APPLICATIONS

Much of the material to follow has been drawn from the literature.[1,7,26] The Illustrative Examples in this section were extracted from numerous sources including problem workbooks prepared under National Science Foundation grants.[9,10,26,27]

Illustrative Example 28.1 Highlight the difference between a wastewater "direct discharger" and an "indirect discharger."

Solution A direct discharger is an industrial plant that discharges its effluent wastewater directly to a surrounding water source with no intermediate means of treatment. An indirect discharger is a plant that first discharges to a publicly owned treatment works (POTW) facility prior to release to the environment.

Illustrative Example 28.2 Identify three water systems (two of which involve a flowing medium) in the order of increasing complexity that are commonly modeled for water quality analysis. What assumptions are made for each system?

Solution

1. Rivers—Can be modeled in one, two and three dimensions.
2. Lakes—Closed system. Will be subject to evaporative effects. Assumes poor mixing within the lake (temperature stratification with depth).
3. Estuaries—Complex mass balance. Boundaries must be defined. Can be defined as a steady-state condition. Assume time averaged and distance averaged conditions with respect to area, flow, and reaction rates.

Illustrative Example 28.3 PALT (Pat Abulencia and Louis Theodore) engineers have been requested to determine the minimum distance downstream from a cement dust emitting source that will be free of cement deposit. The source is equipped with a cyclone located 150 ft above ground level. They assume ambient conditions are at 60°F and 1 atm and neglect meteorological aspects. Additional data are given below:

Particle size range of cement dust is 2.5–50 microns.
Specific gravity of the cement dust is 1.96.
Wind speed is 3.0 miles/hr.

Solution   A particle diameter of 2.5 microns is used to calculate the minimum distance downstream free of dust since the smallest particle will travel the greatest horizontal distance. In order to determine the value of K for the appropriate size of the dust, first calculate the particle density using the specific gravity given and determine the properties of the gas (assume air)

$$\rho_p = SG(62.4) = 1.96(62.4) = 122.3 \, \text{lb/ft}^3$$

$$\rho_{\text{air}} = \frac{PM}{RT} = \frac{(1)(29)}{(0.73)(60 + 460)} = 0.0764 \, \text{lb/ft}^3$$

The viscosity of air (Appendix A, Table A.3), μ, at 60°F is $1.22 \times 10^{-5} \, \text{lb/ft} \cdot \text{s}$. Calculate the value of K.

$$K = d_p \left(\frac{g \rho_p \rho_{\text{air}}}{\mu^2}\right)^{1/3} = \frac{2.5}{(25,400)(12)} \left(\frac{(32.174)(122.3)(0.0764)}{(1.22 \times 10^{-5})^2}\right)^{1/3} = 0.104$$

Stokes' law range applies since $K < 3.3$.
    Use the appropriate terminal settling velocity equation and calculate the terminal settling velocity in ft/s.

$$v = \frac{g d_p^2 \rho_p}{18\mu} = \frac{(32.174)[2.5/(25,400)(12)]^2(122.3)}{18(1.22 \times 10^{-5})}$$

$$= 1.21 \times 10^{-3} \, \text{ft/s}$$

Calculate the approximate time for descent in seconds.

$$t = \frac{h}{v} = \frac{150}{1.21 \times 10^{-3}}$$

$$= 1.24 \times 10^5 \, \text{s}$$

Calculate the horizontal distance traveled in miles

$$x = \frac{t}{v_{\text{wind}}} = \frac{1.24 \times 10^5}{3.0/3600}$$

$$= 103.3 \, \text{miles}$$

Illustrative Example 28.4   It is proposed to install a pulse-jet fabric filter system to clean an airstream containing particulate pollutants. You are asked to select the most appropriate filter bag fabric considering performance and cost. Pertinent design and operating data, as well as fabric information, are given below in Table 28.2.

Volumetric flowrate of polluted airstream $= 10,000$ scfm ($60°F$, 1 atm)
Operating temperature $= 250°F$
Concentration of pollutants $= 4.00$ gr/ft^3
Average $ACR = 2.5$ cfm/ft^2 cloth
Collection efficiency requirement $= 99\%$

Table 28.2   Pulse-jet bag data provided by manufacturer

Filter Bag	A	B	C	D
Tensile strength	Excellent	Above average	Fair	Excellent
Recommended maximum temperature ($°F$)	260	275	260	220
Resistance factor	0.9	1.0	0.5	0.9
Cost per bag, ($)	26	38	10	20
Standard size	8 in × 16 ft	10 in × 16 ft	1 ft × 16 ft	1 ft × 20 ft

Note: No bag has an advantage from the standpoint of durability under the operating conditions for which the bag was designed.

Solution   Bag D is eliminated since its recommended maximum temperature (220°F) is below the operating temperature of 250°F. Bag C is also eliminated since a pulse-jet fabric filter system requires the tensile strength of the bag to be at least above average.

Consider the economics for the two remaining choices. The cost per bag is $26.00 for A and $38.00 for B. The gas flowrate and filtration velocity are

$$q = 10,000 \left(\frac{250 + 460}{60 + 460} \right)$$

$$= 13,654 \text{ acfm}$$

$$v_f = 2.5 \text{ cfm/ft}^2 \text{ cloth}$$

$$= 2.5 \text{ ft/ min}$$

The filtering (bag) area is then

$$S_c = q/v_f$$

$$= 13,654/2.5$$

$$= 5462 \text{ ft}^2$$

For bag A, the area and number, N, of bags are

$$S = \pi DH$$
$$= \pi\left(\tfrac{8}{12}\right)(16)$$
$$= 33.5 \, \text{ft}^2$$
$$N = S_c/S$$
$$= 5462/33.5$$
$$= 163$$

For bag B:

$$S = \pi\left(\tfrac{10}{12}\right)(16)$$
$$= 41.9 \, \text{ft}^2$$
$$N = 5462/41.9$$
$$= 130$$

The total cost (TC) for each bag is as follows:

For bag A:

$$TC = N \text{ (cost per bag)}$$
$$= (163)(26.00)$$
$$= \$4238$$

For bag B:

$$TC = (130)(38.00)$$
$$= \$4940$$

Since the total cost for bag A is less than bag B, select bag A.

Illustrative Example 28.5   You are requested to determine the number of filtering bags required and cleaning frequency for a plant equipped with a fabric system. Operating and design data are given below:

Volumetric flow rate of the gas stream $= 50,000 \, \text{acfm}$
Dust concentration $= 5.0 \, \text{gr/ft}^3$
Efficiency of the fabric filter system $= 98.0\%$
Filtration velocity $= 10 \, \text{ft/min}$
Diameter of filtering bag $= 1.0 \, \text{ft}$
Length of filtering bag $= 15 \, \text{ft}$

The system is designed to begin cleaning when the pressure drop reaches 8.0 in. H_2O. The pressure drop is given by the following empirical equation

$$\Delta P = 0.2v_f + 5\,cv_f^2 t$$

where ΔP is the pressure drop, in. H_2O; v_f is the filtration velocity, ft/min; c is the dust concentration, lb/ft^3; and t is the time since the bags were cleaned, min.

Solution   To calculate N, you again need the total required surface area of the bags and the surface area of each bag.
    Calculate the filtering area

$$S_c = q/v_f$$
$$= 50{,}000/10$$
$$= 5000\,ft^2$$

The area per bag is

$$S = \pi DH$$
$$= (\pi)(1.0)(15)$$
$$= 47.12\,ft^2$$

The number of bags may now be calculated

$$N = S_c/S$$
$$= 5000/47.12$$
$$= 106$$

Since $5.0\,gr/ft^3 = 0.0007143\,lb/ft^3$ and

$$\Delta P = 0.2\,v_f + 5\,cv_t^2 t,$$

Substituting

$$8.0 = (0.2)(10) + (5)(0.0007143)(10)^2 t$$

Solving for t:

$$t = 16.8\,min$$

Illustrative Example 28.6   The primary difference between confined flow in pipes and open channel flow is that in open channel flow, the cross-sectional area of the flow is not predetermined but may be a variable that depends on other factors.
    One of the basic equations for calculating the flow rate, q (ft^3/s), as a function of depth of flow and channel characteristics is the Manning equation. It is given by:

$$q = \frac{1.486}{n} Ar_h^{2/3} S^{1/2}$$

In this equation, n, is a roughness coefficient which may vary from 0.01 for smooth uniform channels to 0.03 or higher for irregular natural river channels, and S is the channel bottom slope (not the cross-sectional area). The cross-sectional area for flow is (A) and r_h is the hydraulic radius (equal to A/P), where P is the wetted perimeter of the cross-section.

Water is passing through a trapezoidal channel whose bottom base, top base (open to the atmosphere), and height are 20 ft, 50 ft, and 7.5 ft, respectively. If $S = 0.0008$ and $n = 0.02$, calculate the volumetric flow rate of the water in ft^3/s.

Solution   Calculate the cross-sectional area available for flow in the channel.

$$A = (l_{base} + l_{top})(h/2) = (20 + 50)(7.5/2) = 262.5 \text{ ft}^2$$

Calculate the wetted perimeter of the trapezoid.

$$P = 20 + [7.5^2 + 15^2]^{1/2} = 36.8 \text{ ft}$$

Calculate the hydraulic radius.

$$r_h = \frac{A}{P} = \frac{262.5}{36.8} = 7.13 \text{ ft}$$

Solve Manning's equation.

$$q = \frac{1.486}{n} A r_h^{2/3} S^{1/2} = \frac{1.486}{0.02}(262.5)(7.13)^{2/3}(0.0008)^{1/2} = 2042 \text{ ft}^3/\text{s}$$

Another equation that can be used to calculate the flow rate in open channels is the Hazen–Williams equation. It is given by

$$v \text{ (ft/s)} = 1.318 \, n r_h^{0.63} S^{0.54}$$

or

$$q \text{ (ft}^3/\text{s)} = 1.318 n A r_h^{0.63} S^{0.54}$$

Illustrative Example 28.7   A watershed has an area of 8 mi^2. On average, rainfall occurs every 3 days at a rate of 0.06 mL/day and for a period of 5 h. Approximately 50% of the rain runoff reaches the sewers and contains an average total nitrogen concentration of 9.0 mg/L. In addition, the city wastewater treatment plant discharges 10 MGD (10^6 gal/day) with a total nitrogen concentration of 35 mg/L. Compare the total nitrogen discharge from runoff from the watershed with that of the city's sewage treatment plant.

Solution   First calculate the total nitrogen discharge, $\dot{m}_w$, from the treatment plant:

$$\dot{m}_w = (10)(35)(8.34)$$
$$= 2919 \, \text{lb/day}$$

The volumetric flow of the runoff, q, is

$$q = (0.5)(0.06)(8)(5280)^2/(3600)(12)$$
$$= 155 \, \text{ft}^3/\text{s}$$

The total nitrogen discharge, $\dot{m}_r$, from runoff is then

$$\dot{m}_r = \left(155 \, \text{ft}^3/\text{s}\right)(9 \, \text{mg/L})\left(10^{-6} \, \text{L/mg}\right)(3600 \times 24 \, \text{s/day})\left(62.4 \, \text{lb/ft}^3\right)$$
$$= 7521 \, \text{lb/day}$$

During rain, the runoff is over 2.5 times that for the treatment plant.

Illustrative Example 28.8   A municipality generates 1000 lb of solids daily. Size an aerobic digester to treat the solids. The following design parameters and information are provided:

Detention time, hydraulic $= t_h = 20$ days, etc.
Detention time, solids $t_s = 20$ days;
Temperature $= 95°$F;
Organic loading (OL) $= 0.2 \, \text{lbVS}/(\text{ft}^3 \cdot \text{day})$;
Volatile solids (VS) $= 78\%$ of total solids;
Percentage solids (TS) entering digester $= 4.4\%$;
VS destruction $= 62\%$.

Solution   Check the design based on the organic load and the hydraulic load. The volume based on the organic load, V_{OL}, is

$$V_{OL} = (1000)(0.78)/(0.2)$$
$$= 3900 \, \text{ft}^3$$

Based on the hydraulic load the volume, V_{HL}, is

$$V_{HL} = \frac{(1000)(20)}{(0.044)(8.33)(7.48)}$$
$$= 7300 \, \text{ft}^3$$

Since $7300 > 3900$, the hydraulic detention time controls and the design volume is $7300\,\text{ft}^3$.

Illustrative Example 28.9 A large, deep cavern (formed from a salt dome), located north of Houston, Texas, has been proposed as an ultimate disposal site for both solid hazardous and municipal wastes. Preliminary geological studies indicate that there is little chance that the wastes and any corresponding leachates will penetrate the cavern walls and contaminate adjacent soil and aquifers. A risk assessment analysis was also conducted during the preliminary study and the results indicate that there was a greater than 99% probability that no hazardous and/or toxic material would "meander" beyond the cavern walls during the next 25 years.

The company preparing the permit application for the Texas Water Pollution Board has provided the following data and information:

Approximate total volume of cavern $= 0.78\,\text{mi}^3$.

Approximate volume of cavern available for solid waste depository $= 75\%$ of total volume.

Proposed maximum waste feed rate to cavern $= 20{,}000\,\text{lb/day}$.

Feed rate schedule $= 6\,\text{days/week}$.

Average bulk density of waste $= 30\,\text{lb/ft}^3$.

Based on the above data, estimate the minimum amount of time it will take to fill the volume of the cavern available for the waste deposition.

Note: The proposed operation could extend well beyond the 25 years upon which the risk assessment analysis was based. The decision whether to grant the permit is somewhat subjective since there is a finite, though extremely low, probability that the cavern walls will be penetrated. Another, more detailed and exhaustive, risk analysis study should be considered.

Solution   The volume of the cavern, V, in cubic miles available for the solid waste is

$$V = (0.75)(0.78)$$

$$= 0.585\,\text{mi}^3$$

This volume can be converted to cubic feet:

$$V = (0.585\,\text{mi}^3)(5280\,\text{ft/mi})^3$$

$$= 8.61 \times 10^{10}\,\text{ft}^3$$

The daily volume rate of solids deposited within the cavern in cubic feet/day, q, is

$$q = (20{,}000\,\text{lb/day})/(30\,\text{lb/ft}^3)$$

$$= 667\,\text{ft}^3/\text{day}$$

The solids volume rate can now be converted to cubic feet/year:

$$q = (667 \text{ ft}^3/\text{day})(6 \text{ days/week})(52 \text{ weeks/yr})$$
$$= 208{,}000 \text{ ft}^3/\text{yr}$$

The time it will take to fill the cavern is therefore

$$t = V/q$$
$$= 8.61 \times 10^{10}/208{,}000$$
$$= 414{,}000 \text{ yr}$$

Deep-well injection is an ultimate disposal method that transfers liquid wastes far underground and away from freshwater sources. Like landfarming, this disposal process has been used for many years by the petroleum industry. It is also used to dispose of saltwater in oil fields. When the method first came into use, the injected brine would often eventually contaminate groundwater and freshwater sands because the site was poorly chosen. The process has since been improved, and laws such as the Safe Drinking Water Act of 1974 ensure that sites for potential wells are better surveyed.

Illustrative Example 28.10   A compliance stack test on a facility yields the results provided below in Table 28.3. Determine whether the incinerator meets the state particulate standard of 0.05 gr/dscf. Estimate the amount of particulate matter escaping the stack, and indicate the molecular weight of the stack gas. Use standard conditions of 70°F and 1 atm pressure.

Table 28.3   Compliance test data

Volume sampled	35 dscf
Diameter of stack	2 ft
Pressure of stack gas	29.6 in Hg
Stack gas temperature	140°F
Mass of particulate collected	0.16 g
% moisture in stack gas	7% (by volume)
% O_2 in stack gas (dry)	7% (by volume)
% CO_2 in stack gas (dry)	14% (by volume)
% N_2 in stack gas (dry)	79% (by volume)
Pitot tube factor (k)	0.85

Pitot tube measurements made at eight points across the diameter of the stack provided values of 0.3, 0.35, 0.4, 0.5, 0.5, 0.4, 0.3, and 0.3 in of H_2O.

Use the following equations for S-type pitot tube velocity, v (m/s), measurements (see also Equation (19.9)):

$$v = C\sqrt{2gH}$$
$$= C\sqrt{2g\frac{\rho_l}{\rho}(0.0254)h}$$

where g = gravitational acceleration (9.81 m/s^2)
H = fluid velocity head, in H$_2$O
ρ_l = density of manometer fluid, 1000 kg/m^3
ρ = density of flue gas, 1.084 kg/m^3
h = mean pitot tube reading, in H$_2$O
C = pitot tube coefficient = 0.85 (dimensionless)

Solution   The particulate concentration in the stack is

$$\text{Particulate concentration} = \frac{0.16 \text{ g collected}}{35 \text{ dscf sampled}} \left(\frac{15.43 \text{ gr}}{\text{g}} \right)$$

$$= 0.0706 \text{ gr/dscf}$$

Since this does exceed the particulate standard of 0.05 gr/dscf, *the facility is not in compliance.*

The actual particulate emission rate is the product of the stack flowrate and the stack flue gas particulate concentration. The stack flowrate is calculated from the velocity measurements provided in the problem statement using the second velocity equation given.

$$v = 0.85 \sqrt{2(9.81 \text{ m/s}^2) \left(\frac{1000 \text{ kg/m}^3}{1.084 \text{ kg/m}^3} \right) 0.0254h}$$

$$= 0.85(21.4)\sqrt{h}$$
$$= 0.85(21.4)(0.6142)$$
$$= 11.2 \text{ m/s} = 36.75 \text{ fps}$$

$$\text{Stack flowrate} = v \text{ (cross-sectional area)}$$

$$v = 36.75 \text{ fps} \left(\frac{\pi}{4} \right) (2 \text{ ft})^2$$

$$= 115.45 \text{ acfs} = 6.924 \text{ acfm}$$
$$\text{Dry volumetric flowrate} = (1-0.07) \times 6924 \text{ acfm} = 6439 \text{ dacfm}$$

Correct to standard conditions of 70°F and 1 atm pressure:

$$\text{Standard volumetric flowrate} = 6439 \text{ dacfm} \left(\frac{530°\text{R}}{600°\text{R}} \right) \left(\frac{29.6 \text{ psi}}{29.9 \text{ psi}} \right)$$

$$= 5631 \text{ dscfm}$$

$$\text{Particulate emission rate} = 0.0706 \, \text{gr}/\text{dscf}(5631 \, \text{dscfm})$$
$$= 398 \, \text{gr}/\text{min}$$
$$= 398 \, \text{gr}/\text{min}\left(\frac{1 \, \text{lb}}{7000 \, \text{gr}}\right) = 0.0569 \, \text{lb}/\text{min}$$
$$= (0.0569 \, \text{lb}/\text{min})(1440 \, \text{min}/\text{day}) = 81.9 \, \text{lb}/\text{day}$$

The molecular weight of flue gas is based on the mole fraction of the flue gas components. The flue gas is 7% water and 93% other components by volume. On a dry basis, the flue gas molecular weight is

$$MW = 0.07 \, O_2(32 \, \text{lb}/\text{lbmol}) + 0.14 \, CO_2(44 \, \text{lb}/\text{lbmol}) + 0.79 \, N_2(28 \, \text{lb}/\text{lbmol})$$
$$= 30.52 \, \text{lb}/\text{lbmol}$$

The average molecular weight of the stack gas on an actual (wet) basis is then

$$MW = 0.07 \, \text{water}(18 \, \text{lb}/\text{lbmol}) + 0.93 \, \text{other components}(30.52 \, \text{lb}/\text{lbmol})$$
$$= 29.64 \, \text{lb}/\text{lbmol}$$

Illustrative Example 28.11   Perchloroethylene (PCE) is utilized in a degreasing operation and is lost from the process via evaporation from the degreasing tank. This process has an emission factor (estimated emission rate/unit measure of production) of 0.78 lb PCE released per lb PCE entering the degreasing operation. The PCE entering the degreaser is made up of recycled PCE from a solvent recovery operation plus a fresh PCE makeup. The solvent recovery system is 75% efficient with the 25% reject going offsite for disposal.

1. Draw a flow diagram for the process.
2. Develop a mass balance around the degreaser.
3. Develop a mass balance around the solvent recovery system.
4. Develop a mass balance around the entire system.
5. Determine the mass of PCE emitted per pound of fresh PCE utilized.

Quantify the impact of the emission factor in the degreasing operation on the flowrates within the solvent recovery unit.

Solution

1. A flow diagram for the system is provided in Fig. 28.1.

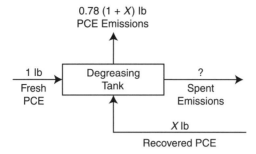

Figure 28.1   PCE mass balance around degreasing tank.

2. Assume a basis of 1 lb of fresh PCE feed. A PCE mass balance around the degreasing tank can now be written for the unit pictured in Fig. 28.1.

$$\text{Input} = \text{Fresh} - \text{Recycled PCE}$$
$$= 1\,\text{lb} + X\,\text{lb}$$
$$\text{Output} = \text{PCE emissions} + \text{Spent PCE}$$
$$= 0.78(1 + X) + \text{Spent PCE}$$

Equating the input with the output gives

$$\text{Spent PCE} = (1 - 0.78)(1 + X)$$
$$= 0.22(1 + X)\,\text{lb PCE}$$

3. A PCE mass balance around the solvent recovery unit is shown in Fig. 28.2.

$$\text{Input} = \text{Spent PCE}$$
$$= 0.22(1 + X)\,\text{lb}$$
$$\text{Output} = \text{Recycle PCE} + \text{Reject PCE}$$

where

$$\text{Recycle PCE} = 75\% \text{ of spent PCE}$$
$$= 0.75[0.22(1 + X)] + \text{Reject PCE}$$

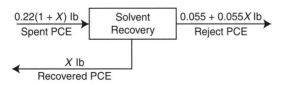

Figure 28.2   PCE mass balance around solvent recovery unit.

Since

$$\text{Input} = \text{Output}$$
$$0.22(1 + X) = 0.75[0.22(1 + X)] + \text{Reject PCE}$$
$$\text{Reject PCE} = (1 - 0.75)[0.22(1 + X)]$$
$$= 0.055 + 0.055X \text{ lb}$$

4. PCE mass balance around the entire system is shown in Fig. 28.3.

$$\text{Input} = 1 \text{ lb PCE}$$
$$\text{Output} = \text{PCE emissions} + \text{Spent PCE}$$
$$= (0.78 + 0.78X) + (0.055 + 0.055X)$$

Since

$$\text{Input} = \text{Output}$$
$$1 \text{ lb PCE} = (0.78 + 0.78X) + (0.055 + 0.055X)$$
$$= 0.835 + 0.835X$$
$$X = 0.165/0.835$$
$$= 0.198 \approx 0.20 \text{ lb PCE}$$

5. From the flow diagram in Fig. 28.3:

$$\text{PCE emissions} = 0.78 + 0.78X$$
$$= 0.78 + 0.78(0.20)$$
$$= 0.94 \text{ lb PCE emitted per lb fresh PCE}$$

If the emission factor were lower, the flowrates to the solvent recovery unit and the recycle stream would be higher. Additionally, there would be less PCE lost from the system. To determine the effect of the emissions factor on the system flow streams,

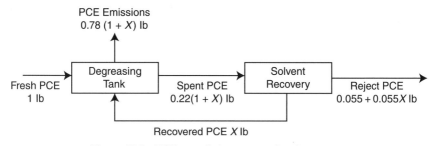

Figure 28.3 PCE mass balance around entire system.

the equations above were solved using three different emission factors: 0.78, 0.60, and 0.40. These results are summarized below in Table 28.4.

Table 28.4 PCE emission results

Emission Factor	0.78	0.60	0.40
Fresh PCE	1.0	1.0	1.0
Recovered PCE	0.198	0.429	0.818
Spent PCE	0.263	0.571	1.091
PCE emissions	0.934	0.857	0.727
Reject PCE	0.066	0.142	0.273

The sum of the recovered and fresh PCE provides a measure of the degreasing capability of the system per kilogram feed. Notice that, as the emission factors decrease, this sum goes up significantly.

REFERENCES

1. L. Stander and L. Theodore, "Environmental Regulatory Calculations Handbook", John Wiley & Sons, Hoboken, NJ, 2008.
2. L. Theodore and A. Buonicore, "Air Pollution Control Equipment for Gaseous and Particulates Pollutants", USEPA Training Manual, RTP, North Carolina, 1981.
3. L. Theodore, "Air Pollution Control Equipment" John Wiley & Sons, Hoboken, NJ, 2009.
4. J. Mycock, J. McKenna, and L. Theodore, "Handbook of Air Pollution Control Technology", ETS, Roanoke, VA, 1994.
5. M. K. Theodore and L. Theodore, "Major Environmental Issues Facing the 21st Century", 2nd edition, CRC Press, Boca Raton, FL, 2009.
6. G. Holmes, R. Singh, and L. Theodore, "Handbook of Environmental Management and Technology", 2nd edition, Wiley-Interscience, Hoboken, NJ, 2000.
7. J. Reynolds, J. Jeris, and L. Theodore, "Handbook of Chemical and Environmental Engineering Calculations", John Wiley & Sons, Hoboken, NJ, 2002.
8. J. Santoleri, J. Reynolds, and L. Theodore, "Introduction to Hazardous Waste Incineration", 2nd edition, John Wiley & Sons, Hoboken, NJ, 2000.
9. L. Theodore, R. Dupont, and J. Reynolds, "Pollution Prevention: Problems and Solutions", Gordon and Breach Publishers, Amsterdam, Holland, 1994.
10. R. Dupont, T. Baxter, and L. Theodore, "Environmental Management: Problems and Solutions", CRC Press, Boca Raton, FL, 1998.
11. ERM, "Pollution Prevention Quarterly", Miami, FL, Winter 1999.
12. R. Dupont, L. Theodore, and K. Ganesan, "Pollution Prevention: The Waste Management Approach for the 21st Century", CRC Press, Boca Raton, FL, 2000.
13. M. K. Theodore, "Pollution Prevention Calendar—Domestic Version", Theodore Tutorials, East Williston, NY, 1999.

14. M. K. Theodore, "Pollution Prevention Calendar—Office Version", Theodore Tutorials, East Williston, NY, 1999.

15. L. Theodore and R. Allen, "Pollution Prevention", Theodore Tutorials, East Williston, NY, 1993.

16. A. Thumann and C. Miller, "Fundamentals of Noise Control Engineering", The Fairmont Press, Englewood Cliffs, NJ, 1990.

17. P. Chermisinoff and P. Cheremisinoff, "Industrial Noise Control Handbook", Ann Arbor Science Publishers, Ann Arbor, MI, 1977.

18. USEPA, "ISO 14000 Resource Directory", Office of Research and Development, Washington, DC, EPA/625/R-97/003, 1997.

19. T. Welch, "Moving Beyond Environmental Compliance, A Handbook for Integrating Pollution Prevention with ISO 14000", Lewis Publishers, Boca Raton, FL, 1998.

20. D. Goleman, "Assessing Risk: Why Fear May Outweigh Harm", New York Times, February, 1, 1994.

21. M. Russell, "Communicating Risk to a Concerned Public", EPA Journal, November, 1989.

22. EPA, "Seven Cardinal Rules of Risk Communication", EPA OPA/8700, April, 1988.

23. C. Main, Inc. "Health Risk Assessment for Air Emissions of Metals and Organic Compounds from the PERC Municipal Waste to Energy Facility", Prepared for Penobscot Energy Recovery Company (PERC), Boston, MA, December 1985.

24. A. Flynn and L. Theodore, "Health, Safety and Accident Management on the Chemical Process Industries," (Marcel Dekker), CRC Press, Boca Raton, FL, 2002.

25. L. Theodore and R. Kunz, "Nanotechnology: Environmental Implications and Solutions", John Wiley & Sons, Hoboken, NJ, 2005.

26. L. Theodore, "Nanotechnology: Basic Calculations for Engineers and Scientists", John Wiley & Sons, Hoboken, NJ, 2006.

27. K. Ganeson, L. Theodore, and J. Reynolds, "Accident and Emergency Management: Problems and Solutions", Gordon and Breach Publishers, Amsterdam, Holland, 1996.

NOTE: Additional problems are available for all readers at www.wiley.com. Follow links for this title.

29

ACCIDENT AND EMERGENCY MANAGEMENT

29.1 INTRODUCTION

Accidents are a fact of life, whether they are a careless mishap at home, an unavoidable collision on the freeway, or a miscalculation at a chemical plant. Even in prehistoric times, long before the advent of technology, a club-wielding caveman might have swung at his prey and inadvertently toppled his friend in what can only be classified as an "accident."

As man progressed, so did the severity of his misfortunes. The "Modern Era" has brought about assembly lines, chemical manufacturers, nuclear power plants, and so on, all carrying the capability of disaster. To keep pace with the changing times, safety precautions must constantly be upgraded. It is no longer sufficient, as with the caveman, to shout the warning, "Watch out with that thing!" Today's problems require more elaborate systems of warnings and controls to minimize the chances of serious accidents.

Industrial accidents occur in many ways—a chemical spill, an explosion, a nuclear power plant melt down, and so on. There are often problems in transport, with trucks overturning, trains derailing, or ships capsizing. There are "acts of God," such as earthquakes and storms. The one common thread through all of these situations is that they are rarely expected and frequently mismanaged.

Most industrial process plants are safe to be around. Plant management, aided by reliable operators, who are in turn backed up by still-more-reliable automatic controls, does its best to keep operations moving along within the limits usually considered reasonably safe to man and machine. Occasionally, however, there is a whoosh or

Fluid Flow for the Practicing Chemical Engineer. By J. Patrick Abulencia and Louis Theodore
Copyright © 2009 John Wiley & Sons, Inc.

a bang that is invariably to the detriment of the operation, endangering investment and human life, and rudely upsetting the plant's loss expectancy.[1]

Accidents have occurred since the birth of civilization. Anyone who crosses a street, rides in a car, or swims in a pool runs the risk of injury through carelessness, poor judgment, ignorance, or other circumstances. This has not changed throughout history. In the following pages, a number of accidents and disasters that took place before the advances of modern technology will be examined.

29.2 LEGISLATION

The concern for emergency planning and response is reflected in the legislation[2-4] summarized in this Section. Although the Clean Air Act does not cover emergency planning and response in a clear and comprehensive manner, certain elements of the act are particularly significant. These include implementation plans and national emission standards for hazardous air pollutants. The Clean Water Act as well as other legislation pertaining to water pollution provides emergency planning and response that is more developed than it is for air. The Resource Conservation and Recovery Act (RCRA) and the Comprehensive Environmental Response, Compensation and Liability Act (CERCLA) are two important pieces of legislation that are concerned with preventing releases, and with the requirements for the cleanup of hazardous and toxic sites. RCRA and CERCLA contain specific sections that address emergency planning and response. The Superfund Amendments and Reauthorization Act (SARA) is another important piece of legislation. SARA deals with the cleanup of hazardous waste sites as well as emergency planning and response. Title III, which is the heart of SARA, establishes requirements for emergency planning and "community right to know" for federal, state and local government, as well as industry. Title III is a major stepping-stone in the protection of the environment, but its principal thrust is to facilitate planning in the event of a catastrophe.

Three other important topics as they relate to the subject of this chapter include the US Environmental Protection Agency's (USEPA's) Risk Management Program, the Occupational Heath and Safety Administration (OSHA), and potential environmental violations.

29.2.1 Comprehensive Environmental Response, Compensation, and Liability Act (CERCLA)

The Comprehensive Environmental Response, Compensation, and Liability Act (CERCLA) of 1980 was the first major response to the problem of abandoned hazardous waste sites throughout the nation. CERCLA was the beginning of the remediation of hazardous waste sites. This program was designed to:

1. Develop a comprehensive program to set priorities for cleaning up the worst existing hazardous waste sites.
2. Make responsible parties pay for these cleanups wherever possible.

3. Set up (initially) a $1.6 billion *Hazardous Waste Trust Fund*, properly known as the "Superfund," for the twofold purpose of performing remedial cleanups when responsible parties could not be held accountable and responding to emergency situations involving hazardous substances.
4. Advance scientific and technological capabilities in all aspects of hazardous waste management, treatment, and disposal.

CERCLA requires the person in charge of a process or facility to notify the *National Response Center* (NRC) immediately when there is a release of a designated hazardous substance in an amount equal to or greater than the reportable quantity. CERCLA establishes the reportable quantity for releases of designated hazardous substances at one pound, unless otherwise specified. Such releases require notification to government officials to ensure that the need for response can be evaluated and any response can be undertaken in a timely fashion.

The development of the emergency planning and response actions under CERCLA is based primarily on a national contingency plan that was developed under the *Clean Water Act*. Although the actions of CERCLA have the capabilities to handle hazardous and toxic releases, the act was primarily directed toward the cleanup of abandoned hazardous waste sites.

Under Section 7003 of the RCRA legislation (1984), private citizens are authorized to bring legal action against companies, governmental entities, or individual citizens if past or present hazardous waste management practices are believed to pose an imminent danger. Section 7003 applies to past generators as well as to situations or sites where past acts or failures to act may have contributed to a present endangerment to human health and the environment. Citizen rights to sue are limited, however: (1) if USEPA or the state government is diligently bringing and prosecuting a related action under Section 7003 of RCRA or Section 106 of CERCLA, or (2) if USEPA or the state has settled a related action by entering into a consent decree. CERCLA was amended by the *Superfund Amendments and Reauthorization Act* (SARA) in 1986.

29.2.2 Superfund Amendments and Reauthorization Act of 1986 (SARA)

The *Superfund Amendments and Reauthorization Act of 1986* renewed the national commitment to correcting problems arising from previous mismanagement of hazardous wastes. While SARA was similar in many respects to the original law (i.e., CERCLA), it also contained new approaches to the program's operation. The 1986 Superfund legislation:[5]

1. Reauthorized the program for 5 more years and increased the size of the cleanup fund from $1.6 billion to $8.5 billion.
2. Set specific cleanup goals and standards, and stressed the achievement of permanent remedies.

3. Expanded the involvement of states and citizens in decision making.

4. Provided for new enforcement authorities and responsibilities.

5. Increased the focus on human health problems caused by hazardous waste sites.

The new law is more specific than the original statute with regard to remedies to be used at Superfund sites, public participation, and accomplishment of cleanup activities. The most important part of SARA with respect to public participation is Title III, which addresses the important issues of community awareness and participation in the event of a chemical release.

As mentioned earlier, Title III of SARA addresses hazardous materials release; its subtitle is the *Emergency Planning and Community Right-to-Know Act of 1986*. Title III establishes requirements for emergency planning, hazardous emissions reporting, emergency notification, and "community right-to-know." The objectives of Title III are to improve local chemical emergency response capabilities, primarily through improved emergency planning and notification, and to provide citizens and local governments with access to information about chemicals in their localities. The major sections of Title III that aid in the development of contingency plans are as follows:

1. Emergency Planning (Sections 301–303).

2. Emergency Notification (Section 304).

3. Community Right To Know Reporting Requirements (Sections 311 and 312).

4. Toxic Chemicals Release Reporting—Emissions Inventory (Section 313).

Title III also developed time frames for the implementation of the Emergency Planning and Community Right-to-Know Act of 1986.

Sections 301–303 of Title III, which are responsible for emergency planning, are designed to develop state and local governments' emergency response and preparedness capabilities through better coordination and planning, especially within local communities.

29.3 HEALTH RISK ASSESSMENT[5–7]

There are many definitions for the word risk. It is a combination of uncertainty and damage; a ratio of hazards to safeguards; a triplet combination of event, probability, and consequences; or even a measure of economic loss or human injury in terms of both the incident likelihood and the magnitude of the loss or injury. People face all kinds of risks everyday, some voluntarily and others involuntarily. Therefore, risk plays a very important role in today's world. Studies on cancer caused a turning point in the world of risk because it opened the eyes of risk scientists and health professionals to the world of risk assessments.

Since 1970, the field of risk assessment has received widespread attention within both the scientific and regulatory committees. It has also attracted the attention of the public. Properly conducted risk assessments have received fairly broad acceptance, in part because they put into perspective the terms toxic, hazard, and risk. Toxicity is an inherent property of all substances. It states that all chemical and physical agents can produce adverse health effects at some dose or under specific exposure conditions. In contrast, exposure to a chemical that has the capacity to produce a particular type of adverse effect, represents a hazard. Risk, however, is the probability or likelihood that an adverse outcome will occur in a person or a group that is exposed to a particular concentration or dose of the hazardous agent. Therefore, risk is generally a function of exposure or dose. Consequently, health risk assessment is defined as the process or procedure used to estimate the likelihood that humans or ecological systems will be adversely affected by a chemical or physical agent under a specific set of conditions.[8]

The term risk assessment is not only used to describe the likelihood of an adverse response to a chemical or physical agent, but it has also been used to describe the likelihood of any unwanted event. This subject is treated in more detail in the next section. These include risks such as: explosions or injuries in the workplace; natural catastrophes; injury or death due to various voluntary activities such as skiing, sky-diving, flying, and bungee jumping; diseases; death due to natural causes; and many others.[9]

Risk assessment and risk management are two different processes, but they are intertwined. Risk assessment and risk management give a framework not only for setting regulatory priorities, but also for making decisions that cut across different environmental areas. Risk management refers to a decision-making process that involves such considerations as risk assessment, technology feasibility, economic information about costs and benefits, statutory requirements, public concerns, and other factors. Therefore, risk assessment supports risk management in that the choices on whether and how much to control future exposure to the suspected hazards may be determined.

Regarding both risk assessment and risk management, this section will primarily address this subject from a health perspective; the next section will primarily address this subject from a safety and accident perspective.

The reader should note that two general types of potential health risk exist. These are classified as:

1. *Acute.* Exposures that occur for relatively short periods of time, generally from minutes to one or two days. Concentrations of (toxic) air contaminants are usually high relative to their protection criteria. In addition to inhalation, airborne substances might directly contact the skin, or liquids and sludges may be splashed on the skin or into the eyes, leading to adverse health effects. This subject area falls, in a general sense, in the domain of hazard risk assessment (HZRA).

2. *Chronic.* Continuous exposure occurring over long periods of time, generally several months to years. Concentrations of inhaled (toxic) contaminants are usually relatively low. This subject area falls in the general domain of health

risk assessment (HRA) and it is this subject that is addressed in this section. Thus, in contrast to the acute (short-term) exposures that predominate in hazard risk assessment, chronic (long-term) exposures are the major concern in health risk assessments.

Health risk assessments provide an orderly, explicit, and consistent way to deal with scientific issues in evaluating whether a hazard exists and what the magnitude of the hazard may be. This evaluation typically involves large uncertainties because the available scientific data are limited, and the mechanisms for adverse health impacts or environmental damage are only imperfectly understood. When one examines risk, how does one decide how safe is safe, or how clean is clean? To begin with, one has to look at both sides of the risk equation—that is, both the toxicity of a pollutant and the extent of public exposure. Information is required at both the current and potential exposure, considering all possible exposure pathways. In addition to human health risks, one needs to look at potential ecological or other environmental effects. In conducting a comprehensive risk assessment, one should remember that there are always uncertainties, and these assumptions must be included in the analysis.

29.3.1 Risk Evaluation Process for Health

In recent years, several guidelines and handbooks have been produced to help explain approaches for doing health risk assessments. As discussed by a special National Academy of Sciences committee convened in 1983, most human or environmental health hazards can be evaluated by dissecting the analysis into four parts: hazard identification, dose-response assessment or hazard assessment, exposure assessment, and risk characterization (see Fig. 29.1). For some perceived hazards, the risk assessment might stop with the first step, hazard identification, if no adverse effect is identified or if an agency elects to take regulatory action without further analysis.[8] Regarding hazard identification, a hazard is defined as a toxic agent or a set of conditions that has the potential to cause adverse effects to human health or the environment. Hazard identification involves an evaluation of various forms of information in order to identify the different hazards. Dose-response or toxicity assessment is required in an overall assessment: responses/effects can vary widely since all chemicals and contaminants vary in their capacity to cause adverse effects. This step frequently requires that assumptions be made to relate experimental data for animals and humans. Exposure assessment is the determination of the magnitude, frequency, duration, and routes of exposure of human populations and ecosystems. Finally, in risk characterization, toxicology and exposure data/information are combined to obtain a qualitative or quantitative expression of risk.

Risk assessment involves the integration of the information and analysis associated with the above four steps to provide a complete characterization of the nature and magnitude of risk and the degree of confidence associated with this characterization. A critical component of the assessment is a full elucidation of the uncertainties associated with each of the major steps. Under this broad concept of risk assessment

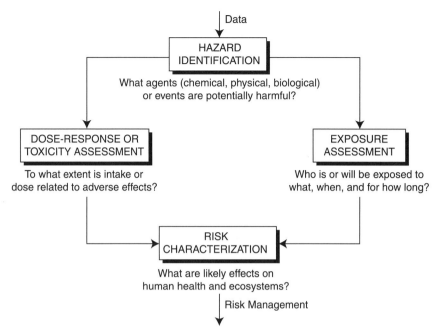

Figure 29.1   The health risk evaluation process.

are encompassed all of the essential problems of toxicology. Risk assessment takes into account all of the available dose-response data. It should treat uncertainty not by the application of arbitrary safety factors, but by stating them in quantitatively and qualitatively explicit terms, so that they are not hidden from decision-makers. Risk assessment, defined in this broad way, forces an assessor to confront all the scientific uncertainties and to set forth in explicit terms the means used in specific cases to deal with these uncertainties.[10]

29.4 HAZARD RISK ASSESSMENT[4-7]

Risk evaluation of accidents serves a dual purpose. It estimates the probability that an accident will occur and also assesses the severity of the consequences of an accident. Consequences may include damage to the surrounding environment, financial loss, or injury to life. This section is primarily concerned with the methods used to identify hazards and the causes and consequences of accidents. Issues dealing with health risks have been explored in the previous chapter. Risk assessment of accidents provides an effective way to help ensure either that a mishap does not occur or that the likelihood of an accident is reduced. The result of the risk assessment allows concerned parties to take precautions to prevent an accident before it happens.

Regarding definitions, the first thing an individual needs to know is what exactly is an accident. An accident is an unexpected event that has undesirable consequences. The causes of accidents have to be identified in order to help prevent accidents from occurring. Any situation or characteristic of a system, plant, or process that has the potential to cause damage to life, property, or the environment is considered a hazard. A hazard can also be defined as any characteristic that has the potential to cause an accident. The severity of a hazard plays a large part in the potential amount of damage a hazard can cause if it occurs. Risk is the probability that human injury, damage to property, damage to the environment, or financial loss will occur. An acceptable risk is a risk whose probability is unlikely to occur during the lifetime of the plant or process. An acceptable risk can also be defined as an accident that has a high probability of occurring, with negligible consequences. Risks can be ranked qualitatively in categories of high, medium, and low. Risk can also be ranked quantitatively as an annual number of fatalities per million affected individuals. This is normally denoted as a number times one millionth, for example, 3×10^{-6}. This number indicates that on average three workers will die every year out of one million individuals. Another quantitative approach that has become popular in industry is the Fatal Accident Rate (FAR) concept. This determines or estimates the number of fatalities over the lifetime of 1000 workers. The lifetime of a worker is defined as 10^5 hours, which is based on a 40-hour work week for 50 years. A reasonable FAR for a chemical plant is 3.0 with 4.0 usually taken as a maximum. A FAR of 3.0 means that there are 3 deaths for every 1000 workers over a 50-year period. Interestingly, the FAR for an individual at home is approximately 3.0. Some of the Illustrative Examples in Section 29.5 compliment many of the concepts described below with technical calculations and elaborations.

29.4.1 Risk Evaluation Process for Accidents

As with Health Risk Assessment (HRA), there are four key steps involved in a Hazardous Risk Assessment (HZRA). These are presented below in Fig. 29.2.

A more detailed flowchart is presented in Fig. 29.3, if the system in question is a chemical plant. These steps are detailed below:

1. A brief description of the equipment and chemicals used in the plant is needed
2. Any hazard in the system has to be identified. Hazards that may occur in a chemical plant include:
   a. Fire.
   b. Toxic vapor release.
   c. Slippage.
   d. Corrosion.
   e. Explosions.
   f. Rupture of pressurized vessel.
   g. Runaway reactions.

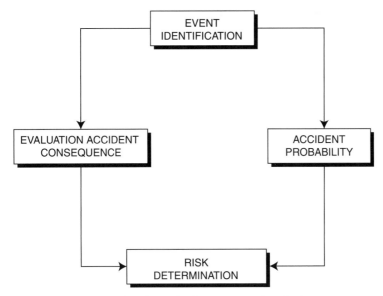

Figure 29.2 Hazard risk assessment flowchart.

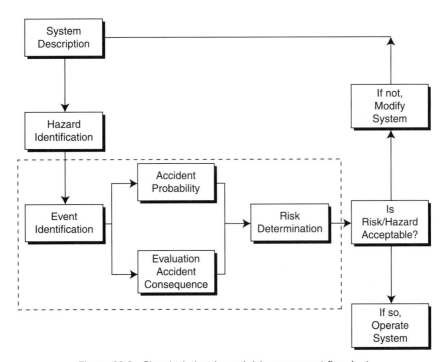

Figure 29.3 Chemical plant hazard risk assessment flowchart.

3. The event or series of events that will initiate an accident has to be identified. An event could be a failure to follow correct safety procedures, improperly repaired equipment, or a safety mechanism.

4. The probability that the accident will occur has to be determined. For example, if a chemical plant has a given life, what is the probability that a pump will fail? The probability can be ranked from low to high. A low probability means that it is unlikely for the event to occur in the life of the plant. A medium probability suggests that there is a possibility that the event will occur. A high probability means that the event will probably occur during the life of the plant.

5. The severity of the consequences of the accident must be determined.

6. The information from (4) and (5) are combined. If the probability of the accident and the severity of its consequences are low, then the risk is usually deemed acceptable and the plant should be allowed to operate. If the probability of occurrence is too high or the damage to the surroundings is too great, then the risk is usually unacceptable and the system needs to be modified to minimize these effects.

The heart of the hazard risk assessment algorithm provided is enclosed in the dashed box of Fig. 29.3. The algorithm allows for reevaluation of the process if the risk is deemed unacceptable (the process is repeated starting with either step one or two).

As evident in the lessons from past accidents, it is essential for industry to abide by stringent safety procedures. The more knowledgeable the personnel, from the management to the operators of a plant, and the more information that is available to them, the less likely a serious incident will occur. The new regulations, and especially Title III of 1986, help to ensure that safety practices are up to standard. However, these regulations should only provide a minimum standard. It should be up to the companies, and specifically the plants, to see that every possible measure is taken to ensure the safety and well-being of the community and the environment in the surrounding area. It is also up to the community itself, under Title III, to be aware of what goes on inside local industry, and to prepare for any problems that might arise.

29.5 ILLUSTRATIVE EXAMPLES

The remainder of this chapter is devoted to Illustrative Examples, many of which contain technical development material. A good number of applications have been drawn from National Science Foundation (NSF) literature,[11-16] and two other sources.[6,17]

Illustrative Example 29.1 Consider the release of a toxic gas from a storage tank. List and discuss possible causes for the release.

Solution   Some possible causes for a toxic gas release from a storage tank are:

1. Rupture in storage tank.
2. Fire in tank farm; explosion of storage tank.
3. Collapse of tank due to earthquake.
4. Rupture in main line.
5. Leak in line or from tank.

Illustrative Example 29.2   Consider the probability distribution of the number of defectives in a sample of five pumps drawn with replacement from a lot of 1000 pumps, 50 of which are defective. Associate "success" with drawing a defective pump from the lot. Then the result of each drawing can be classified as success (defective pump) or failure (non-defective pump). The sample of pumps is drawn with replacement (i.e., each item in the sample is returned before the next is drawn from the lot; therefore, the probability of success remains constant at 0.05). Calculate the probability that the sample contains exactly three defective pumps.

Solution   The probability distribution of x, the number of successes in n performances of the random experiment, is the binomial distribution, with probability distribution function (pdf) specified by[16]

$$P(x) = \frac{n!}{x!(n! - x!)} p^x q^{n-x}; \quad x = 0, 1, \ldots, n$$

where $P(x)$ is the probability of x successes in n performances.
Substituting the values $n = 5$, $p = 0.05$, and $q = 0.95$ into the pdf,

$$P(x) = \frac{n!}{x!(n! - x!)} p^x q^{n-x} = \frac{5!}{x!(5! - x!)} (0.05)^x (0.95)^{5-x}$$

Therefore,

$$P(x = 3) = \frac{5!}{3!(5! - 3!)} (0.05)^3 (0.95)^{5-3} = 0.11\%$$

Illustrative Example 29.3   An iron foundry has four work stations that are connected to a single duct. In order to reduce the possibility of an accident, each work station has a hood that transports 3000 acfm of air flow. The duct length is 400 feet and the pressure loss at the hood entrance is 0.5 in H_2O. There also is a cyclone air cleaner that creates 3.5 in H_2O pressure drop. Determine the diameter of the duct to ensure adequate transport of the dust. Also determine the power required for a combined blower/motor efficiency of 40%.

Solution   Determine the minimum air velocity required for general foundry dust

$$v_{air} = 4000\,\text{ft/min}$$
$$= 66.67\,\text{ft/s}$$

Calculate the total air flow required in acfm.

$$q_{air} = (3000)(4)$$
$$= 12,000\,\text{acfm}$$

Calculate required cross-sectional area in ft².

$$A = q_{air}/v_{air}$$
$$= (12,000)/(4000)$$
$$= 3\,\text{ft}^2$$

Calculate the duct diameter.

$$D = (4A/\pi)^{1/2}$$
$$= [(4)(3)/3.14]^{1/2}$$
$$= 1.9544\,\text{ft}$$
$$= 24\,\text{in}$$

In order to determine power requirements, the pressure drop across the system needs to be calculated first. Calculate the Reynolds number for the above duct.

$$\text{Re} = Dv\rho/\mu$$
$$= (1.9544)(66.67)(0.075)/(1.21 \times 10^{-5})$$
$$= 8.08 \times 10^5$$

The pressure drop in the duct in lb_f/ft^2 is then (for $f = 0.003$ since Re > 20,000).

$$\Delta P_{duct} = 4\,fLv^2\rho/2g_cD$$
$$= (4)(0.003)(400)(66.67)^2(0.075)/[(2)(32.2)(1.9544)]$$
$$= 12.7\,lb_f/ft^2$$

Calculate the total system pressure drop.

$$\Delta P_{tot} = \Delta P_{duct} + \Delta P_{hood} + \Delta P_{cyc}$$
$$= 12.7 + (0.5)(5.2\,lb_f/ft.^2\text{-in } H_2O) + (3.5)(5.2\,lb_f/ft.^2\text{-in } H_2O)$$
$$= 33.5\,lb_f/ft^2$$

Finally, calculate the power required in hp.

$$hp = 3.03 \times 10^{-5}\,\Delta P_{tot}q_{air}/\eta$$
$$= 3.03 \times 10^{-5}(33.5)(12,000)/0.4$$
$$= 30.5\,hp$$

Illustrative Example 29.4 Discuss the HAZOP (Hazard and Operability) procedure.

Solution Specific details regarding this procedure are available in the literature.[4,17] The overall HAZOP method, however, is summarized in the following steps:

1. Define objective(s).
2. Define plant limits.
3. Appoint and train a team.
4. Obtain complete preparative work.
5. Conduct examination meetings in order to:
   a. Select a manageable portion of the process.
   b. Review the flowsheet and operating instructions.
   c. Agree on how the process is intended to operate.
   d. State and record the intention.
   e. Search for possible ways to deviate from the intention, utilizing the HAZOP "guide" words.
   f. Determine possible causes for the deviation.
   g. Determine possible consequences of the deviation.
   h. Recommend action(s) to be taken.
6. Issue meeting report.
7. Follow up on recommendations.

After the serious hazards have been identified with a HAZOP study or some other type of qualitative approach, a quantitative examination should be performed. Hazard quantification or hazard analysis (HAZAN) involves the estimation of the expected frequencies or probabilities of events with adverse or potentially adverse consequences. It logically ties together historical occurrences, experience, and imagination. To analyze the sequence of events that lead to an accident or failure, event and fault trees are used to represent the possible failure sequences.

Illustrative Example 29.5 Consider a water pumping system consisting of two pumps (A and B), where A is the pump ordinarily operating and B is a standby unit that automatically takes over if A fails. A control valve in both cases regulates flow of water through the pump. Suppose that the top event is no water flow, resulting from the following basic events: failure of pump A and failure of pump B, or failure of control valve. Prepare a fault tree diagram for this system.[4,17]

Solution Generally, a fault tree may be viewed as a diagram that shows the path that a specific accident takes. Fault tree analysis (FTA) begins with the ultimate consequence and works backward to the possible causes and failures. It is based on the most likely or most credible events that lead to the accident. FTA demonstrates the mitigating or reducing effects, and can include causes stemming from human error as well as equipment failure. The task of constructing a fault tree is tedious and requires a probability background to handle common mode failures, dependent events, and time constraints.

Fault tree analysis seeks to relate the occurrence of an undesired event, the "top event," to one or more antecedent events, called "basic events." The top event may be, and usually is, related to the basic events via certain intermediate events. A fault tree diagram exhibits the casual chain linking the basic events to the intermediate events and the latter to the top event. In this chain, the logical connection between events is indicated by so-called "logic gates." The principal logic gates are the AND gate, symbolized on the fault tree by AND, and the OR gate symbolized by OR.

The fault tree symbols and their definitions are presented in Table 29.1. The construction of the fault tree for a tank overflow example is demonstrated in Fig. 29.4.

Details on event trees are available in the literature.[16,17]

Illustrative Example 29.6 A baghouse has been used to clean a particulate gas stream for nearly 30 years. There are 600, 8-inch diameter bags in the unit. 50,000 acfm of dirty gas at 250°F enters the baghouse with a loading of 5.0 grains/ft^3. The outlet loading is 0.3 grains/ft^3. Local EPA regulations state that the maximum allowable outlet loading is 0.4 grains/ft^3. If the system operates at a pressure drop of 6 in. H$_2$O, how many bags can fail before the unit is out of compliance? The Theodore–Reynolds equation (see below) applies and all the contaminated gas emitted through the broken bags may be assumed the same as that passing through the tube sheet thimble.

Table 29.1 Fault tree symbols[a]

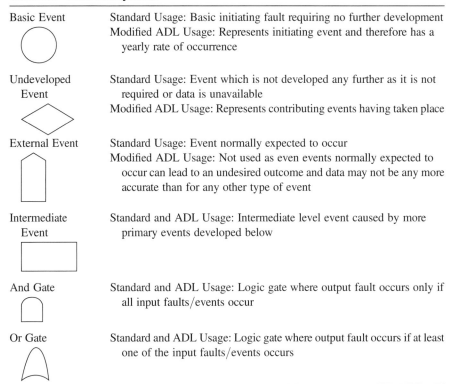

Basic Event	Standard Usage: Basic initiating fault requiring no further development Modified ADL Usage: Represents initiating event and therefore has a yearly rate of occurrence
Undeveloped Event	Standard Usage: Event which is not developed any further as it is not required or data is unavailable Modified ADL Usage: Represents contributing events having taken place
External Event	Standard Usage: Event normally expected to occur Modified ADL Usage: Not used as even events normally expected to occur can lead to an undesired outcome and data may not be any more accurate than for any other type of event
Intermediate Event	Standard and ADL Usage: Intermediate level event caused by more primary events developed below
And Gate	Standard and ADL Usage: Logic gate where output fault occurs only if all input faults/events occur
Or Gate	Standard and ADL Usage: Logic gate where output fault occurs if at least one of the input faults/events occurs

[a]ADL = Alternate Digital Logic.

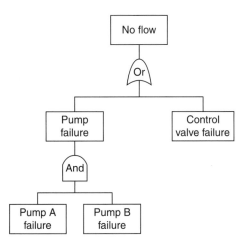

Figure 29.4 Fault tree diagram for a water pumping system consisting of two pumps (A and B).

The effect of bag failure on baghouse fractional penetration (or efficiency) can be described by the following equation:

$$P_t^* = P_t + P_{tc}$$

$$P_{tc} = \frac{0.582(\Delta P)^{1/2}}{\phi}$$

$$\phi = \frac{q}{[LD^2(T + 460)^{1/2}]}$$

where P_t^* = penetration after bag failure
P_t = penetration before bag failure
P_{tc} = penetration correction term; contribution of broken bags to P_t^*
ΔP = pressure drop, in. H_2O
ϕ = dimensionless parameter
q = volumetric flow rate of contaminated gas, acfm
L = number of broken bags
D = bag diameter, inches
T = temperature, °F

Note: $P = 1 - E$; E = fractional efficiency.
A detailed development of the above equation is provided in the literature.[18]

Solution   Calculate the efficiency, E, and penetration, P, before bag failure(s):

$$E = (\text{inlet loading} - \text{outlet loading})/(\text{inlet loading})$$
$$= (5.0 - 0.03)/(5.0)$$
$$= 0.9940 = 99.40\%$$

$$P_t = 1 - 0.9940$$
$$= 0.0060 = 0.60\%$$

Calculate the efficiency and penetration, P_t^*, based on regulatory conditions:

$$E = (5.0 - 0.4)/(5.0)$$
$$= 0.9200 = 92.00\%$$

$$P_t^* = 1 - 0.9200$$
$$= 0.0800 = 8.00\%$$

Calculate the penetration term, P_{tc}, associated with the failed bags:

$$P_{tc} = 0.0800 - 0.0060$$
$$= 0.0740$$

Write the equation(s) for P_{tc} in terms of the number of failed bags, L:

$$P_{tc} = \frac{0.582(\Delta P)^{1/2}}{\phi}$$

where

$$\phi = \frac{q}{\left[LD^2(T+460)^{1/2}\right]}$$

Calculate the number of bag failures that the system can tolerate and still remain in compliance:

$$L = \frac{qP_{tc}}{(0.582)(\Delta P)^{0.5}(D)^2(T+460)^{0.5}}$$
$$= (50{,}000)(0.074)/(0.582)(6)^{0.5}(8)^2(250+460)^{0.5}$$
$$= 1.52$$

Thus, if two bags fail, the baghouse is out of compliance.

Illustrative Example 29.7   A reactor is located in a relatively large laboratory with a volume of 1100 m³ at 22°C and 1 atm. The reactor can emit as much as 0.75 gmol of hydrocarbon (HC) into the room if a safety valve ruptures. A hydrocarbon mole fraction in the air greater than 425 parts per billion (ppb) constitutes a health and safety hazard.

Suppose the reactor valve ruptures and the maximum amount of HC is released instantaneously. Assume the air flow in the room is sufficient to cause the room to behave as a continuously stirred tank reactor (CSTR), i.e., the air composition is spatially uniform. Calculate the ppb of hydrocarbon in the room. Is there a health risk? From a treatment point-of-view, what can be done to decrease the environmental hazard or to improve the safety of the reactor?

Solution   Calculate the total number of gmols of air in the room, n_{air}. Assuming that air is an ideal gas, 1 gmol of air occupies 22.4 liters (0.0224 m³) at standard temperature and pressure (273K, 1 atm). Since the room temperature is not 273K,

$$n_{air} = (1100\,\text{m}^3)\left(\frac{1\,\text{gmol}}{0.0224\,\text{STP m}^3}\right)\left(\frac{273\text{K}}{295\text{K}}\right)$$
$$= 45{,}445\,\text{gmol}$$

Note: STP m³ indicates the volume (in m³) that the gas would have at standard temperature and pressure.

The mole fraction of hydrocarbon in the room, x_{HC}, is

$$x_{HC} = \frac{0.75 \, \text{gmol HC}}{45{,}445 \, \text{gmol air} + 0.75 \, \text{gmol HC}} = 16.5 \, \text{ppm} = 16{,}500 \, \text{ppb}$$

Since $16{,}500 \, \text{ppb} \gg 850 \, \text{ppb}$, the hazard presents a significant health risk.

To implement safety measures, the potential rupture area should be vented directly into a hood or a duct to capture any leakage in the event of a rupture. Another alternative is input substitution, a source reduction measure. Input substitution is the replacement of the material in the reactor with material with a lower vapor pressure.

REFERENCES

1. H. deHeer, "Calculating How Much Safety is Enough", Chemical Engineering, New York, February 19, 1973.

2. L. Theodore, M. Hyland, Y. McGuinn, L. Schoen, and F. Taylor, USEPA Manual, "Principles of Accident and Emergency Management", Air Pollution Training Institute, RTP, NC, 1988.

3. Personal notes:A. M. Flynn and L. Theodore, 1999.

4. A. M. Flynn and L. Theodore, "Health, Safety and Accident Management in the Chemical Process Industries", Marcel Dekker, New York (acquired by Taylor & Francis, Boca Raton, FL), 2002.

5. G. Burke, B. Singh, and L. Theodore, "Handbook of Environmental Management and Technology", 2nd edition, John Wiley & Sons, Hoboken, NJ, 2001.

6. L. Stander and L. Theodore, "Environmental Regulatory Calculations Handbook," John Wiley & Sons, Hoboken, NJ, 2008.

7. M. K. Theodore and L. Theodore, "Major Environmental Issues Facing the 21st Century," 2nd edition, CRC Press, Boca Raton, FL, 2009.

8. D. Paustenbach, "The Risk Assessment of Environmental and Human Health Hazards: A Textbook of Case Studies", John Wiley & Sons, Hoboken, NJ, 1989.

9. Manual of Industrial Hazard Assessment Techniques, Office of Environmental and Scientific Affairs, The World Bank, October, 1985, London.

10. J. Rodricks and R. Tardiff, "Assessment and Management of Chemical Risks", ACS, Washington DC, 1984.

11. J. Reynolds, R. Dupont, and L. Theodore, "Hazardous Waste Incineration Calculations: Problems and Software", John Wiley & Sons, Hoboken, NJ, 1991.

12. R. Dupont, L. Theodore, and J. Reynolds, "Accident and Emergency Management: Problems and Solutions", VCH Publishers, New York, 1991.

13. L. Theodore, R. Dupont, and J. Reynolds, "Pollution Prevention: Problems and Solutions", Gordon and Breach Publishers, Amsterdam, Holland, 1994.

14. K. Ganeson, L. Theodore, and J. Reynolds, "Air Toxics—Problems and Solutions", Gordon and Breach Publishers, Amsterdam, Holland, 1996.

15. R. Dupont, T. Baxter, and L. Theodore, "Environmental Management: Problems and Solutions", CRC Press, Boca Raton, FL, 1998.

16. S. Shaefer and L. Theodore, "Probability and Statistics for Environmental Scientists", CRC Press, Boca Raton, FL, 2007.

17. J. Reynolds, J. Jeris, and L. Theodore, "Handbook of Chemical and Environmental Engineering Calculations", John Wiley & Sons, Hoboken, NJ, 2002.

18. L. Theodore and J. Reynolds, "Effect of Baghouse Failure and Baghouse Outlet Loading", *J. Air Poll. Control.*, 29, 870–872, 1979.

NOTE: Additional problems are available for all readers at www.wiley.com. Follow links for this title.

30

ETHICS

30.1 INTRODUCTION

This chapter is concerned with ethics. The topics reviewed include:

Teaching Ethics
Case Study Approach
Integrity
Moral Issues
Guardianship
Engineering and Environmental Ethics

The remainder of the chapter consists of two additional sections:

Applications
References

The Applications section primarily addresses topics concerned (in part) with fluid flow issues.

The reader should note that the bulk of the material in this chapter has been drawn from the work of Wilcox and Theodore.[1]

Fluid Flow for the Practicing Chemical Engineer. By J. Patrick Abulencia and Louis Theodore
Copyright © 2009 John Wiley & Sons, Inc.

30.2 TEACHING ETHICS

Professionals are often skeptical about the value or practicality of discussing ethics in the workplace. When students hear that they are required to take an ethics course or if they opt for one as an elective in their schedules, they frequently wonder whether ethics can be taught. They share the skepticism of the practitioners about such discussion. Of course, both groups are usually thinking of ethics as instruction in goodness, and they are rightly skeptical, given their own wealth of experience with or knowledge of moral problems. They have seen enough already to know that you cannot change a person's way of doing things simply by teaching about correct behavior.

The teaching of ethics is not a challenge if ethics is understood *only* as a philosophical system. Parks notes that teaching ethics is important, but "if we are concerned with the teaching of ethics is understood as the practice of accountability to a profession vital to the common good, the underlying and more profound challenge before all professional schools [and other organizations] is located in the question, How do we foster the formation of leadership characterized, in part, by practice of moral courage?"[2]

Moral courage requires knowing *and* acting. College and university educators, as well as those charged with ethics training in the private sector, develop a sense of uneasiness when topics such as "fostering leadership formation," "moral courage," or "knowing *and* acting on that knowledge" are mentioned. Such terms resurrect images of theological indoctrination, Sunday school recitations, or pulpit sermonizing. These images contrast sharply with what the present-day professor envisions as the groves of academic freedom and dispassionate analysis. Perhaps out of fear of disrespecting the dignity of students and devaluing their critical reasoning powers or their ability to understand where the truth lies, faculty will take a dim view of academic goals that go beyond those strictly cognitive. The consequence of such values among the professoriate is the further erosion of a moral commons where an agreed-upon set of values and beliefs allows for discourse on ethics. Of course, the erosion has continued steadily from the inception of the Enlightenment Project in the seventeenth century until the present day wherever industrialized and postindustrialized societies have been subject to rapid cultural, economic, political, and technological change. It is not simply an erosion in the realm of higher education. Practitioners in the engineering and scientific communities experience the same erosion of the moral commons taking place in society as a whole.

The authors are certainly in agreement with their colleagues in higher education and those who do ethics training in the private sector, that individuals are not to be manipulated or indoctrinated. However, they are also convinced that not only should students and other participants in ethics analysis have a body of knowledge, but that they have a responsibility for the civic life of society. Such responsibility requires leadership, moral courage, and action. Of course, none of these characteristics can be demanded or forced, only elicited. That is the great, yet delicate challenge facing the professoriate and all those charged with ethics training in other sectors. Eliciting a sense of civic responsibility as a goal of ethics analysis can be realized only as a derivative of cognitive processes and not as a direct goal. In sum, the formation of personal character and the practice of virtue are not to be subject to external control and the diminution of individual freedom through manipulation or indoctrination.

30.3 CASE STUDY APPROACH

The authors believe that the case study method is a valuable way to take seriously Parks's response to the question "Can ethics be taught?" He also considers the method to be an important tool in investigating the relationship between assumptions, values, and moral life, as well as an ethical reflection on those three aspects of life. The authors are convinced that the case study method is one of the most useful ways of teaching ethics and of achieving the goals of ethics education outlined by the Hastings Center.[3]

1. *Stimulating the Moral Imagination.* The concreteness of the case study appeals very much to the learning style of most people. While a certain amount of ambiguity is essential to evoke interest and discussion, it is also a stimulus for enlivening knowledge. Hopefully, the participant will begin to appreciate the moral complexity of a situation, which in the past might have been thought of only as a technical or managerial problem. Practice in the art of case discussion has the larger intent of leading the individual to bring an ethical frame of reference to bear on the variety of problems faced in the discipline studied. Stimulating the moral imagination is similar to putting on a pair of glasses that are tinted. The result is that the whole world is seen through that tint. As a consequence of the case study method, the authors of the cases hope that each individual will see his or her field of study through the interpretive glasses of engineering and environmental ethics. He or she would then routinely ask: "What is the moral issue here?"

2. *Recognizing Ethical Issues.* The case analyst should not be content with a good "imagination." The further challenge is the recognition of specific moral problems and how they differ from one another in terms of immediacy or urgency. Concreteness is an important asset of the case study and clearly assists in achieving this second goal. Comparing and contrasting a variety of cases through discussion is essential to recognition and leads to achievement of the next goal.

3. *Developing Analytical Skills.* Differentiation, comparison, contrasts—all of these must be related to an enhanced ability to solve the problem. To achieve this goal, the student of ethics is taught to bring the skills developed in his or her major field of study to bear on the ambiguous situation, the moral dilemma, or the competing values that must be addressed. Analytic skills are best honed through the use of examples or cases. The technical ability to analyze all dimensions of an environmental spill will have an impact on how the moral aspect of the problem is understood in terms of resolving the problem. Of course, ethical systems that emphasize the importance of consequences, the obligations inherent in a duty-based ethic, as well as theories of

justice or virtue will enhance the ability to use technical or discipline-based analytic skills in resolving the problem. Knowing, however, is related to acting. This leads to the fourth goal.

4. *Eliciting a Sense of Moral Obligation and Personal Responsibility.* Much has already been said about the importance of this goal. However, it should be clear that a sense of moral obligation does not mean that there is one set of absolute answers. Dictating a solution is quite different from an internalization process whereby the individual commits himself or herself to be a "seeker," one who takes personal responsibility for addressing and resolving the moral problems facing engineers or scientists. Both professions constitute the "guardians of the system" in the technical community. They are the first line of response to the problems and dilemmas facing the professions as such. To point to the Environmental Protection Agency, the Occupational Safety and Health Administration, the Federal Bureau of Investigation, congressional formulators of public policy, or other sovereign countries as the parties responsible for resolving acute problems is to abnegate one's moral responsibility as a professional person. To say this is not to dictate solutions, but to alert individuals to their personal responsibility for the integrity of the respective field. Eliciting a sense of responsibility depends on an assessment of the assumptions or "images at the core of one's heart." Assessment of ethical systems or normative frames of reference must be connected to the actual assumptions or images that constitute a person's worldview. Challenging the individual to examine that worldview in relation to a case and ethical systems is the first step in joining doing to knowing. Closely related to the achievement of this goal is the following one.

5. *Tolerating—and Resisting—Disagreement and Ambiguity.* An essential component of case discussion is the willingness to listen carefully to the points-of-view held by others. Cases, by their nature, are ambiguous. They are bare-boned affairs meant more to be provocative than to lead to a clear-cut jury decision. The purpose of the case is to stimulate discussion and learning among individuals. As a result, there will be much disagreement surrounding the ethical issues in the case and the best option for resolving it. Toleration does not mean "putting up with people with whom I disagree." Respect for the inherent dignity of the person and a willingness to understand not only another position but also a person's reasons for or interest in that point-of-view, should be part of the case discussion. Toleration does not mean all opinions must be of equal value and worth. It is true that respect for and listening to another person's argument may lead one to change a position. However, a careful description and discussion of the other person's position may also lead to a greater conviction that one's own position is correct. What is clearly of central concern is the belief that the free flow of ideas and carefully wrought arguments, presented from all sides without fear of control, manipulation, threat, or disdain, is at the core of human understanding and development. This hallowed concept of academic freedom is the catalyst, that allows human communities to be committed to the search for truth, without at the same time declaring absolute possession of the truth.

30.4 INTEGRITY

Scenarios are, for the most part, designed to reflect ambiguity in work situations. The ethicist hopes to get his or her hands dirty, dealing with the bottom-line motives of survival, competitiveness, and profitability as well as the mixed motives of self-interest, respect for the rights of others, and altruism. Obtaining an ethical solution to a difficult moral problem or dilemma is based on much more than choosing the correct ethical framework with its normative frame of reference. One must also be ready to examine fundamental assumptions and the values to which the assumptions give rise. Carter has made this point recently in a discussion of "integrity."[4]

1. *Honesty in Relation to Integrity.* Carter explores integrity in relation to the value society places on honesty. On this subject, one of the best-known and most popular ethics books of the last few decades is Sissela Bok's *Lying: Moral Choice in Public and Private Life* (Bok, 1978). Without taking away from the merits of *Lying*, Carter notes: "Plainly, one cannot have integrity without being honest (although, as we shall see, the matter gets complicated), but one can certainly be honest and yet have little integrity." Honesty is far easier to practice than the tough work of figuring out what it takes to have integrity in a situation. Integrity requires a high degree of moral reflectiveness. Honesty may result in harm to another person. Furthermore, "if forthrightness is not preceded by discernment, it may result in the expression of an incorrect moral judgment".[4] The racist may be transparently honest, Carter declares, but he certainly lacks integrity because his beliefs, deeply held as they might be, are wrong. He has not engaged in the hard work of examining his fundamental assumptions, values, and beliefs.

2. *Personal Integrity Without Public Responsibility?* It would appear that one cannot have integrity without responsibility, since any consideration of integrity addresses the effects of ones conduct on other people. In work life and community life, individuals have public responsibilities for their clients and fellow citizens. That is the nature of public life. It demands civic virtue of all. In this light, consider an example supplied by Carter:

Having been taught all his life that women are not as smart as men, a manager gives the women on his staff less-challenging assignments than he gives the men. He does this, he believes, for their own benefit: he does not want them to fail, and he believes that they will if he gives them tougher assignments. Moreover, when one of the women on his staff does poor work, he does not berate her as harshly as he would a man, because he expects nothing more. And he claims to be acting with integrity because he is acting according to his own deepest beliefs.

The manager has the most basic test of integrity. The question is not whether his actions are consistent with what he most deeply believes but whether he has done the hard work of discerning whether what he most deeply believes is right. The manager has not taken this harder step.

> Moreover, even within the universe that the manager has constructed for himself, he is not acting with integrity. Although he is obviously wrong to think that the women on his staff are not as good as the men, even were he right, that would not justify applying different standards to their work. By so doing he betrays both his obligation to the institution that employs him and his duty as a manager to evaluate his employees.[4]

Carter's reasoning concerning the hard work leading to integrity must be applied to the cases in this text. Answers to problems or dilemmas are not easily arrived at and require a willingness to examine one's fundamental assumptions, values, and beliefs. The theme of integrity plays itself out in a somewhat different fashion in Stark's provocative essay, "What's the Matter with Business Ethics?"[5] He notes that managers are not getting the needed help from business ethicists in addressing two types of ethical challenges:

> first, identifying ethical courses of action in difficult gray-area situations and second, navigating those situations where the right course is clear, but real-world competitive and institutional pressures lead even well-intentioned managers astray.[5]

Much as Carter faults those who opt for the easy road of "honesty," Stark faults business, and by extension engineering or environmental ethicists, for offering "a kind of ethical absolutism that avoids many of the difficult [and most interesting] questions." Such absolutism devalues the bottom-line interest and marketplace success. Ethicists of the absolutist persuasion would rather see the corporation sink than compromise idealism. Stark takes as the starting point the existence of the corporation and managers who "still lack solutions for the basic problem of how to balance ethical demands and economic realities when they do in fact conflict".[5] The litmus test for all applied ethics, then, is whether it is an ethics of practice, a "dirty-hands ethics." A practitioner of such ethics "must help managers do the arduous, conceptual balancing required in difficult cases where every alternative has both moral and financial costs".[5] Furthermore, these ethicists must address the complexity of personal motivations. "The fact is, most people's motives are a confusing mix of self-interest, altruism, and other influences".[5]

The new business ethic—and, by extension, engineering ethics—may be identified in the following way:

> Moderation, pragmatism, minimalism: these are new words for business ethicists. In each of these new approaches, what is important is ... the commitment to converse with real managers in a language relevant to the world they inhabit and the problems they face. That is an understanding of business ethics worthy of managers' attention.[5]

30.5 MORAL ISSUES[6]

The conflict of interest between Chief Seattle (and Native Americans in general) and President Pierce (and the European–American expansion) provides a perfect example of how ethics and the resulting codes of behavior they engender can differ drastically from culture to culture, from religion to religion, and even from person to person. This enigma, too, is noted again and again by Seattle[7]:

> I do not know. Our ways are different from your ways . . . . But perhaps it is because the red man is a savage and does not understand . . . . The air is precious to the red man, for all things share the same breath . . . the white man does not seem to notice the air he breathes . . . . I am a savage and do not understand any other way. I have seen a thousand rotting buffaloes on the prairie, left by the white man who shot them from a passing train. I am a savage and I do not understand how the smoking iron horse can be more important than the buffalo we kill only to stay alive.

Chief Seattle sarcastically uses the European word "savage" and all its connotations throughout his address. When one finishes reading the work it becomes obvious which viewpoint (President Pierce's or his own) Chief Seattle feels is the savage one. What his culture holds dearest (the wilderness) the whites see as untamed, dangerous, and savage. What the whites hold in highest regard (utilization of the Earth and technological advancement) the Native Americans see as irreverent of all other living things. Each culture maintains a distinct and conflicting standard for the welfare of the world. Opposing viewpoints and moralities such as these are prevalent throughout the world and have never ceased to present a challenge to international, national, state, community, and interpersonal peace.

It is generally accepted, however, that any historical ethic can be found to focus on one of four different underlying moral concepts:

1. *Utilitarianism* focuses on good consequences for all.
2. *Duties Ethics* focus on one's duties.
3. *Rights Ethics* focus on human rights.
4. *Virtue Ethics* focus on virtuous behavior.

Utilitarians hold that the most basic reason why actions are morally right is that they lead to the greatest good for the greatest number. "Good and bad consequences are the only relevant considerations, and, hence all moral principles reduce to one: 'We ought to maximize utility' ".[7]

Duties Ethicists concentrate on an action itself rather than the consequences of that action. To these ethicists there are certain principles of duty such as "Do not deceive" and "Protect innocent life" that should be fulfilled even if the most good does not result. The list and hierarchy of duties differs from culture to culture, religion to religion. For Judeo-Christians, the Ten Commandments provide an ordered list of duties imposed by their religion.[7]

Often considered to be linked with Duties Ethics, Rights Ethics also assesses the act itself rather than its consequences. Rights Ethicists emphasize the rights of the people affected by an act rather than the duty of the person(s) performing the act. For example, because a person has a *right* to life, murder is morally wrong. Rights Ethicists propose that duties actually stem from a corresponding right. Since each person has a *right* to life, it is everyone's *duty* to not kill. It is because of this link and their common emphasis on the actions themselves that Rights Ethics and Duty Ethics are often grouped under the common heading Deontological Ethics.[8]

The display of virtuous behavior is the central principle governing Virtue Ethics. An action would be wrong if it expressed or developed vices—for example, bad character traits. *Virtue Ethicists*, therefore, focus upon becoming a morally good person.

To display the different ways that these moral theories view the same situation, one can explore their approach to the following scenario that Martin and Schinzinger present:[7]

> On a midnight shift, a botched solution of sodium cyanide, a reactant in organic synthesis, is temporarily stored in drums for reprocessing. Two weeks later, the day shift foreperson cannot find the drums. Roy, the plant manager, finds out that the batch has been illegally dumped into the sanitary sewer. He severely disciplines the night shift foreperson. Upon making discreet inquiries, he finds out that no apparent harm has resulted from the dumping. Should Roy inform government authorities, as is required by law in this kind of situation?

If a representative of each of the four different theories on ethics just mentioned were presented with this dilemma, their decision-making process would focus on different principles.

The Utilitarian Roy would assess the consequences of his options. If he told the government, his company might suffer immediately under any fines administered and later (perhaps more seriously) due to exposure of the incident by the media. If he chose not to inform authorities, he risks heavier fines (and perhaps even worse press) in the event that someone discovers the cover-up. Consequences are the utilitarian Roy's only consideration in his decision-making process.

The Duties Ethicist Roy would weigh his duties and his decision would probably be more clear-cut than his utilitarian counterpart. He is obliged foremost by his duty to obey the law and must inform the government.

The Rights Ethicist mindframe would lead Roy to the same course of action as the duties ethicist—not necessarily because he has a duty to obey the law but because the people in the community have the right to informed consent. Even though Roy's inquiries informed him that no harm resulted from the spill, he knows that the public around the plant have the right to be informed of how the plant is operating.

Vices and virtues would be weighed by the Virtue Ethicist Roy. The course of his thought process would be determined by his own subjective definition of what things are virtuous, what things would make him a morally good person. Most likely, he would consider both honesty and obeying the law virtuous, and withholding information from the government and public as virtueless and would therefore, tell the authorities.

30.6 GUARDIANSHIP

Despite the great teaching advantage that comes with case use, there are two important questions that case discussants must keep in mind when they assess the ethical problem:

1. *Who are the Guardians of the System?* This question addresses the issue of who, among engineering or science professionals, is responsible for the ethical standards in the organization. If professionals point the finger at senior management, the legal department, the Environmental Protection Agency, or the Department of Justice, they have indeed misunderstood the nature of a professional calling. The first line of defense is the willingness of professionals themselves to maintain and enhance the integrity of the engineering or scientific profession through their own personal adherence to the highest standards of conduct and to assume responsibility for commitment to these standards within the companies where they work. Moreover, ethics is a positive task, not a list of dos and don'ts. To achieve excellence in one's work presumes a commitment to the client's contract, public safety, and environmental integrity, among several factors that are all too often thought of as "management" issues. They are, in reality, the ethical standards of the work itself. Thus, the ethical engineer or scientist is the one who identifies with the profession and all that is involved in the work assigned or contracted.

2. *Who Gives Support to the Guardians of the System?* This second issue goes to the heart of the assessment problem, but also has an impact on the first issue. Unless the organization backs those who assume positive responsibility for the ethical tenor of the group, very little will change. Why would someone risk ostracism or retaliation by confronting a person engaging in unethical behavior or illegal behavior if there is no institutional support for the one assuming responsibility?

Effective guardianship is facilitated if:

1. There are clear-cut standards of behavior and high expectations of the membership.
2. The standards are brought to the attention of the members through a well-developed training program.
3. The standards are taken seriously by the senior leadership team of the firm. They must demonstrate that seriousness by taking an active role in the training, without, at the same time, creating a chilly climate stifling discussion and participation in the training. The ethics program must be seen not as frosting on the cake or as a value added on to forestall legal problems through better compliance. The CEO needs to demonstrate a commitment to the values and principles that drive the business. Ethics training is no add-on. Ethics is what *drives* the organization: trust, integrity, fidelity to the client.

4. It is evident that the leadership "walks the talk" in all aspects of its decision making and actions.

5. There are mechanisms in place to address the concerns of the members, mechanisms such as an ombudsperson, a hotline.

6. Those who adversely affect the integrity of the business are effectively and fairly disciplined.

Another way of addressing the question of who supports the guardians is to emphasize the importance of organizational or corporate culture. A positive response to the six points just raised has a great impact on the culture of the organization. Unless there is what is sometimes called a "thick" culture, wherein respect for and adherence to guardianship and the tenets of integrity and trust are palpable, individually ethical persons can do very little to raise the moral climate. An organization is more than the sum total of the individuals who constitute the membership. The attitudes conveyed, values expressed, and ways of doing business in an organization profoundly affect the perceptions of the members therein that is set the tone of the company. Having a positive impact on culture is a great challenge that is not easily achieved. Culture is so subtle that one often does not even realize or understand its dimensions until a significantly different culture is experienced.

An organization will not have effective guardianship of the system unless there is a concerted attempt to create, enhance, or reinforce a culture where values and ethics are clear and fully supported. There is little doubt, however, that the twin issues of guardianship and culture are much more difficult to address than the institutionalization of the ethics program itself.

30.7 ENGINEERING AND ENVIRONMENTAL ETHICS[9]

In the ethical theories presented here, established hierarchies of duties, rights, virtues, and desired consequences exist so that situations where no single course of action satisfies all of the maxims can still be resolved. The entry of environmentalism into the realm of ethics raises questions concerning where it falls in this hierarchy. Much debate continues over these questions of how much weight the natural environment should be given in ethical dilemmas, particularly in those where ecological responsibility seems to oppose economic profitability and technological advances. Those wrapped up in this technology/economy/ecology debate can generally be divided into three groups:

1. Environmental extremists.

2. Technologists to whom ecology is acceptable provided it does not inhibit technological or economic growth.

3. Those who feel technology should be checked with ecological responsibility.

Each is briefly discussed below.

After his year-and-a-half of simple living on the shores of Walden Pond, Henry David Thoreau rejected the pursuit of technology and industrialization. While most would agree with his vision of nature as being inspirational, few would choose his way of life. Even so, the movement rejecting technological advances in favor of simple, sustainable, and self-sufficient living is being embraced by more and more people who see technology as nothing but a threat to the purity and balance of nature. Often called environmental extremists by other groups, they even disregard "environmental" technologies that attempt to correct pollution and irresponsibilities, past and present. They see all technology as manipulative and uncontrollable and choose to separate themselves from it. To them, the environment is at the top of the heirarchy.

On the other extreme are the pure technologists. They view the natural world as a thing to be subdued and manipulated in the interest of progress—technological and economic. This is not to say one will not find technologists wandering in a national park admiring the scenery. They do not necessarily deny the beauty of the natural environment, but they see themselves as separate from it. They believe that technology is the key to freedom, liberation, and a higher standard of living. It is viewed, therefore, as inherently good. They see the environmental extremists as unreasonable and hold that even the undeniably negative side effects of certain technologies are best handled by more technological advance. The technologists place environmental responsibility at the bottom of their ethical heirarchy.

Somewhere in the middle of the road travels the third group. While they reap the benefits of technology, they are concerned much more deeply than the technologists with the environmental costs associated with industrialization. It is in this group that most environmental engineers find themselves. They are unlike the environmental extremists since, as engineers, they inherently study and design technological devices and have faith in the ability of such devices to have a positive effect on the condition of the environment. They also differ from the technologists. They scrutinize the effects of technologies much more closely and critically. While they may see a brief, dilute leak of a barely toxic chemical as an unacceptable side effect of the production of a consumer product, the technologists may have to observe destruction—the magnitude of that caused by Chernobyl—before they consider rethinking a technology they view as economically and socially beneficial. In general, this group sees the good in technology but stresses that it cannot be reaped if technological growth goes on unchecked.

The ethical behavior of engineers is more important today than at any time in the history of the profession. The engineers' ability to direct and control the technologies they master has never been stronger. In the wrong hands, the scientific advances and technologies of today's engineer could become the worst form of corruption, manipulation, and exploitation. Engineers, however, *are* bound by a code of ethics that carry certain obligations associated with the profession. Some of these obligations include:

1. Support ones professional society.
2. Guard privileged information.
3. Accept responsibility for one's actions.

4. Employ proper use of authority.
5. Maintain one's expertise in a state-of-the-art world.
6. Build and maintain public confidence.
7. Avoid improper gift exchange.
8. Practice conservation of resources and pollution prevention.
9. Avoid conflict of interest.
10. Apply equal opportunity employment.
11. Practice health, safety, and accident prevention.
12. Maintain honesty in dealing with employers and clients.

There are many codes of ethics that have appeared in the literature. The preamble for one of these codes is provided below:[7]

Engineers in general, in the pursuit of their profession, affect the quality of life for all people in our society. Therefore, an engineer, in humility and with the need for divine guidance, shall participate in none but honest enterprises. When needed, skill and knowledge shall be given without reservation for the public good. In the perform-ance of duty and in fidelity to the profession, engineers shall give utmost.

30.8   APPLICATIONS

The three Illustrative Examples below have been drawn from the work of Wilcox and Theodore,[1] keying primarily on fluid flow issues. Each example is presented in case-study format, containing both a fact pattern and finally, questions for discussion. These questions are by no means definitive. While they will help individuals focus on the case, the issues raised will make the most sense if they lead to a wide-ranging dis-cussion among all readers. Analysis of ethics cases comes alive in group work. Answering the questions individually is a helpful first step, but one's understanding of ethical problems and dilemmas improves dramatically in group discussion.

Illustrative Example 30.1   Fact Pattern   Michael's company is currently in the process of designing a chemical plant. Michael has been given the job of designing the emergency pressure relief system for one of the plant's reactors, which operates at high pressure as well as at high temperature in order to achieve a high single-pass conversion. The design requires that two high-pressure valves be used to vent the gases in the reactor should the pressure exceed the upper design limits.

The engineering company contracts UP, a company that markets safety valves. The valve company has had some problems in the past with their line of high-pressure valves, but they assure Michael that their valves have been tested and passed. Michael realizes that if a reactor were to proceed uncontrolled and if the pressure relief valve did not function properly, the result would be disastrous. But redesigning the system to use several lower-pressure valves would push back the completion date of the plant as well as cost the company more in terms of capital and maintenance costs.

Michael decides to use the high-pressure system. A week after the plant is started, the reactor pressure exceeds the upper limits. The valves fail to open, and the resulting explosion kills a man. After an official investigation, it is determined that the explosion was due to operator error, and no company benefits will be paid to the victim's family.

Questions for Discussion

1. What are the facts in this case?
2. What is the ethical problem facing Michael?
3. Do you think Michael or the valve company is to blame?
4. If there was no accident, would the morality of the decision change?

Illustrative Example 30.2 Fact Pattern Laura is an engineer working in a chemical plant. She has recently received a job offer from another company, which she accepts because she knows that the new job could be a big step in her career.

Laura is responsible for one of the production lines in the plant she will soon be leaving. She has always been a reliable worker and an effective manager. However, having handed in her letter of resignation, she has been less attentive to her work over the past couple of weeks. She figures that there is no need to worry about this job anymore; she has to concentrate on her future.

On Laura's next-to-last day of work at the plant, Harry, a coworker on the same production line, finds out that there is a problem with the purity of the product: The level of impurities is a little higher than acceptable. Harry decides to consult Laura.

He says, "The product coming out is below the required purity. I think you should investigate it so we can solve this problem."

Laura replies, "I would love to help you, Harry, but tomorrow is my last day here. I don't want to start dealing with this problem; it could take a while to solve. Let my replacement worry about it."

Harry answers, "Laura, if we let this problem go, we'll continue to have a product that doesn't meet regulation. The problem could also get *worse*. You are the expert here, so you could easily fix this mess."

"Harry, you're a friend of mine. Please don't ask me to get involved in this problem; it's not my concern anymore. I just want to relax during my last two days at work," pleads Laura. "It's not like the plant will blow up. Wait for two days. You can pretend that you didn't notice anything until then."

Reluctantly, Harry agrees. "I know you're really looking forward to your new job. It's just that I'll feel guilty knowing that something is wrong, and I'm not doing anything about it. But I guess I can wait for two days."

"Harry, don't worry. Take it easy for a couple of days. Just think of it as a minor delay," replies Laura.

Questions for Discussion

1. What are the facts in this case?
2. Do you think Laura should stay focused on her current job?

3. Should Laura handle the problem?

4. Do you think it's okay for Harry to ignore the problem for the next two days?

5. Should Harry consult someone else now that Laura has refused to deal with the problem?

Illustrative Example 30.3   Fact Pattern   Tom is preparing for his final exam in fluid flow, a course that he has been struggling with all semester long. He desperately needs to pass this exam because as it stands he has a D in the course; not passing this final exam means he will have to repeat the class.

One night Tom sees a janitor that he is acquainted with, and they begin to talk about his course difficulties. The janitor, whose name is Mike, likes Tom and thinks he's a good kid. Mike offers to help Tom with his fluid flow exam.

The student, bewildered as to how the janitor can help, asks, "Did you take fluids while you were in school, Mike?"

"No, Tom, I went only as far as high school, but I know for a fact that your professor keeps all of his exams on his desk; I saw them when I was cleaning his office one night."

Mike says he can let Tom into his professor's office to look around for the final exam. Tom is excited about the idea of getting it ahead of time and feels a sense of relief. On the other hand, Tom realizes that if he gets caught, he can get thrown out of school, and that would not be good. He has two options: to study hard for the exam (and possibly still not do well) or obtain the exam (at the risk of getting caught). Which is the best alternative?

Questions for Discussion

1. What are the facts in this case?

2. What is the ethical problem with what Tom is doing?

3. Do you think Tom should try to get the exam ahead of time?

REFERENCES

1. J. Wilcox and L. Theodore, "Engineering and Environmental Ethics: A Case Study Approach," John Wiley & Sons, Hoboken, NJ, 1998.

2. S. Parks—Professional Ethics, Moral Courage, and the Limits of Personal Virtue, "Can Virtue be Taught," edited by B. Darling-Smith, University of Notre Dame Press, 1993.

3. "The Teaching of Ethics in Higher Education," The Hastings Center, Hastings on the Hudson, NY, 1980.

4. S. Carter, "The Insufficiency of Honesty," *Atlanta Monthly*, 1996.

5. A. Stark, "What's the Matter with Business Ethics?" *Harvard Business Review*, 38–48, May–June 1993.

6. M. K. Theodore and L. Theodore, "Introduction to Environmental Management," 2nd edition, CRC Press, Boca Raton, FL, 2009.

7. M. Martin and R. Schinzinger, "Ethics in Engineering," McGraw-Hill, New York, 1989.

8. I. Barbour, "Ethics in an Age of Technology," Harper, San Francisco, 1993.

9. G. Burke, B. Singh, and L. Theodore, "Handbook of Environmental Management and Technology," 2nd edition, John Wiley & Sons, Hoboken, NJ, 2000.

NOTE: Additional problems are available for all readers at www.wiley.com. Follow links for this title.

31

NUMERICAL METHODS

31.1 INTRODUCTION

Chapter 31 is concerned with Numerical Methods. This subject was taught in the past as a means of providing engineers with ways to solve complicated mathematical expressions that they could not solve otherwise. However, with the advent of computers, these solutions have become readily obtainable.

A brief overview of Numerical Methods is given to provide the practicing engineer with some insight into what many of the currently used software packages (MathCad, Mathematica, MatLab, etc.) are actually doing. The authors have not attempted to cover all the topics of Numerical Methods. There are several excellent texts in the literature that deal with this subject matter in more detail.[1,2]

Ordinarily, discussion of the following eight numerical methods would be included in this chapter:

1. Simultaneous linear algebraic equations.
2. Nonlinear algebraic equations.
3. Numerical integration.
4. Numerical differentiation.
5. Ordinary differential equations.
6. Partial differential equations.
7. Regression analysis.
8. Optimization.

Fluid Flow for the Practicing Chemical Engineer. By J. Patrick Abulencia and Louis Theodore
Copyright © 2009 John Wiley & Sons, Inc.

However, because of the breadth of the subject matter, the reader should note that only three numerical methods receive treatment in the chapter. They are the first three topics listed above. The remaining five methods are to be found in the literature.[3,4] It should be noted that the problems section contains fluid flow material dealing with all eight subject topics with solutions available for those who adopt the text for classroom/training purposes.

31.2 EARLY HISTORY

Early in one's career, the engineer/scientist learns how to use equations and mathematical methods to obtain exact answers to a large range of relatively simple problems. Unfortunately, these techniques are often inadequate for solving real-world problems. The reader should note that one rarely needs exact answers in technical practice. Most real-world applications are usually inexact because they have been generated from data or parameters that are measured, and hence represent only approximations. What one is likely to require in a realistic situation is not an exact answer but rather one having reasonable accuracy from an engineering point of view.

The solution to an engineering or scientific problem usually requires an answer to an equation or equations, and the answer(s) may be approximate or exact. Obviously an exact answer is preferred but because of the complexity of some equations, often representing a system or process, exact solutions may not be attainable. For this condition, one may resort to another method that has come to be defined as a numerical method. Unlike the exact solution, which is continuous and in closed form, numerical methods provide an inexact (but reasonably accurate) solution. The numerical method leads to discrete answers that are almost always acceptable.

The numerical methods referred to above provide a step-by-step procedure that ultimately leads to an answer and a solution to a particular problem. The method usually requires a large number of calculations and is therefore ideally suited for digital computation.

High-speed computing equipment has had a tremendous impact on engineering design, scientific computation, and data processing. The ability of computers to handle large quantities of data and to perform mathematical operations at tremendous speeds permits the examination of many more cases and more engineering variables than could possibly be handled on the slide rule—the trademark of engineers of yesteryear. Scientific calculations previously estimated in lifetimes of computation time are currently generated in seconds and, in many instances, microseconds.[5]

A procedure-oriented language (POL) is a way of expressing commands to a computer in a form somewhat similar to such natural languages as English and mathematics. The instructions that make up a program written in a POL are called the source code. Because the computer understands only machine language or object code, a translator program must be run to translate the source code into an object code. In terms of input, processing, and output, the source code is the input to the translator program, which processes (translates) the code. The output is the object code. It is the object code that is actually executed in order to process data and information.

The first POL to be widely used was FORTRAN, an acronym that was coined from the words "FORmula TRANslation." FORTRAN was designed initially for use on problems of a mathematical nature and it is still used for solving some problems in mathematics, engineering, and science.

PASCAL is a POL designed by Niklaus Wirth in 1968. The motivation behind its design was to provide a language that encouraged the programmer to write programs according to the principles of structured programming. An important aspect of the PASCAL design philosophy is that it is a "small" language. The purpose of this is to provide the programmer with a language that can be easily learned and retained. PASCAL provides a variety of data structuring that enable programmers to easily define new data types. PASCAL is also most commonly used in mathematics, engineering and science.

BASIC is an acronym for Beginner's All-Purpose Symbolic Instruction Code. J. G. Kemeny and T. E. Kurtz developed BASIC in 1967 to give students a simple language for learning programming. BASIC is an interactive language, that is, the programmer sees an error or output as soon as it occurs. The simplicity of BASIC makes it easy to learn and use. Many versions of BASIC have been written since the late 1960s. BASIC can be used effectively for a variety of business and scientific applications.

Two types of translator programs—compilers and interpreters—are used to convert program statements to a machine-readable format. A compiler first translates the entire program to machine language. If any syntax or translation errors are encountered, a complete listing of each error and the incorrect statement is given to the programmer. After the programmer corrects the errors, the program is compiled again. When no errors are detected, the compiled code (object code) can be executed. The machine-language version can then be saved separately so that the compiling step need not be repeated each time the program is executed unless the original program is changed. Compiled programs run much faster than the interpreted ones. An interpreter translates and executes one source code instruction of a program at a time. Each time an instruction is executed, the interpreter uses the key words in the source code to call pre-written machine-language routines that perform the functions specified in the source code. The disadvantage of an interpreter is that the program must be translated each time it is executed.

Today, many powerful commercial mathematical applications are available and widely used in academia and industry. Some of these programs include MathCad, Matlab, Mathematica, etc. These user-friendly programs allow engineers and scientists to perform mathematical calculations without knowing any programming. In addition, new programs, for example, Visual Basic.NET, JAVA, C++, etc., are constantly evolving.

31.3 SIMULTANEOUS LINEAR ALGEBRAIC EQUATIONS

The engineer often encounters problems that not only contain more than two or three simultaneous algebraic equations but also those that are sometimes nonlinear as well. There is therefore, an obvious need for systematic methods of solving simultaneous linear and simultaneous nonlinear equations.[1] This section will address the linear sets of equations; information on nonlinear sets is available in the literature.[6]

Consider the following set of n equations:

$$a_{11}x_1 + a_{12}x_2 + \cdots + a_{1n}x_n = c_1$$
$$a_{21}x_1 + a_{22}x_2 + \cdots + a_{2n}x_n = c_2$$
$$\cdots \tag{31.1}$$
$$\cdots$$
$$a_{n1}x_1 + a_{n2}x_2 + \cdots + a_{nn}x_n = c_n$$

where a is the coefficient of the variable x and c is a constant. The above set is considered to be linear as long as none of the x-terms are nonlinear, for example, x_2^2 or $\ln x_1$. Thus, a linear system requires that all terms in x be linear.

A system of linear algebraic equations may be set in matrix form:

$$
\begin{bmatrix}
a_{11} & a_{12} & \cdots & a_{1n} \\
a_{21} & a_{22} & \cdots & a_{2n} \\
\cdots & \cdots & \cdots & \cdots \\
a_{n1} & a_{n2} & \cdots & a_{nn}
\end{bmatrix}
\begin{bmatrix}
x_1 \\
x_2 \\
\cdots \\
x_n
\end{bmatrix}
=
\begin{bmatrix}
c_1 \\
c_2 \\
\cdots \\
c_n
\end{bmatrix}
\tag{31.2}
$$

However, it is often more convenient to represent Equation (31.2) in the *augmented matrix* provided in Equation (31.3)

$$
\begin{bmatrix}
a_{11} & a_{12} & \cdots & a_{1n} & c_1 \\
a_{21} & a_{22} & \cdots & a_{2n} & c_2 \\
\cdots & \cdots & \cdots & \cdots & \cdots \\
a_{n1} & a_{n2} & \cdots & a_{nn} & c_n
\end{bmatrix}
\tag{31.3}
$$

Methods of solution available for solving these linear sets of equations include:

1. Gauss–Jordan reduction.
2. Gauss elimination.
3. Gauss–Seidel.
4. Cramer's rule.
5. Cholesky's methods.

Only the first three methods are discussed in this section.

31.3.1 Gauss–Jordan Reduction

Carnahan and Wilkes[1] solved the following two simultaneous equations using the Gauss–Jordan reduction method

$$3x_1 + 4x_2 = 29 \tag{31.4}$$
$$6x_1 + 10x_2 = 68 \tag{31.5}$$

The four step procedure is provided below.
Step 1. Divide Equation (31.4) through by the coefficient of x_1:

$$x_1 + \frac{4}{3}x_2 = \frac{29}{3} \tag{31.6}$$

Step 2. Subtract a suitable multiple (6, in this case) of Equation (31.6) from Equation (31.5), so that x_1 is "eliminated". Equation (31.6) remains intact so that what remains is:

$$x_1 + \frac{4}{3}x_2 = \frac{29}{3} \tag{31.7}$$
$$2x_2 = 10 \tag{31.8}$$

Step 3. Divide Equation (31.8) by the coefficient of x_2, that is, solve Equation (31.8).

$$x_2 = 5 \tag{31.9}$$

Step 4. Subtract a suitable factor of Equation (31.9) from Equation (31.7) so that x_2 is eliminated. When $(4/3)x_2 = 20/3$ is subtracted from Equation (31.7), one obtains

$$x_1 = 9 \tag{31.10}$$

31.3.2 Gauss Elimination

Gauss elimination is another method used to solve linear sets of equations. This method utilizes the augmented matrix described in Equation (31.3). The goal with Gauss elimination is to rearrange the augmented matrix into a "triangle form" where all the elements below the diagonal are zero. This is accomplished in much the same way as in Gauss–Jordan reduction. The procedure employed follows. Start with the first equation in the set. This is known as the pivot equation and will not change throughout the procedure. Once the matrix is in triangle form, back substitution can be used to solve for the variables. The Gauss elimination algorithm can be found in Figs. 31.1–31.3.

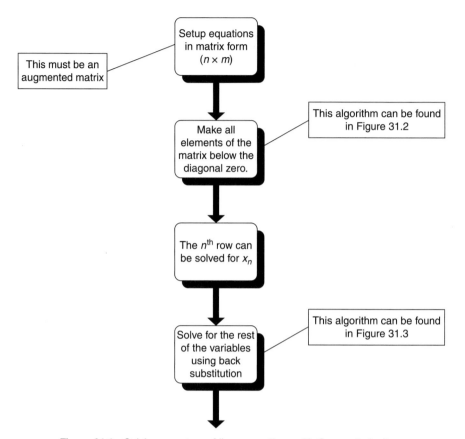

Figure 31.1 Solving a system of linear equations with Gauss elimination.

Illustrative Example 31.1 Solve the following set of linear algebraic equations using Gauss elimination.

$$3x_1 - 2x_2 + 1x_3 = 7$$
$$x_1 + 4x_2 - 2x_3 = 21$$
$$2x_1 - 3x_2 - 4x_3 = 9$$

Solution First, setup the augmented matrix

$$\begin{bmatrix} 3 & -2 & 1 & 7 \\ 1 & 4 & -2 & 21 \\ 2 & -3 & -4 & 9 \end{bmatrix}$$

In order to convert the first column of elements below the diagonal to be zero, start by setting the first row as the pivot row. For the remaining rows, multiply each element

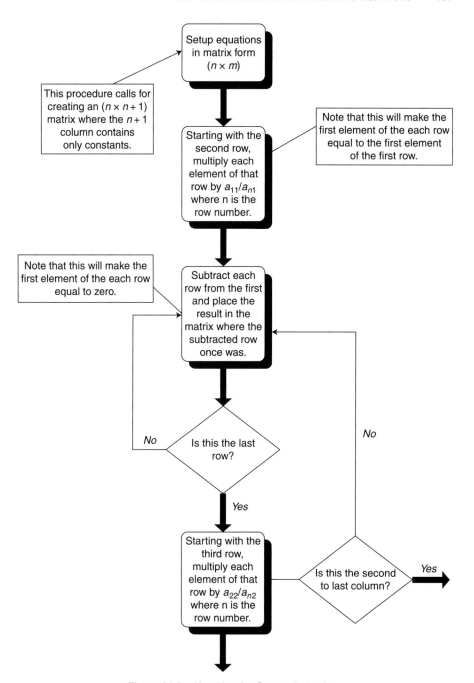

Figure 31.2 Algorithm for Gauss elimination.

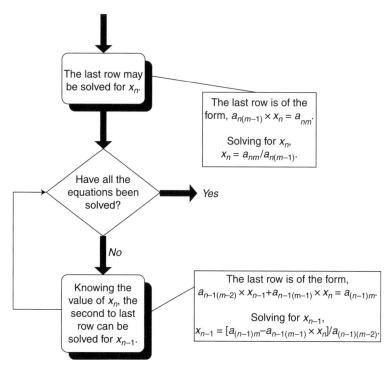

Figure 31.3 Algorithm for Gauss elimination back substitution.

by a_{11}/a_{n1} and subtract each row from the pivot row. Thus:

Row 2:

$$-(3)(1) + 3 = 0$$
$$-(3)(4) - 2 = -14$$
$$-(3)(-2) + 1 = 7$$
$$-(3)(21) + 7 = -56$$

Row 3:

$$-\frac{(3)(2)}{2} + 3 = 0$$
$$-\frac{(3)(-3)}{2} - 2 = 2.5$$
$$-\frac{(3)(-4)}{2} + 1 = 7$$
$$-\frac{(3)(9)}{2} + 7 = -6.5$$

The resulting matrix is:

$$\begin{bmatrix} 3 & -2 & 1 & 7 \\ 0 & -14 & 7 & -56 \\ 0 & 2.5 & 7 & -6.5 \end{bmatrix}$$

Now starting with the second column and the second row, perform the same procedure again, with the second row as the pivot row. Thus:

Row 3:

$$-\frac{(3)(2)}{2} + 3 = 0$$

$$-\frac{(3)(-3)}{2} - 2 = 2.5$$

$$-\frac{(3)(-4)}{2} + 1 = 7$$

$$-\frac{(3)(9)}{2} + 7 = -6.5$$

This produces the following matrix:

$$\begin{bmatrix} 3 & -2 & 1 & 7 \\ 0 & -14 & 7 & -56 \\ 0 & 0 & 46.2 & -92.4 \end{bmatrix}$$

At this point, back substitution may be employed. The following are the results:

$$46.2x_3 = -92.4; \quad x_3 = -2$$
$$-14x_2 + (-2)(7) = -56; \quad x_2 = 3$$
$$3x_1 + (-2)(3) + (1)(-2) = 7; \quad x_1 = 5$$

Gauss elimination is useful for systems that contain fewer than 30 equations. Systems larger than 30 equations become subject to round-off error where numbers are truncated by computers performing the calculations.

31.3.3 Gauss–Seidel

Another approach to solving an equation or series/sets of equations is to make an "informed" or "educated" guess. If the first assumed value(s) does not work, the value is updated. By carefully noting the influence of these guesses on each variable, these answers or correct set of values for a system of equations can be approached. The reader should note that when this type of iterative procedure is employed, a poor guess does not prevent the correct solution from ultimately being obtained.

Ketter and Prawel[2] provide the following example. Consider the equations below

$$4x_1 + 2x_2 + 0x_3 = +2$$
$$2x_1 + 10x_2 + 4x_3 = +6 \qquad (31.11)$$
$$0x_1 + 4x_2 + 5x_3 = +5$$

The reader may choose to assume, as a starting point, $x_1 = x_2 = x_3 = 0$. Solving each equation for the underlined terms (found on the diagonal), one obtains

$$x_1 = 0.50; \quad x_2 = 0.60; \quad x_3 = 1.00 \qquad (31.12)$$

Using these computed values, an updated set of xs can be obtained. Thus,

$$x_1 = \underline{x_1} - \frac{2}{4}x_2 + \frac{0}{4}x_3 = 0.50 - \frac{2}{4}(0.60) + \frac{0}{4}(1.00) - 0.50 = 0.30$$
$$x_2 = \underline{x_2} - \frac{2}{10}x_1 + \frac{4}{10}x_3 = 0.60 - \frac{2}{10}(0.50) + \frac{4}{10}(1.00) = 0.10 \qquad (31.13)$$
$$x_3 = \underline{x_3} - \frac{0}{5}x_1 + \frac{4}{5}x_2 = 1.00 - \frac{0}{5}(0.50) + \frac{4}{5}(0.60) = 0.52$$

The right-hand side of the equations may be viewed as residuals. The procedure is repeated until convergence (the "residuals" approach zero) is obtained. More rapid convergence techniques are available in the literature.[2]

31.4 NONLINEAR ALGEBRAIC EQUATIONS

The subject of the solution to a nonlinear algebraic equation is considered in this section. Although several algorithms are available, the presentation will key on the Newton–Raphson method of evaluating the root(s) of a nonlinear algebraic equation.

The solution to the equation

$$f(x) = 0 \qquad (31.14)$$

is obtained by guessing a value for x (x_{old}) that will satisfy the above equation. This value is continuously updated (x_{new}) using the equation

$$x_{new} = x_{old} - \frac{f(x_{old})}{f'(x_{new})} \qquad (31.15)$$

until either little or no change in ($x_{new} - x_{old}$) is obtained. One can express this operation graphically (see Fig. 31.4). Noting that

$$f'(x_{old}) = \frac{df(x)}{dx} \approx \frac{\Delta f(x)}{\Delta x} = \frac{f(x_{old}) - 0}{x_{old} - x_{new}} \qquad (31.16)$$

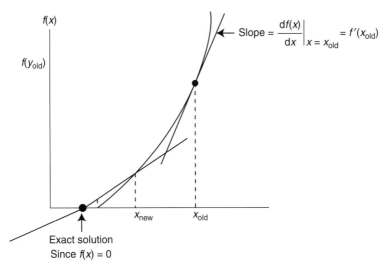

Figure 31.4   Newton–Raphson method.

one may rearrange Equation (31.16) to yield Equation (31.17). The x_{new} then becomes the x_{old} in the next calculation.

This method is also referred to as Newton's Method of Tangents and is a widely used method for improving a first approximation to a root to the aforementioned equation of the form $f(x) = 0$. The above development can be rewritten in subscripted form to (perhaps) better accommodate a computer calculation. Thus

$$f'(x_n) = \frac{f(x_n)}{x_n - x_{n+1}}$$ (31.17)

from which

$$x_{n+1} = x_n - \frac{f(x_n)}{f'(x_n)}$$ (31.18)

where x_{n+1} is again the improved estimate of x_n, the solution to the equation $f(x) = 0$. The value of the function and the value of the derivative of the function are determined at $x = x_n$, for this procedure, and the new approximation to the root, x_{n+1}, is obtained. The same procedure is repeated, with the new approximation, to obtain a still better approximation of the root. This continues until successive values for the approximate root differ by less than a prescribed small value, ε, which controls the allowable error (or tolerance) in the root. Relative to the previous estimate, ε is given by

$$\varepsilon = \frac{x_{n+1} - x_n}{x_n}$$ (31.19)

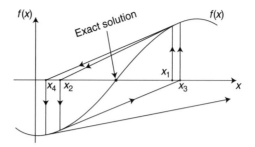

Figure 31.5   Failure of the Newton–Raphson method.

Despite its popularity, the method suffers for two reasons. First, an analytical expression for the derivative, that is, $f'(x_n)$ is required. In addition to the problem of having to compute an analytical derivative value at each iteration, one would expect Newton's method to converge fairly rapidly to a root in the majority of cases. However, as is common with most numerical methods, it may fail occasionally in certain instances. A possible initial oscillation followed by a displacement away from a root is illustrated in Fig. 31.5. Note, however, that the method would have converged (in this case) if the initial guess had been somewhat closer to the exact root. Thus, it can be seen that the initial guess may be critical to the success of the calculation.

Illustrative Example 31.2   The vapor pressure, p', for a new synthetic chemical at a given temperature has been determined to take the form:

$$p' = T^3 - 2T^2 + 2T; \ T = \text{K}, \ P' = \text{atm}$$

If $p' = 1$, one may then write

$$f(T) = T^3 - 2T^2 + 2T - 1 = 0$$

where the actual temperature, t (in K), is given by

$$t = 10^3 T$$

Solve the above equation for the actual temperature in K for $p' = 1$. Earlier studies indicate that t is in the $1000 - 1200\text{K}$ range.

Solution   Assume an initial temperature t_1. Set $t_1 = 1100$, so that

$$T_1 = (1100)(10^{-3}) = 1.1$$

Obtain the analytical derivative, $f'(T)$

$$f'(T) = 3T^2 - 4T + 2$$

Calculate $f(T_1)$ and $f'(T_1)$

$$f(1.1) = T^3 - 2T^2 + 2T - 1 = (1.1)^3 - 2(1.1)^2 + 2(1.1) - 1 = 0.111$$

$$f'(1.1) = 3T^2 - 4T + 2 = 3(1.1)^2 - 4(1.1) + 2 = 1.23$$

Use the Newton–Raphson method to estimate T_2. Employ Equation (31.15):

$$T_2 = T_1 - \frac{f(T_1)}{f'(T_1)}$$

Substituting,

$$T_2 = 1.1 - \frac{0.111}{1.23} = 1.0098$$

Calculate T_3.

$$f(T_2) = 0.0099$$
$$f'(T_2) = 1.0198$$
$$T_3 = 1.0001$$

Finally, calculate the best estimate (based on two iterations) of t.

$$t = 1000.1\text{K}$$

Other methods that may be employed include:

1. Wegstein's method.
2. False-position.
3. Half-interval.
4. Second-order Newton–Raphson.

Details are available in the literature.[2,6]

Illustrative Example 31.3   The friction factor for smooth tubes can be approximated by

$$f = 0.079\,\text{Re}^{-(1/4)}$$

if $2100 < \text{Re} < 2 \times 10^5$. It can be shown that the average velocity in the system shown in Fig. 31.6, involving the flow of water at $60°$F, is given by

$$v = \sqrt{\frac{2180}{213.5\,\text{Re}^{-(1/4)} + 10}}$$

For water at 60°F, Re $= 12,168v$. Calculate the average velocity, v (ft/s), using the Newton–Raphson method of solution.[7]

Solution   Substitute the expression of Reynolds number as a function of velocity into the velocity equation

$$v = \sqrt{\frac{2180}{213.5 \text{ Re}^{-(1/4)} + 10}} = \sqrt{\frac{2180}{213.5(12,168v)^{-(1/4)} + 10}}$$

$$= \sqrt{\frac{2180}{[213.5/(12,168v)^{(1/4)}] + 10}}$$

Manipulate the above equation into one that is easier to differentiate. Squaring both sides gives

$$v^2 = \frac{2180}{[213.5/(12,168v)^{(1/4)}] + 10} = \frac{2180(12,168v)^{(1/4)}}{213.5 + 10(12,168v)^{(1/4)}}$$

Cross-multiplying leads to

$$213.5v^2 + 10(12,168v)^{(1/4)}v^2 - 2180(12,168v)^{(1/4)} = 0$$

or

$$f(v) = 213.5v^2 + 105.03v^{2.25} - 22,896.08v^{0.25} = 0$$

The analytical derivative of $f(v)$ is

$$f'(v) = 427v + 236.313v^{1.25} - 5724.02v^{-0.75}$$

Make an initial guess of $v = 5$ ft/s and substitute into the above equations:

$$f_1(v) = f_1(5) = -24,973.8$$

Applying the initial guess,

$$f_1'(v) = f_1'(5) = 2189.97$$

Calculate the next guess using Equation (31.18).

$$v_2 = v_1 - \frac{f_1(v)}{f_1'(v)} = 5 - \frac{-24,973.8}{2189.97} = 16.40 \text{ ft/s}$$

Solve for v_3 etc., until the result converges

$$v_3 = 11.56\,\text{ft/s}$$
$$v_4 = 10.22\,\text{ft/s}$$
$$v_5 = 10.09\,\text{ft/s}$$
$$v_6 = 10.09\,\text{ft/s}$$

The average velocity is therefore $10.09\,\text{ft/s}$.

31.5 NUMERICAL INTEGRATION

Numerous engineering and science problems require the solution of integral equations. In a general sense, the problem is to evaluate the function

$$I = \int_a^b f(x)\,dx \tag{31.20}$$

where I is the value of the integral. There are two key methods employed in their solution: analytical and numerical. If $f(x)$ is a simple function, it may be integrated analytically. For example, if $f(x) = x^2$

$$I = \int_a^b x^2\,dx = \frac{1}{3}(b^3 - a^3) \tag{31.21}$$

If, however, $f(x)$ is a function too complex to integrate analytically (e.g., $\ln[\sinh(e^{x^2-2})]$), one may resort to any of the many numerical methods available. Two simple numerical integration methods that are commonly employed are the trapezoidal rule and the Simpon's rule. These are detailed below.

31.5.1 Trapezoidal Rule

In order to use the trapezoidal rule to evaluate the integral with I given by Equation (31.20) as

$$I = \int_a^b f(x)\,dx$$

use the equation

$$I = \frac{h}{2}[y_0 + 2y_1 + 2y_2 + \cdots + 2y_{n-1} + y_n] \tag{31.22}$$

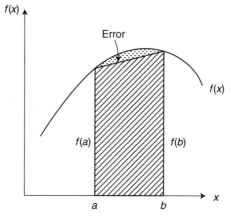

Figure 31.7 Trapezoidal rule error.

where h is the incremental change in x, i.e., Δx, and y_i are the values of $f(x)$ at x_i (i.e., $f(x_i)$). Thus,

$$y_0 = f(x_0) = f(x = a)$$
$$y_n = f(x_n) = f(x = b)$$
$$h = (b - a)/n$$

This method is known as the trapezoidal rule because it approximates the area under the function $f(x)$—which is generally curved—with a 2-point trapezoidal rule calculation. The error associated with this rule is illustrated in Fig. 31.7.

There is an alternative available for improving the accuracy of this calculation. The interval $(a - b)$ can be subdivided into smaller intervals. The trapezoidal rule can be applied repeatedly in turn over each subdivision.

31.5.2 Simpson's Rule

A higher degree interpolating polynomial scheme can be employed for more accurate results. One of the more popular integration approaches is Simpson's rule. For Simpson's 3-point (or one-third) rule, one may use the equation

$$I = \frac{h}{3}[y_a + 4y_{(b+a)/2} + y_b] \tag{31.23}$$

For the general form of Simpson's rule (n is an even integer), the equation is

$$I = \frac{h}{3}[y_0 + 4y_1 + 4y_2 + \cdots + 4y_{n-1} + y_n] \tag{31.24}$$

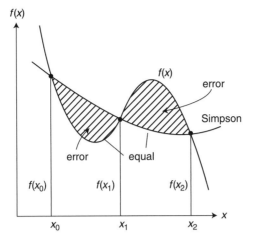

Figure 31.8 Simpson's rule error.

This method also generates an error, although it is usually smaller than that associated with the trapezoidal rule. A diagrammatic representation of the error for a 3-point calculation is provided in Fig. 31.8.

Illustrative Example 31.4   Evaluate the integral below using Simpson's 3-point rule. The term I in this application represents the volume requirement for a tubular flow reactor.

$$I = \int_{0}^{0.468} \left[\frac{(1 - 0.4x)^2}{(1 - x)(1 - 0.4x) - 1.19x^2} \right] dx$$

Solution   Write the 3-point rule. See Equation (31.23).

$$I = \frac{h}{3}[y_a + 4y_{(b+a)/2} + y_b]$$

$$= \frac{h}{3}[f(x = a) + 4f(x = (b + a)/2) + f(x = b)]$$

Evaluate h.

$$h = \frac{0.468}{2} = 0.234$$

Calculate y_a, $y_{(b+a)/2}$ and y_b. For $x = a = 0$,

$$y(0) = \frac{(1 - 0.4(0))^2}{(1 - (0))(1 - 0.4(0)) - 1.19(0)^2} = 1$$

Similarly,

$$y(x = 0.234) = 1.548$$
$$y(x = 0.468) = 3.80$$

Finally, calculate the integral I.

$$I = \frac{h}{3}[y_a + 4y_{(b+a)/2} + y_b] = \frac{0.234}{3}[1 + 4(1.548) + 3.80] = 0.857$$

Other numerical integration methods include:

1. Romberg's method.
2. Composite formulas.
3. Gregory's formulas.
4. Taylor's theorem.
5. Method of undetermined coefficients.
6. Richardson's extrapolation.

Details are available in the literature.[3,6,7] Some useful analytical integrals are also provided in mathematical reference handbooks. Note some integrals are indefinite, i.e., the upper and lower limits are not specified.

In closing, the reader should realize that analytical approaches yield closed form and/or exact solutions. Numerical methods provide discrete and/or inexact answers. Thus, the analytical approach should always be attempted first even though numerical methods have become the preferred choice.

REFERENCES

1. B. Carnahan and J. Wilkes, "Digital Computing and Numerical Methods," John Wiley & Sons, Hoboken, NJ, 1973.
2. R. L. Ketter and S. P. Prawler, "Modern Methods of Engineering Computations," McGraw-Hill, New York, NY, 1969.
3. L. Theodore, "Heat Transfer for the Practicing Engineer" (in preparation), John Wiley & Sons, Hoboken, NJ, 2009.
4. L. Theodore, "Mass Transfer for the Practicing Engineer," John Wiley & Sons, Hoboken, NJ, (in preparation).
5. M. Moyle, "Introduction to Computers for Engineers," John Wiley & Sons, Hoboken, NJ, 1967.
6. J. Reynolds, class notes (with permission), Manhattan College, Bronx, NY, 2001.
7. J. Famularo, class notes (with permission), Manhattan College, Bronx, NY, 1981.

NOTE: Additional problems are available for all readers at www.wiley.com. Follow links for this title.

32

ECONOMICS AND FINANCE

32.1 INTRODUCTION

Chapter 32 is concerned with Economics and Finance. These two topics can ultimately dictate the decisions made by the practicing engineer and their company. For example, a company may decide that due to the rising price of their raw materials, they will explore the possibility of producing a raw material instead of purchasing it. A decision will then be based on whether it makes sense economically in the short- and long-term. Furthermore, economic evaluations are a major part of process and plant design.

This chapter provides introductory material to this vast field within engineering. The next section is devoted to definitions. This is followed with an overview of accounting principles. The chapter concludes with seven Illustrative Examples in the Applications section.

Both the qualitative and quantitative viewpoint is emphasized in this chapter although it is realized that the broad subject of engineering economics cannot be fitted into any rigid set of formulas. The material presented falls into roughly three parts: namely, general principles, practical information, and applications. The presentation starts with simple situations and proceeds to more complicated formulations and techniques that may be employed if there are sufficient data available. Other texts in the literature provide further details on the subject.

Fluid Flow for the Practicing Chemical Engineer. By J. Patrick Abulencia and Louis Theodore
Copyright © 2009 John Wiley & Sons, Inc.

32.2 THE NEED FOR ECONOMIC ANALYSES

A company or individual hoping to increase its profitability must carefully assess a range of investment opportunities and select the most profitable options from those available. Increasing competitiveness also requires that efforts be expended to reducing costs of existing processes. In order to accomplish this, engineers should be fully aware of not only technical factors but also economic factors, particularly those that have the largest effect on profitability.

In earlier years, engineers concentrated on the technical side of projects and left the financial studies to the economist. In effect, engineers involved in making estimates of the capital and operating costs have often left the overall economic analysis and investment decision-making to others. This approach is no longer acceptable.

Some engineers are not equipped to perform a financial and/or economic analysis. Furthermore, many engineers already working for companies have never taken courses in this area. This shortsighted attitude is surprising in a group of people who normally go to great lengths to get all the available technical data before making an assessment of a project or study. The attitude is even more surprising when one notes that data are readily available to enable an engineer to assess the prospects of both his/her own company and those of his/her particular industry.[1]

As noted above, the purpose of this chapter is to provide a working tool to assist the student or engineer in not only understanding economics and finance but also in applying technical information to the economic design and operation of processes and plants. The material to follow will often focus on industrial and/or plant applications. Hopefully, this approach will provide the reader with a better understanding of some of the fundamentals and principles.

Bridging the gap between theory and practice is often a matter of experience acquired over a number of years. Even then, methods developed from experience all too often must be re-evaluated in the light of changing economic conditions if optimum designs are to result. The approach presented here therefore represents an attempt to provide a consistent and reasonably concise method for the solution of these problems involving economic alternatives.[2]

The term "economic analysis" in engineering problems generally refers to calculations made to determine the conditions for realizing maximum financial return for a design or operation. The same general principles apply, whether one is interested in the choice of alternatives for completing projects, in the design of plants so that the various components are economically proportioned, or in the technique of economical operation of existing plants. General considerations that form the framework on which sound decisions must be made are often simple. Sometimes their application to the problems encountered in the development of a commercial enterprise involves too many intangibles to allow exact analysis, in which case judgment must be intuitive. Often, however, such calculations may be made with a considerable degree of exactness. This chapter will attempt to develop a relatively concise method for applying these principles.

Finally, concern with maximum financial return implies that the criterion for judging projects involved is profit. While this is usually true, there are many important objectives which, though aimed at ultimate profit increase, cannot be immediately evaluated in quantitative terms. Perhaps the most significant of these is increased concern with environmental degradation and sustainability. Thus, there has been some tendency in recent years to regard management of commercial organizations as a profession with social obligations and responsibilities; considerations other than the profit motive may govern business decisions. However, these additional social objectives are for the most part often not inconsistent with the economic goal of satisfying human wants with the minimum effort. In fact, even in the operation of primarily nonprofit organizations, it is still important to determine the effect of various policies on profit.[2]

The next section is devoted to definitions. This is followed with an overview of accounting principles and applications. This chapter concludes with applications.

32.3 DEFINITIONS

Before proceeding to the applications, it would be wise to provide the reader with certain key definitions in the field. Fourteen concepts that often come into play in an economic analysis are given below. The definitions have been drawn from the literature.[3]

32.3.1 Simple Interest

The term interest can be defined as the money paid for the use of money. It is also referred to as the value or worth of money. Two terms of concern are simple interest and compound interest. Simple interest is always computed on the original principal. The basic formula to employ in simple interest calculations is:

$$S = P(1 + ni) \tag{32.1}$$

where $P =$ original principal
  $n =$ time in years
  $i =$ annual interest rate
  $S =$ sum of interest and principal after n years

Normally, the interest period is one year, in which case i is referred to as the effective interest rate.

32.3.2 Compound Interest

Unlike simple interest, with compound interest, interest is added periodically to the original principal. The term conversion or compounding of interest simply refers to the addition of interest to the principal. The interest period or conversion period in compound interest calculations is the time interval between successive conversions of interest and the interest period is the ratio of the stated annual rate to this number of interest periods in one year. Thus, if the given interest rate is 10% compounded semiannually, the interest period is 6 months and the interest rate per interest period is 5%. Alternately, if the given interest rate is 10% compounded quarterly, then the interest period is 3 months and the interest rate per interest period is 2.5%. One should always assume the interest is compounded annually unless otherwise stated. The basic formula to employ for compound interest is:

$$S = P(1 + i)^n \qquad (32.2)$$

If interest payments become due m times per year at compound interest, $(m)(n)$ payments are required in n years. A nominal annual interest rate, i', may be defined by:

$$S = P\left(1 + \frac{i'}{m}\right)^{mn} \qquad (32.3)$$

In this case, the effective annual interest, i, is:

$$i = \left(1 + \frac{i'}{m}\right)^m - 1 \qquad (32.4)$$

In the limit (as m approaches infinity), such payments may be considered to be required at infinitesimally short intervals, in which case, the interest is said to be compounded continuously. Numerically, the difference between continuous and annual compounding is small. However, annual compounding may be significant when applied to very large sums of money.

32.3.3 Present Worth

The present worth is the current value of a sum of money due at time n and at interest rate i. This equation is the compound interest equation solved for the present worth term, P

$$P = S(1 + i)^{-n} \qquad (32.5)$$

32.3.4 Evaluation of Sums of Money

The value of a sum of money changes with time because of interest considerations. $1000 today, $1000 ten years from now, and $1000 ten years ago all have different meanings when interest is taken into account. $1000 today would be worth more ten years from now because of the interest that could be accumulated in the interim. On the other hand, $1000 today would have been worth less ten years ago because a smaller sum of money could have been invested then so as to yield $1000 today. Therefore, one must refer to the date as well as the sum of money.

Summarizing, evaluating single sums of money requires multiplying by $(1 + i)^n$ if the required date of evaluation is after the date associated with the obligation or multiplying by $(1 + i)^{-n}$ if the required date of evaluation is before the date associated with the obligation. The term n is always the time in periods between the date associated with the obligation and the date of evaluation.

The evaluation of sums of money may be applied to the evaluation of a uniform series of payments. A uniform series is a series of equal payments made at equal intervals. Suppose R is invested at the end of every interest period for n periods. The total value of all these payments, S, as of the date of the last payment, may be calculated from the equation

$$S = R[(1 + i)^n - 1]/i \qquad (32.6)$$

The term S is then called the amount of the uniform series.

32.3.5 Depreciation

The term depreciation refers to the decrease in the value of an asset. Two approaches that can be employed are the straight line and sinking fund method. In the straight line method of depreciation, the value of the asset is decreased each year by a constant amount. The annual depreciation amount, D, is given by

$$D = (\text{Original cost} - \text{Salvage value})/(\text{Estimated life in years}) \qquad (32.7)$$

In the sinking fund method of depreciation, the value of the asset is determined by first assuming that a sinking fund consisting of uniform annual payments had been set up for the purpose of replacing the asset at the end of its estimated life. The uniform annual payment (UAP) may be calculated from the equation

$$UAP = (\text{Original cost} - \text{Salvage value})(\text{SFDF})$$

where SFDF is the sinking fund deposit factor and is given by

$$SFDF = i/[(1 + i)^n - 1] \qquad (32.8)$$

The value of the asset at any time is estimated to be the difference between the original cost and the amount that would have accumulated in the sinking fund. The amount accumulated in the sinking fund is obtained by multiplying the SFDF by the compound amount factor (CAF) where

$$CAF = [(1 + i)^n - 1]/i \qquad (32.9)$$

32.3.6 Fabricated Equipment Cost Index

A simple process is available to estimate the equipment cost from past cost data. The method consists of adjusting the earlier cost data to present values using factors that correct for inflation. A number of such indices are available; one of the most commonly used is the fabricated equipment cost index (FECI)

$$\text{Cost}_{\text{year B}} = \text{Cost}_{\text{year A}} \left(\frac{\text{FECI}_{\text{year B}}}{\text{FECI}_{\text{year A}}} \right) \qquad (32.10)$$

Given the cost and FECI for year A, as well as the FECI for year B, the cost of the equipment in year B can be estimated.

32.3.7 Capital Recovery Factor

In comparing alternative processes or different options for a particular process from an economic point-of-view, one recommended procedure to follow is that the total capital cost can be converted to an annual basis by distributing it over the projected lifetime of the facility (or the equivalent). The sum of both the annualized capital cost (ACC), including installation, and the annual operating cost (AOC), is called the total annualized cost (TAC) for the project or facility. The economic merit of the proposed facility, process, or scheme can be examined once the total annual cost is available.

The conversion of the total capital cost (TCC) to an ACC requires the determination of an economic parameter known as the capital recovery factor (CRF). This parameter can be found in any standard economics textbook or calculated directly from the following equation:

$$\text{CRF} = i(1 + i)^n / [(1 + i)^n - 1] \qquad (32.11)$$

where $n =$ projected lifetime of the system,
$\quad\quad\quad i =$ annual interest rate (as a fraction).

The CRF is a positive, fractional number. Once this factor has been determined, the ACC can be calculated from the following equation:

$$\text{ACC} = (\text{TCC})(\text{CRF}) \qquad (32.12)$$

The annualized capital cost reflects the cost associated with recovering the initial capital expenditure over the depreciable life of the system.

32.3.8 Present Net Worth

There are various approaches that may be employed in the economic selection of the best of several alternatives. For each alternative in the present net worth (PNW) method of economic selection, the single sum is calculated that would provide for all expenditures over a common time period. The alternative having the least PNW of expenditures is selected as the most economical. The equation to employ is

$$\text{PNW} = \text{CC} + \text{PN} + \text{PWD} - \text{PWS} \qquad (32.13)$$

where   CC = Capital cost,
       PN = Future renewals,
    PWD = Other disbursements,
    PWS = Salvage value.

If the estimated lifetimes differ for the various alternatives, employ a period of time equal to the least common multiple of the different lifetimes for renewal purposes.

32.3.9 Perpetual Life

Capitalized cost can be viewed as present worth under the assumption of perpetual life. Computing capitalized cost involves, in a very real sense, finding the present worth of an infinite series of payments. To obtain the present worth of an infinite series of payments of $R at the end of each interest period forever, one needs simply to divide R by i, where i is the interest rate per interest period. Thus, to determine what sum of money, P, would have to be invested at 8.0% to provide payments of $100,000 at the end of each year forever, P, would have to be such that the interest on it each period would be $100,000. Withdrawal of the interest at the end of each period would leave the original sum intact to again draw $100,000 interest at the end of the next period. For this example,

$$P = 100,000/0.08$$
$$= \$1,250,000$$

The $1,250,000 would be the present worth of an infinite series of payments of $100,000 at the end of each year forever, assuming money is worth 8%.

To determine the present worth of an infinite series of payments of $R at the end of each n periods forever, first multiply by the SFDF to convert to an equivalent single period payment and then divide by i to obtain the present worth.

32.3.10 Break-Even Point

From an economic point-of-view, the break-even point of a process operation is defined as that condition when the costs (C) exactly balance the income (I). The profit (P) is therefore,

$$P = I - C \tag{32.14}$$

At break-even, the profit is zero.

32.3.11 Approximate Rate of Return

Rate of return can be viewed as the interest that will make the present worth of net receipts equal to the investment. The approximate rate of return (ARR), denoted by some as p, may be estimated from the equation below:

$$p = \text{ARR} = \text{Average annual profit or earnings/Initial total investment} \quad (32.15)$$

To determine the average annual profit, simply divide the difference between the total money receipts (income) and the total money disbursements (expenses) by the number of years in the period of the investment.

32.3.12 Exact Rate of Return

Using the approximate rate of return as a guide, one can generate the exact rate of return (ERR). This is usually obtained by trial-and-error and interpolation calculations of the rate of interest that makes the present worth of net receipts equal to the investment. The approximate rate of return will tend to overestimate the exact rate of return when all or a large part of the receipts occur at the end of a period of investment. The approximate rate will tend to underestimate the exact rate when the salvage value is zero and also when the salvage value is a high percentage of the investment.

32.3.13 Bonds

A bond is a written promise to pay both a certain sum of money (redemption price) at a future date (redemption date) and equal interest payments at equal intervals in the interim. The holder of a $1000, 5% bond, redeemable at 105 (bond prices are listed without the last zero) in 10 years, with interest payable semiannually would be entitled to semiannual payments of $1000 (0.25) or $25 for 10 years and 105% of $1000, that is $1050, at the end of 10 years when the bond is redeemed.

The interest payment on a bond is found by multiplying the face value of the bond by the bond interest rate per period. From above, the face value is $1000 and the bond interest rate per period is 0.025. Therefore, the periodic interest payment is $25. Redeemable at 105 means that the redemption price is 105% of the face value of the bond.

The purchase price of a bond depends on the yield rate, that is the actual rate of return on the investment represented by the bond purchase. Therefore, the purchase price of a bond is the present worth of the redemption price plus the present worth of future interest payments, all computed at the yield rate. The bond purchase price formula is:

$$V = C(1 + i)^{-n} + R[1 - (1 + i)^{-n}]/i \quad (32.16)$$

where V = purchase price,
   C = redemption price,
   R = periodic interest payment,
   n = time in periods to maturity,
   i = yield rate.

32.3.14 Incremental Cost

By definition, the average unit increment cost is the increase in cost divided by the increase in production. Only those cost factors which vary with production can affect the average unit increment cost. In problems involving decisions as to whether to stay in production or (temporarily) shut down, the average unit increment cost may be compared with the unit increment cost or the unit selling price.

32.4 PRINCIPLES OF ACCOUNTING[3]

Accounting is the science of recording business transactions in a systematic manner. Financial statements are both the basis for and the result of management decisions. Such statements can tell a manager or an engineer a great deal about a company, provided that one can interpret the information correctly.

Since a fair allocation of costs requires considerable technical knowledge of operations in the chemical process industries, a close liaison between the senior process engineers and the accountants in a company is desirable. Indeed, the success of a company depends on a combination of financial, technical and managerial skills.

Accounting is also the language of business and the different departments of management use it to communicate within a broad context of financial and cost terms. The engineer who does not take the trouble to learn the language of accountancy denies himself the most important means available for communicating with top management. He may be thought by them to lack business acumen. Some engineers have only themselves to blame for their lowly status within the company hierarchy since they seem determined to hide themselves from business realities behind the screen of their specialized technical expertise. However, more and more engineers are becoming involved in decisions that are business related.

Engineers involved in feasibility studies and detailed process evaluations are dependent on financial information from the company accountants, especially information regarding the way the company intends to allocate its overhead costs. It is vital that the engineer should correctly interpret such information and that he/she can, if necessary, make the accountant understand the effect of the chosen method of allocation.

The method of allocating overheads can seriously affect the assigned costs of a project and hence the apparent cash flow for that project. Since these cash flows are often used to assess profitability by such methods as PNW, unfair allocation of overhead costs can result in a wrong choice between alternative projects.

In addition to understanding the principles of accountancy and obtaining a working knowledge of its practical techniques, the engineer should be aware of possible inaccuracies of accounting information in the same way that he/she allows for errors in any technical data.

At first acquaintance, the language of accountancy appears illogical to most engineers. Although the accountant normally expresses information in tabular form, the basis of all practice can be simply expressed by:

$$\text{Capital} = \text{Assets} - \text{Liabilities} \qquad (32.17)$$

or

$$\text{Assets} = \text{Capital} + \text{Liabilities} \qquad (32.18)$$

Capital, often referred to as net worth, is the money value of the business, since assets are the money values of things the business owns while liabilities are the money value of the things the business owes.

Most engineers have great difficulty in thinking of capital (also known as ownership) as a liability. This is easily overcome once it is realized that a business is a legal entity in its own right, owing money to the individuals who own it. This realization is absolutely essential when considering large companies with stockholders, and is used for consistency even for sole ownerships and partnerships. If a person (say LT) puts up $10,000 capital to start a business, then that business has a liability to repay $10,000 to that person.

It is even more difficult to think of profit as being a liability. Profit is the increase in money available for distribution to the owners, and effectively represents the interest obtained on the capital. If the profit is not distributed, it represents an increase in capital by the normal concept of compound interest. Thus, if the business makes a profit of $5000, the liability is increased to $15,000. With this concept in mind Equation (32.18) can be expanded to:

$$\text{Assets} = \text{Capital} + \text{Liabilities} + \text{Profit} \qquad (32.19)$$

where the capital is considered as the cash investment in the business and is distinguished from the resultant profit in the same way that principal and interest are separated.

Profit (as referred to above) is the difference between the total cash revenue from sales and the total of all costs and other expenses incurred in making those sales. With this definition, Equation (32.19) can be further expanded to:

$$\text{Assets} + \text{Expenses} = \text{Capital} + \text{Liabilities} + \text{Profit}$$
$$+ \text{Revenue from sales} \qquad (32.20)$$

Some engineers have the greatest difficulty in regarding an expense as being equivalent to an asset, as is implied by Equation (32.20). However, consider LT's earnings. During the period in which he made a profit of $5000, his total expenses excluding his earnings were $8000. If he assessed the worth of his labor to the business at $12,000, then the revenue required from sales would be $25,000. Effectively, LT has made a personal income of $17,000 in the year but he has apportioned it to the business as $12,000 expense for his labor and $5000 return on his capital. In larger businesses, there will also be those who receive salaries but do not hold stock and therefore, receive no profits, and stockholders who receive profits but no salaries. Thus, the difference between expenses and profits is very practical.

The period covered by the published accounts of a company is usually one year, but the details from which these accounts are compiled are often entered daily in a journal. The journal is a chronological listing of every transaction of the business, with details of the corresponding income or expenditure. For the smallest businesses, this may provide sufficient documentation but, in most cases, the unsystematic nature of the journal can lead to computational errors. Therefore, the usual practice is to keep accounts that are listings of transactions related to a specific topic such as "Purchase of Oil Account." This account would list the cost of each purchase of oil, together with the date of purchase, as extracted from the journal.

The traditional work of accountants has been to prepare balance sheets and income statements. Nowadays, accountants are becoming increasingly concerned with forward planning. Modern accountancy can roughly be divided into two branches: financial accountancy and management or cost accountancy.

Financial accountancy is concerned with stewardship. This involves the preparation of balance sheets and income statements that represent the interest of stockholders and are consistent with the existing legal requirements. Taxation is an important element of financial accounting.

Management accounting is concerned with decision-making and control. This is the branch of accountancy closest to the interest of most (process) engineers. Management accounting is concerned with standard costing, budgetary control, and investment decisions.

Accounting statements only present facts that can be expressed in financial terms. They do not indicate whether a company is developing new products that will ensure a sound business future. A company may have impressive current financial statements and yet may be heading for bankruptcy in a few years' time if provision is not being made for the introduction of sufficient new products or services.

32.5 APPLICATIONS

The remainder of the chapter is devoted to Illustrative Examples, many of which contain technical developmental material. A good number of fluid flow related applications have been drawn from the National Science Foundation (NSF) literature[4–8] and two other key sources.[9,10]

Illustrative Example 32.1   List the major fixed capital costs for the chemical process industry.

Solution

1. Major process equipment (i.e., reactors, tanks, pumps, filters, distillation columns, etc.).
2. Installation of major process equipment.
3. Process piping.
4. Insulation.
5. Instrumentation.

6. Auxiliary facilities (i.e., power substations, transformers, boiler houses, fire-control equipment, etc.).

7. Outside lines (i.e., piping external to buildings, supports and posts for overhead piping, electric feeders from power substations, etc.).

8. Land and site improvements.

9. Building and structures.

10. Consultant fees.

11. Engineering and construction (design and engineering fees plus supervision of plant erection).

12. Contractors' fees (administrative).

Illustrative Example 32.2   List the major working capital costs for the chemical process industry.

Solution

1. Raw materials for plant startup.

2. Raw material, intermediate and finished product inventories.

3. Cost of handling and transportation of materials to and from sites.

4. Cost inventory control, warehouse, associated insurance, security arrangements, etc.

5. Money to carry accounts receivable (i.e., credit extended to customers) less accounts payable (i.e., credit extended by suppliers).

6. Money to meet payrolls when starting up.

7. Readily available cash for emergency.

8. Any additional cash required to operate the process or business.

9. Expenses associated with new hirees.

10. Startup consultant fees.

Illustrative Example 32.3   Answer the following three questions:

1. Define the straight-line method of analysis that is employed in calculating depreciation allowances.

2. Define the double-declining balance (DDB) method of analysis.

3. Define the sum-of-the-year's digits (SYD) method of analysis.

Solution   The straight-line rate of depreciation is a constant equal to $1/r$, where r is the life of the facility for tax purposes. Thus, if the life of the plant is 10 yr, the straight-line rate of depreciation is 0.1. This rate, applied over each of the 10 yr, will result in a depreciation reserve equal to the initial investment.

A declining balance rate is obtained by first computing the straight-line rate and then applying some multiple of that rate to each year's unrecovered cost rather than to the original investment. Under the double-declining balance method, twice the straight-line rate is applied to each year's remaining unrecovered cost. Thus, if the life of a facility is 10 years, the straight-line rate will be 0.1, and the first year's double-declining balance will be 0.2. If the original investment is I, the depreciation allowance the first year will be $0.2I$. For the second year, it will be $0.16I$, or 0.2 of the unrecovered cost of $0.8I$. The depreciation allowances for the remaining years are calculated in a similar manner until the tenth year has been completed. Since this method involves taking a fraction of an unrecovered cost each year, it will never result in the complete recovery of the investment. To overcome this objection, the U.S. Internal Revenue Service (IRS) allowed the taxpayer in the past to shift from the DDB depreciation method to the straight-line method any time after the start of the project.

The rate of depreciation for the sum-of-the-year's digits method is a fraction. The numerator of this fraction is the remaining useful life of the property at the beginning of the tax year, while the denominator is the sum of the individual digits corresponding to the total years of life of the project. Thus, with a project life, r, of 10 years, the sum of the year's digits will be $10 + 9 + 8 + 7 + 6 + 5 + 4 + 3 + 2 + 1 = 55$. The depreciation rate the first year will be $10/55 = 0.182$. If the initial cost of the facility is I, the depreciation for the first year will be $0.182I$, $9/55 = 0.164I$ for the second year, and so on until the last year. The SYD method will recover 100% of the investment at the end of r years. A shift from SYD to straight-line depreciation cannot be made once the SYD method has been started.

Illustrative Example 32.4   Compare the results of the three methods discussed in Illustrative Example 32.3.

Solution   A tabular summary of the results of depreciation according to the straight-line, double-declining, and sum-of-the-year's digits methods are shown in Table 32.1.

Table 32.1 Comparative methods of analysis

Year	Straight-Line	Double-Declining	Sum-of-the-Year's Digits
0	1.000	1.000	1.000
1	0.900	0.800	0.818
2	0.800	0.640	0.655
3	0.700	0.512	0.510
4	0.600	0.410	0.383
5	0.500	0.328	0.274
6	0.400	0.262	0.183
7	0.300	0.210	0.110
8	0.200	0.168	0.056
9	0.100	0.134	0.018
10	0.000	0.108	0.000

Illustrative Example 32.5   A fluid is to be transported 4 miles under turbulent flow conditions. An engineer is confronted with two choices in designing the system:

A. Employ a 2 inch ID pipe at a cost of $1/foot.
B. Employ a 4 inch ID pipe at a cost of $6/foot.

Pressure drop costs for the 2 inch ID pipe are $20,000/yr. Assume only that the operating cost is the pressure drop and the only capital cost is the pipe. The capital recovery factor (CRF) for either pipe system is 0.1. Estimate the operating cost for the 4 inch ID pipe. Also determine which is the more economical pipe system?

Solution   The pressure drop is (approximately) proportional to the velocity squared for turbulent flow. The velocity of the fluid is lower for the 4-inch ID pipe. Since the area ratio is 4 (diameter squared), the 4-inch ID pipe velocity is one quarter of the velocity of the fluid in the 2-inch ID pipe. Therefore, the pressure drop for 4-inch ID pipe is approximately one sixteenth of the pressure drop for 2-inch ID pipe. The operating cost associated with the pressure drop cost is equal to:

$$\text{Operating cost} = \frac{\$20,000/\text{yr}}{16} = \$1250/\text{yr}; \quad \text{2-inch ID pipe}$$

To select the more economic pipe system, calculate the total cost

$$\text{Total cost} = \text{Operating cost} + \text{Capital cost}$$

The operating cost for a 2-inch pipe system and a 4-inch pipe system are given in the problem statement and calculated above, respectively. The annual capital cost is calculated as follows:

$$\text{Capital cost} = (\text{Distance})(\text{Cost})(\text{Capital recovery factor})$$
$$\text{Capital cost (2 inch)} = (4 \text{ miles})(5280 \text{ ft/mile})(\$1/\text{ft})(0.1) = \$2110$$
$$\text{Capital cost (4 inch)} = (4 \text{ miles})(5280 \text{ ft/mile})(\$6/\text{ft})(0.1) = \$12,700$$

Annual data is summarized in Table 32.2. Obviously, the 4-inch pipe is more economical.

Table 32.2   Cost results for two different sized pipes

	2 inch	4 inch
Operating cost	$20,000	$1250
Capital cost	$2110	$12,700
Total cost	$22,110	$13,950

Illustrative Example 32.6   A process emits 50,000 acfm of gas containing a dust (it may be considered ash and/or metal) at a loading of 2.0 gr/ft^3. A particulate control device is employed for particle capture and the dust captured from the unit is worth

$0.03/lb of dust. Experimental data have shown that the collection efficiency, E, is related to the system pressure drop, ΔP, by the formula:

$$E = \frac{\Delta P}{\Delta P + 15.0}$$

where $E =$ fractional collection efficiency,
$\Delta P =$ pressure drop, lb_f/ft^2.

If the fan is 55% efficient (overall) and electric power costs $0.18/kW · h, at what collection efficiency is the cost of power equal to the value of the recovered material? What is the pressure drop in inches of water (in H_2O) at this condition?

Solution   The value of the recovered material (RV) may be expressed in terms of the fractional collection efficiency E, the volumetric flowrate q, the inlet dust loading c, and the value of the dust (DV):

$$RV = (q)(c)(DV)(E)$$

Substituting yields

$$RV = \left(\frac{50{,}000\,ft^3}{min}\right)\left(\frac{2.0\,gr}{ft^3}\right)\left(\frac{1\,lb}{7000\,gr}\right)\left(\frac{0.03\$}{lb}\right)(E) = 0.429E\,\$/min$$

The recovered value can be expressed in terms of pressure drop, that is, replace E by ΔP:

$$RV = \frac{(0.429)(\Delta P)}{\Delta P + 15.0}\,\$/min$$

The cost of power (CP) in terms of ΔP, q, the cost of electricity (CE) and the fan fractional efficiency, E_f, is

$$CP = (q)(\Delta P)(CE)/(E_f)$$

Substitution yields

$$CP = \left(\frac{50{,}000\,ft^3}{min}\right)\left(\frac{\Delta P\,lb_f}{ft^2}\right)\left(\frac{0.18\$}{kW \cdot h}\right)\left(\frac{1\,min \cdot kW}{44{,}200\,ft \cdot lb_f}\right)\left(\frac{1}{0.55}\right)\left(\frac{1\,h}{60\,min}\right)$$

$$= 0.006\Delta P\$/min$$

The pressure drop at which the cost of power is equal to the value of the recovered material is found by equating RV with CP:

$$RV = CP$$

$$\Delta P = 66.5 \, \text{lb}_f/\text{ft}^2$$

$$= 12.8 \, \text{in. } H_2O$$

Figure 32.1 shows the variation of RV, CP, and profit with pressure drop. The collection efficiency corresponding to the above calculated ΔP is

$$E = \frac{\Delta P}{\Delta P + 15.0}$$

$$= \frac{66.5}{66.5 + 15.0}$$

$$= 0.82$$

$$= 82.0\%$$

The reader should note that operating below this efficiency (or the corresponding pressure drop) will produce a profit; operating above this value leads to a loss.

The operating condition for maximum profit can be estimated from Fig. 32.1. Calculating this value is left as an exercise for the reader. [*Hint:* Set the first derivative of the profit (i.e., $RV - CP$) with respect to ΔP equal to zero. The answer is 13.9 lb_f/ft^2.]

Illustrative Example 32.8   A filter press costing $60,000 has an estimated lifetime of 9 years and a salvage value of $500. What uniform annual payment must be made

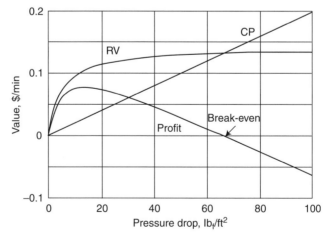

Figure 32.1   Profit as a function of pressure drop.

into a fund at the end of the year to replace the press if the fund earns 3.375%? What would be the appraisal value of the press at the end of the fifth year based on straight line depreciation?

Solution   Write the equation for the uniform annual payment (UAP) in terms of the cost (P) and salvage value (L), using a sinking fund model. See Equation (32.8).

$$\text{UAP} = (P - L)(\text{SFDF})$$

Calculate the sinking fund depreciation factor, SFDF

$$\text{SFDF} = \frac{i}{(1 + i)^n - 1} = \frac{0.03375}{(1 + 0.03375)^9 - 1} = 0.0969$$

Thus,

$$\text{UAP} = (\$60,000 - \$500)(0.0969) = \$5765$$

In determining the appraisal value where the straight line method of depreciation is used, the following equation applies:

$$B = P - \left(\frac{P - L}{n}\right)x$$

The term n refers to the years to the end of life, and x refers to any time from the present before the end of usable life. Employ this equation for the appraisal value and solve for B_5 after 5 years

$$B_5 = \$60,000 - \left(\frac{\$60,000 - \$500}{9}\right)(5) = \$26,945$$

This problem assumed that the depreciation of the filter press followed a sinking fund method, while the appraisal value of the press followed a straight line depreciation trend. For the depreciation calculation, it is assumed that the press will remain in operation for all of its 9 years of usable life. For this reason, the depreciable amount of the press may be thought of as being deposited into a sinking fund to be applied toward the replacement of the press after nine years.

The appraisal value of the press after the fifth year is calculated as part of the appraisal calculation. This value takes into account the fact that the press, even one year after it is purchased, is no longer worth what was paid for it. Since the appraisal had little to do with the fund for its replacement, the press was assumed to follow a straight line depreciation model.

Illustrative Example 32.9   The annual operation costs of an outdated environmental control device is $75,000. Under a proposed emission reduction plan, the installation of a new fan system will require an initial cost of $150,000 and an annual operating cost of $15,000 for the first 5 years. Determine the annualized cost for the new processing system by assuming the system has only 5 years (n)

operational life. The interest rate (i) is 7%. The capital recovery factor (CRF) or annual payment of a capital investment can be calculated as follows:

$$CRF = \left(\frac{A}{P}\right)_{i,n} = \frac{i(1+1)^n}{(1+i)^n - 1}$$

where A is the annual cost and P is the present worth.

Compare the costs for both the outdated and proposed operations.

Solution   The annualized cost for the new fan is determined based on the following input data:

Capital cost = $150,000

Interest, $i = 7\%$

Term, $n = 5$ yr

For $i = 0.07$ and $n = 5$, the CRF is

$$CRF = \frac{0.07(1+0.07)^5}{(1+0.07)^5 - 1}$$

$$= 0.2439$$

The total annualized cost for the fan is then

Annualized cost = Installation cost + Operation cost

= (0.2439)($150,000) + $15,000 = $51,585

Since this cost is lower than the annual cost of $75,000 for the old process, the proposed plan should be substituted.

REFERENCES

1. F. Holland, F. Watson, and J. Wilkinson, "Financing Assets by Equity and Debt," Chemical Engineering, September 2, 1974.
2. J. Happel, "Chemical Process Economics," John Wiley & Sons, Hoboken, NJ, 1958.
3. F. Holland, F. Watson, and J. Wilkinson, (adapted from) "Financing Principles of Accounting," Chemical Engineering, July 8, 1974.
4. J. Reynolds, R. Dupont, and L. Theodore, "Hazardous Waste Incineration Calculations: Problems and Software," John Wiley & Sons, Hoboken, NJ, 1991.
5. R. Dupont, L. Theodore, and J. Reynolds, "Accident and Emergency Management: Problems and Solutions," VCH Publishers, New York, NY, 1991.
6. L. Theodore, R. Dupont, and J. Reynolds, "Pollution Prevention: Problems and Solutions," Gordon and Breach Publishers, Amsterdam, Holland, 1994.

7. K. Ganeson, L. Theodore, and J. Reynolds, "Air Toxics: Problems and Solutions," Gordon and Breach Publishers, Amsterdam, Holland, 1996.

8. R. Dupont, T. Baxter, and L. Theodore, "Environmental Management: Problems and Solutions," CRC Press, Boca Raton, FL, 1998.

9. L. Theodore and K. Neuser, "Engineering Economics and Finance," a Theodore Tutorial, Theodore Tutorials, East Williston, NY, 1996.

10. J. Reynolds, J. Jeris, and L. Theodore, "Handbook of Chemical and Environmental Engineering Calculations," John Wiley & Sons, Hoboken, NJ, 2002.

NOTE: Additional problems are available for all readers at www.wiley.com. Follow links for this title.

33

BIOMEDICAL ENGINEERING

33.1 INTRODUCTION

Biomedical Engineering (BME) is a relatively new discipline in the engineering profession and, as one might expect, it has come to mean different things to different people. Terms such as biomedical engineering, biochemical engineering, bioengineering, biotechnology, biological engineering, genetic engineering, and so on, have been used interchangeably by many in the technical community. To date, standard definitions have not been created to distinguish these genres. Consequently, we will lump them all using the term BME for the sake of simplicity. What one may conclude from all of the above is that BME involves applying the concepts, knowledge, basic fundamentals, and approaches of virtually all engineering disciplines (not only chemical engineering) to solve specific health and health-care related problems; the opportunities for interaction between engineers and health-care professionals are therefore many and varied.

On a personal note, the authors view BME as the application of engineering, mathematics, and physical sciences to principles in biology and medicine. The terms biophysics and bioengineering either involve the interaction of physics or engineering with either biology or medicine.

Because of the broad nature of this subject, this introductory chapter addresses only the application of BME to the anatomy of humans, particularly the cardiovascular system, and attempts to relate four key anatomy topics to fluid flow. The reader is referred to three excellent references in the literature for an extensive comprehensive treatment of this new discipline.[1-3]

Fluid Flow for the Practicing Chemical Engineer. By J. Patrick Abulencia and Louis Theodore
Copyright © 2009 John Wiley & Sons, Inc.

The four key cardiovascular "parts" to be discussed in subsequent sections are:

1. Blood.
2. Blood vessels.
3. Heart.
4. Plasma–cell flow.

The relation of these topics to the general subject of fluid flow is provided in Table 33.1.[4] Thus, from a fluid flow and engineering perspective, the cardiovascular system is comprised of a pump (heart) that generates a pressure difference driving force that involves the flow of a fluid (blood), where the fluid involves the transport of a two-phase medium (plasma and cells), through a complex network of pipes (blood vessels). As noted earlier, BME may therefore be viewed as an interdisciplinary branch of technology that is based on both engineering and the sciences.

Table 33.1   Fluid flow analogies

Topic	Fluid Flow	Biomedical Engineering
Fluid flow	Fluid	Blood
Conduit	Pipe	Blood vessels
Prime mover	Pump	Heart
Two-phase flow	Fluid-particle dynamics	Plasma-cell flow

Following a section on definitions, each of the aforementioned topics receives a qualitative treatment from a fluid flow perspective. The chapter concludes with an abbreviated description of job opportunities and activities in the biomedical engineering field while briefly alluding to regulatory issues.

33.2   DEFINITIONS

The definition of a host of BME and BME-related terms is provided below. A one-sentence (in most instances) description/explanation of each word/phase is provided. As one might suppose, the decision of what to include (as well as what to omit) was somewhat difficult.

1. Anatomy: The structure of an organism or body; its dissection to determine body parts.
2. Aorta: The key artery of the body that carries blood from the left ventricle of the heart to organs.
3. Artery: Any one of the thick-walled tubes that carry blood from the heart to the principal parts of the body.
4. Atrium: Either the left or right upper chamber of the heart.

5. Autonomic nervous system: The functional division of the nervous system that innervates most glands, the heart, and smooth muscle tissue in order to maintain the internal environment of the body.

6. Capillary: Any of the extremely small blood vessels connecting the arteries with veins.

7. Cardiac muscle: Involuntary muscle possessing much of the anatomic attributes of skeletal voluntary muscle and some of the physiologic attributes of involuntary smooth muscle tissue; sinoatrial node induced contraction of its interconnected network of fibers allows the heart to expel blood during systole.

8. Cell: An extremely small complex unit of protoplasm, usually with a nucleus, cytoplasm, and an enclosing membrane; the semielastic, selectively permeable cell membrane controls the transport of molecules into and out of a cell.

9. Chronotropic: Affecting the periodicity of a recurring action such as the slowing (bradycardia) or speeding up (tachycardia) of the heartbeat that results from extrinsic control of the sinoatrial node.

10. Circulatory system: The course taken by the blood through the arteries, capillaries, and veins and back to the heart.

11. Clot: A clot consists primarily of red corpuscles enmeshed in a network of fine fibrils or threads, composed of a substance called fibrin.

12. Corpuscle: An extremely small particle, especially any of the erythrocytes or leukocytes that are carried and/or float in the blood.

13. Cytoplasm: The protoplasm outside the nucleus of a cell.

14. Endocrine system: The system of ductless glands and organs secreting substances directly into the blood to produce a specific response from another "target" organ or body part.

15. Endothelium: Flat cells that line the innermost surfaces of blood and lymphatic vessels and the heart.

16. Erythrocytes: Red corpuscles.

17. Gland: Any organ or group of cells that separates certain elements from the blood and secretes them in a form for the body to use or discard.

18. Heart: The organ that receives blood from the veins and pumps it through the arteries by alternate dilation and contraction.

19. Homeostasis: A tendency to uniformity or stability in an organism by maintaining within narrow limits certain variables that are critical to life.

20. Inotropic: Affecting the contractility of muscular tissue such as the increase in cardiac power that results from extrinsic control of the myocardial musculature.

21. Leukocytes: White corpuscles.

22. Nucleoplasm: The protoplasm that composes the nucleus of a cell.

23. Plasma: The fluid part of blood, as distinguished from corpuscles; its principal component is water.

24. Precapillary sphincters: Rings of smooth muscle surrounding the entrance to capillaries where they branch off from upstream metarterioles. Contraction and realization of these sphincters close and open the access to downstream blood vessels, thus controlling the irrigation of different capillary networks.

25. Protoplasm: A semifluid viscous colloidal that is the living matter of humans and is differentiated into nucleoplasm and cytoplasm.

26. Pulse: The expansion and contraction of the arterial walls that can be felt in all the arteries near the surface of the skin.

27. Stem Cells: A generalized parent cell spawning descendants that become individually specialized.

28. Vein: Any blood vessel that carries blood from some part of the body back to the heart; in a loose sense any blood vessel.

29. Ventricle: Either of the two lower chambers of the heart that receive blood from the atria and pump it into the arteries.

Since the difference between biomedical engineering and biochemical engineering/biotechnology are somewhat overlapping, several additional definitions in the latter subject are listed below[5]:

1. Algae: Algae are a very diverse group of photosynthetic organisms that range from microscopic size to lengths that can exceed the height of a human.

2. Bacteria: The bacteria are tiny single-cell organisms ranging from $0.5-20\,\mu m$ in size although some may be smaller, and a few exceed $100\,\mu m$ in length.

3. Fungi: As a group, fungi are characterized by simple vegetative bodies from which reproductive structures are elaborated; all fungal cells possess distinct nuclei and produce spores in specialized fruiting bodies at some stage in their life cycle.

4. Microorganism: A class of organisms that includes bacteria, protozoan, viruses, and so on.

5. Photosynthesis: All living cells synthesize Adenosine 5′-triphosphate (ATP), but only green plants and a few photosynthetic (or phototrophic) microorganisms can drive biochemical reactions to form ATP with radiant energy though the process of photosynthesis.

6. Protozoan: A single cell or a group of essentially identical cells living primarily in water.

7. Viruses: Viruses are particles of a size below the resolution of the light microscope and are composed mainly of nucleic acid, either DNA or RNA, surrounded by a protein sheath.

8. Yeasts: Yeasts are a kind of fungi; they are unicellular organisms surrounded by a cell wall and possessing a distinct nucleus.

33.3 BLOOD

Blood. It has been justifiably described as the "river of life." It is the transport medium that serves as a dispenser and collector of nutrients, gases, and waste that allows life to be sustained.

In terms of composition, blood is primarily composed of water. The average human contains about 5 L (5000 mL) of blood. This 5 L volume of blood contains in numbers (approximately):

5×10^{16} red corpuscles (erythrocytes)

1×10^{14} white corpuscles (leukocytes)

2.5×10^{15} platelets (thrombocytes)

The red blood cells are typically round disks, concave on two sides, and approximately 7.5 μm in diameter. They are made in the bone marrow and have a relatively robust lifespan of 120 days. These cells take up oxygen as blood passes through the lungs, and subsequently release the oxygen in the capillaries of tissues. From a mechanics aspect, red blood cells have the ability to deform. This is important because erythrocytes often have to navigate through irregular shapes within the vasculature, as well as squeeze through small diameters such as those encountered in capillaries.

In contrast, white blood cells serve a wider variety of functions. These cells can be classified into two categories based on the type of granule within their cytoplasm, and the shape of the nucleus. More specifically, the two categories are: 1) Granulocytes or Polymorphonuclear (PMN) leukocytes, and 2) Mononuclear leukocytes. Examples of PMNs include neutrophils, eosinophils, and basophils, while those of mononuclear leukocytes include lymphocytes and monocytes.

Platelets are small, round, non-nucleated disks with a diameter about one-third that of red blood cells. They are formed from megakaryocytes in the bone marrow, and have a 7 to 10 day lifespan within the vasculature. The main function of platelets is to arrest bleeding.

The final component of blood is plasma, and is the fluid in which the aforementioned cells remain in suspension. It is comprised primarily of water (90%), and the rest plasma protein (7%), inorganic salts (1%), and organic molecules such as amino acids, hormones, and lipoproteins (2%). The primary functions of plasma are to allow exchange of chemical messages between distant parts of the body, and maintenance of body temperature and osmotic balance.[6]

In terms of physical properties, the density of blood (as one might suppose) is approximately that of water, that is, 1.0 g/cm^3 or 62.4 lb/ft^3. The other key property is viscosity and it is approximately 50% greater than that for water. The reader is no doubt familiar with the expression "blood is thicker than water." Well, it turns out not to be that much "thicker" than water since its viscosity is only moderately higher. In addition, the viscosity is a strong function of temperature; it has been reported that the viscosity increases by 50% over the 20–40°C range. Although plasma (the main constituent of blood) is Newtonian, blood (with the added blood cells) is non-Newtonian.

Illustrative Example 33.1   The viscosity of plasma has been estimated to be 1.25 cP at room temperature. Convert this to English units.

Solution   Convert viscosity from centipoises to $lb/ft \cdot s$.

$$(1.25 \text{ cP}) \left(\frac{6.72 \times 10^{-4} \text{ lb}}{\text{ft} \cdot \text{s} \cdot \text{cP}} \right) = 8.4 \times 10^{-4} \text{ lb/ft} \cdot \text{s}$$

$$= 0.00084 \text{ lb/ft} \cdot \text{s}$$

33.4   BLOOD VESSELS

The study of the motion of blood is defined as hemodynamics. Part of the cardiovascular system involves the flow of blood through a complex network of blood vessels. Blood flows through organs and tissues either to nourish and sanitize them or to be itself processed in some sense, e.g., to be oxygenated (pulmonary circulation), filtered of dilapidated red blood cells (splenic circulation), and so on.[3] The aforementioned "river of life" flows through the piping network, which is made up of blood vessels, by the action of two pump stations arranged in series. This complex network also consists of thousands of miles of blood vessels. The network also consists of various complex branching configurations.

Regarding the branching, the aorta divides the discharge from the heart into a number of main branches, which in turn divide into smaller ones until the entire body is supplied by an elaborately branching series of blood vessels. The smallest arteries divide into a fine network of still more minute vessels, defined as capillaries, which have extremely thin walls; thus, the blood is enabled to come into close relation with the fluids and tissues of the body.

In the capillaries, the blood performs three functions:

1. It releases oxygen to the tissues.
2. It furnishes the nutrients and other essential substances that it carries to the body cells.
3. It takes up waste products from the tissues.

The capillaries then unite to form small veins. The veins, in turn, unite with each other to form larger veins until the blood is finally collected into the venal cavae from where it goes to the heart, thus completing the blood vessel circuit (see next section for additional details). This complex network is designed to bring blood to within a capillary size of each and every one of more than 10^{14} cells of the body. Which cells receive blood at any given time, how much blood they get, the composition of the fluid flowing by them, and related physiologic considerations are all matters that are not left up to chance.[1]

Information on average radius and number of each of the various vessels has been provided by LaBarbara.[7] His data/information is given in Table 33.2. The average diameter of blood vessels in humans has not been determined since there is a wide

Table 33.2 The average radii and total numbers of conventional categories of vessels of the human circulatory system

Vessel	Average Radius (mm)	Number
Aorta	12.5	1
Arteries	2	159
Arterioles	0.03	1.4×10^7
Capillaries	0.006	3.9×10^9
Venules	0.02	3.2×10^8
Veins	2.5	200
Vena cava	15	1

distribution of sizes, but based on the data in Table 33.2, one may assume the value of 150 μm for the smaller vessels.

Illustrative Example 33.2 The usual units employed for pressure in cardiovascular studies is mmHg. Convert 80 mm Hg to in Hg, ft H_2O, in H_2O, psia, psfa, and N/m^2.

Solution Expressed in various units, the standard atmosphere is equal to:

1.0	Atmospheres (atm)
33.91	Feet of water (ft H_2O)
14.7	Pounds-force per square inch absolute (psia)
2116	Pounds-force per square foot absolute (psfa)
29.92	Inches of mercury (in Hg)
760	Millimeters of mercury (mm Hg)
1.013×10^5	Newtons per square meter (N/m^2)

Applying the conversion factors to 80 mm Hg from above leads to:

$$P = 80(29.92/760)$$
$$= 3.15 \text{ in Hg}$$
$$P = 80(33.91/760)$$
$$= 3.57 \text{ ft } H_2O$$
$$P = (3.57)(12)$$
$$= 42.8 \text{ in. } H_2O$$
$$P = 80(14.7/760)$$
$$= 1.55 \text{ psia}$$
$$P = 80(2116/760)$$
$$= 223 \text{ psfa}$$
$$P = 80(1.013 \times 10^5/760)$$
$$= 1.07 \times 10^4 \text{ N/m}^2$$

Illustrative Example 33.3   Describe the role arteries play on a heart attack.

Solution   Coronary arteries are very small in diameter and can become narrow as fatty deposits (plaque) build up. As the vessels narrow, less blood can get through to the heart when it needs it. If the heart is not getting enough oxygen to meet its needs, one may experience pain (angina). When the heart muscle does not receive enough oxygen-rich blood, it can become injured since oxygen is important to help keep the heart functioning properly. If blood flow is completely cut off for more than a few minutes, muscle cells die, causing a heart attack.

Illustrative Example 33.4   Discuss why the flow of blood through blood vessels can be safely assumed to be laminar flow.

Solution   For flow in circular conduits, the flow is laminar if the Reynolds number

$$\text{Re} = Dv\rho/\mu$$

is less than 2100. In nearly all biomedical applications, the numerical values for D and v are extremely small while the viscosity is large.

From an engineering perspective, one should note that vessel flow is laminar and the equations presented in Chapter 13 may be assumed to apply. For example, the velocity profile of the blood in the vessel may be assumed to be parabolic with the average velocity equal to one-half the maximum velocity. Equation 13.7 applies so that

$$\Delta P = \frac{4fLv^2}{2g_cD} \tag{33.1}$$

For laminar flow, see Equation (13.6)

$$f = 16/\text{Re} \tag{33.2}$$

so that

$$\Delta P = \frac{32\mu gvL}{g_cD^2} \tag{33.3}$$

This equation may also be expressed in terms of the volumetric flow rate (setting $g/g_c = 1.0$)

$$\Delta P = \frac{128q\mu L}{\pi D^4} \tag{33.4}$$

Equation (33.4) may also be written:

$$q = \frac{\pi D^4(\Delta P)}{128\mu L} \tag{33.5}$$

Since the flow is laminar, the relative irregularity of the inner wall of the blood vessel has a negligible effect (see Chapter 14) on the volumetric flow-pressure drop relationship.

The blood vessel branching discussed above may be viewed as flow through a number of pipes or conduits, an application that often arises in engineering practice. If flow originates from the same source and exits at the same location, the pressure drop across each conduit must be the same. Thus, for flow through conduits 1, 2, and 3, one may write:

$$\Delta P_1 = \Delta P_2 = \Delta P_3 \tag{33.6}$$

From a fluid flow perspective, the presence of blood vessel branching enhances the performance of the cardiovascular circulatory system. When branching involves flow into smaller diameter vessels, the oxygen has a shorter distance and shorter residence time requirement to reach the tissues that require oxygen. In addition, if the velocity through the branched vessels remains unchanged, the pressure drop correspondingly decreases. This drop is linearly related to the diameter though equation (33.3).

Illustrative Example 33.5   If an artery branches into two smaller equal area arteries so that the velocity through the three arteries is the same, determine the ratio of the inlet and discharge artery diameters.

Solution   Based on the problem statement and continuity

$$q = q_1 + q_2$$

Since

$$v = q/S$$

and

$$q_1 = q_2 = q/2$$

Therefore

$$S_1 = S_2 = S/2$$

and

$$\frac{\pi D_1{}^2}{4} = \frac{\pi D^2}{4(2)}$$

$$\left(\frac{D}{D_1}\right)^2 = 2$$

$$\frac{D}{D_1} = 2^{0.5} = 1.414$$

Alternately

$$\frac{D_1}{D} = 0.707$$

Illustrative Example 33.6 A blood vessel branches into three openings. Information on the system is provided in Table 33.3. Determine the magnitude velocity v_3.

Table 33.3   Flow velocity-area data for Illustrative Example 33.6

Vessel	Flow Area (m^2)	Velocity (mm/s)
Inlet	0.2	5
1	0.08	7
2	0.025	12
3	0.031	?

Solution   Calculate the volumetric flow rate through the inlet Sections 1 and 2:

$$q = (0.2)(5) = 1 \,(\text{mm})^3/\text{s}$$

$$q_1 = (0.08)(7) = 0.56 \,(\text{mm})^3/\text{s}$$

$$q_2 = (0.025)(12) = 0.3 \,(\text{mm})^3/\text{s}$$

Since the blood may be assumed to be of constant density, the continuity equation may be applied on a volume rate basis:

$$q = q_1 + q_2 + q_3$$
$$q_3 = q - q_1 - q_2$$
$$= 1 - 0.56 - 0.3$$
$$= 0.14 \,(\text{mm})^3/\text{s}$$

The velocity v_3 is therefore:

$$v_3 = \frac{0.14}{0.031}$$
$$= 4.52 \,\text{mm}/\text{s}$$

The equations developed for laminar flow between two or more parallel plates may also be assumed to apply. The flow is still parabolic but the maximum velocity is only 50% greater than the average value. For the interested reader, there are several biological applications that involve closely spaced and parallel plates.[4]

Finally, it should be noted that the:

1. Flow is pulsating.
2. Fluid is non-Newtonian.

3. Blood vessels vary in cross-sectional area.

4. Blood vessels vary in shape.

5. Terms (3) and (4) vary with time.

Despite the above five limitations, efforts abound that have attempted to model the flow of the "river of life" in the cardiovascular complex blood vessel network.

33.5 HEART

The "river of life" flows through the cardiovascular circulatory system by the action of the heart, which essentially provides two pumps that are arranged in series. (See Chapter 17, Section on flow in pumps arranged in series.)

In simple terms, the heart is an organ that receives blood from veins and propels it into and through the arteries. It is primarily held in place by its attachment to the aforementioned arteries and veins, and by its confinement, via a double-walled sac with one layer enveloping the heart and the other attached to the breastbone. Furthermore, the heart consists of two parallel independent systems, each consisting of an atrium and a ventricle that have been referred to as the right heart and the left heart.

The following is a description of the cardiovascular circulatory system. The heart is divided into four chambers through which blood flows. These chambers are separated by valves that help keep the blood moving in the right direction. The chambers on the right side of the heart take blood from the body and push it through the lungs to pick up oxygen. The left side takes blood from the lungs and pumps it out the aorta. The blood squeezes through a system of arteries, capillaries, and veins that reach every part of the body. The cycle is completed when the blood is returned to the heart by the superior and inferior vena cava and enters the atrium (or chambers) on the upper right side of the heart.

Regarding details of the action of the heart, blood is drawn into the right ventricle by a partial vacuum when the lower chamber relaxes after a beat. On the next contraction of the heart muscle, the blood is squeezed into the pulmonary arteries that carry it to the left and right lungs. The blood receives a fresh supply of oxygen in the lungs and is pumped into the heart by way of the pulmonary veins, entering at the left atrium. The blood is first drawn into the left ventricle and then pumped out again through the aorta (the major artery), which connects with smaller arteries and capillaries reaching all parts of the body. Thus, the cardiovascular circulatory system consists of the heart (a pump), the arteries (pipes) that transport blood from the heart, and the capillaries and veins (pipes) that transport the blood back to the heart. These form a complete recycle process. An expanded discussion of the process is provided in the next paragraph.

A line diagram of the recycle circulatory system is given in Fig. 33.1.[5] The cycle begins at point 1. Oxygenated blood is pumped from the left lower ventricle (LLV) at an elevated pressure through the aorta and discharge oxygen to various parts of the body. Deoxygenated blood then enters the right upper atrium (RUA), passes

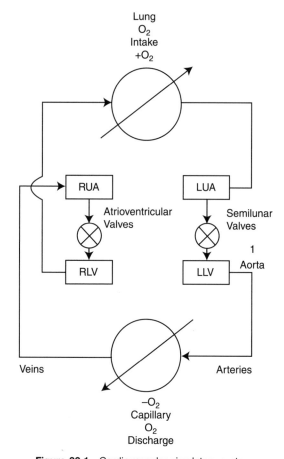

Figure 33.1 Cardiovascular circulatory system.

through the right lower ventricle (RLV) and then enters the lungs where its oxygen supply is replenished. The oxygenated discharge from the lungs enters the left upper atrium (LUA) and returns to the lower left ventricle, completing the cycle.

As noted above, the concepts regarding recycle, bypass and purge presented in Chapter 7 readily apply to the discussion above. In addition to blood being recycled through the circulatory system, it is also being reused. Bypass occurs when part of the blood is bypassed to the liver for cleansing purposes. The purging process may be viewed as occurring during the oxygen transfer to tissues as well as the aforementioned cleansing process.

The typical discharge with each heart beat is 70 mL through the 2.4 cm diameter pulmonary vessel and the 2.25 cm diameter aortic vessel. The return of the 70 mL (from the continuity equation) of blood to the heart enters from the 3.8 cm diameter right atrium and the 3.1 cm diameter left atrium. This discharge occurs approximately every 0.75 s, or the equivalent of 80 heartbeats/min. This translates to a volumetric flow rate of

$$q = 70/0.75$$
$$= 93.3 \, \text{mL/s}$$

This corresponds to a circulation rate of slightly more than 5 L/min.

Illustrative Example 33.7   Calculate the average velocity of blood flowing through the aorta. Assume the diameter of the aorta is 2.5 cm.

Solution   The area for flow in the aorta is

$$S = (\pi/4)(2.5)^2$$
$$= 4.9 \, \text{cm}^2$$

Since the volumetric flow rate is 93.3 mL/s (93.3 cm^3/s), the velocity is

$$v = q/S$$
$$= 93.3/4.9$$
$$= 19 \, \text{cm/s}$$

Illustrative Example 33.8   One of the authors of this book is 74 years old (at the time of the preparation of this manuscript). Based on the data provided above, calculate the number of times (*T*) that the author's heart has beat to date.

Solution

$$T = [(74)(365)(24)(60)](80)$$
$$= 3,110,000,000$$
$$= 3.11 \times 10^9$$

Thus, the author's heart has already beaten approximately 3 billion times over his 74 year lifespan.

Illustrative Example 33.9   Refer to Example 33.8. Calculate the volume of blood that has circulated through the author's system over his lifetime.

Solution   Again, being careful to be dimensionally consistent, the volume of blood (*V*) is

$$V = (3.11 \times 10^9)(70)$$
$$= 2.18 \times 10^{11} \, \text{mL}$$
$$= 2.18 \times 10^8 \, \text{L}$$

This represents approximately 200 million liters of blood.

To summarize, from a fluid flow/engineering perspective, the systemic circulation carries blood to the neighborhood of each cell in the body and then returns it to the right side of the heart low in oxygen and rich in carbon dioxide. The pulmonary circulation carries the blood to the lungs where its oxygen supply is replenished and its carbon dioxide content is purged before it returns to the left side of the heart to repeat the cycle. The driving force for flow arises from the pressure difference between the high pressure left side of the heart (systemic) to the lower pressure right side (pulmonary). This pressure difference provides the impetus for the "river of life" to flow. Thus, the heart may be viewed as an engine pump that has many of the characteristics of a centrifugal pump.

Regarding physical properties of the heart, it is approximately the size of a clenched fist. It is inverted, conically-shaped, measuring 12 to 13 cm from base (top) to apex (bottom) and 7 to 8 cm at its widest point and weighing just under 0.75 lb (less than 0.5% of a human's body weight) and located between the third and sixth ribs. It rests between the lower part of the two lungs.[3]

Illustrative Example 33.10   Dr. Abs, one of the foremost authorities in the biomedical field has concluded, based on many years of a theoretical modeling study that the flow of blood from the aorta to the atrium could be physically represented by a 0.3-mile long 2.5 cm diameter vessel. Is the model a reasonable one if the blood velocity is 19 cm/s (see Illustrative Example 33.7)?

Solution   At a minimum, the pressure drop (ΔP) calculation across the heart should approximately be

$$\Delta P = 120 - 80$$
$$= 40 \, \text{mmHg}$$

Apply Equation (33.3) and assume that the density and viscosity of the blood are 62.4 lb/ft^3 and 1.1 cP, respectively

$$\Delta P = \frac{32 \mu g v L}{\rho g_c D^2} \tag{33.3}$$

Substituting,

$$\Delta P = \frac{32(1.1)(6.72 \times 10^{-4})(19/30.48)(5280)(0.3)}{(62.4)(2.53/30.48)^2(32.2)}$$

$$= 1.729 \, (\text{ft})(\text{lb}_f)/\text{lb}$$

$$= 107.8 \, \text{lb}_f/\text{ft}^2$$

$$= 38.7 \, \text{mmHg}$$

The model is reasonable from a fluid dynamics perspective.

Illustrative Example 33.11   A new blood vessel has been implanted in a patient's circulatory system. Which of the following three options, with dimensions specified, would provide the least resistance to flow of a given quantity of blood?

1. $D = 2.5$ cm, $L = 5$ cm
2. $D = 1.25$ cm, $L = 2.5$ cm
3. $D = 5.0$ cm, $L = 10$ cm

Solution   Assuming laminar flow, apply Equation (33.1). Assign (1) as the base case value

$$\Delta P_1 = \frac{32\mu g v L}{\rho g_c D^2}$$

$$= (K)(Lv/D^2); \quad K = 32\,\mu g/eg_c$$

For (2),

$$\Delta P_2 = K(0.5L)(4v)/(0.5D^2)$$

$$= K(8)(Lv/D^2) = 8(K)(Lv/D^2)$$

For (3),

$$\Delta P_3 = K(2\,L)(0.25v)/(2D^2)$$

$$= K(0.125)(Lv/D^2) = 0.125(K)(Lv/D^2)$$

As expected, (3) provides the lowest pressure drop while (2) provides the highest resistance.

With reference to Fig. 17.2 in Chapter 17, note the relationship between the volume rate of flow and the pressure difference generated by the pump. The pump can operate anywhere within the circular curve, which is referred to as the pump curve. The line from the origin represents the system curve and intersects the pump curve at the maximum output of the pump for those conditions. Notice that for low-pressure drop conditions, the maximum flow rate is not overly sensitive to the pressure drop with which the pump must contend. As can be seen from Fig. 17.2, the pressure increase developed by a pump is a function of the discharge rate. As with a real pump, when the flow rate increases, the pressure increase delivered by the pump decreases. In fact, the pump's power is given by the product of the two terms, that is, the pressure drop and volumetric flow rate

$$\text{hp} = q\Delta P \tag{33.7}$$

For the heart, hp represents the power required to maintain the recycle process that constitutes the circulatory flow of the blood in the cardiovascular system.

Illustrative Example 33.12 Estimate the power generated by the human heart. Assume the pressure drop in the circulatory system is 60 mm Hg.

Solution

$$hp = q\Delta P \tag{33.7}$$

As noted earlier, $q = 93.3$ mL/s.
Converting,

$$q = 0.0033 \, \text{ft}^3/\text{s}$$

Substitution gives

$$hp = (0.0033)(60)(14.7) * (144)/760$$
$$= 10 \times 10^{-4} \, hp$$

33.6 PLASMA/CELL FLOW

As described earlier, blood is comprised of a fluid called plasma and suspended cells that primarily include erythrocytes (red blood cells), leukocytes (white blood cells), and platelets. From a fluid dynamics perspective, blood motion can be viewed as a fluid–particle application involving a two-phase flow. As noted in both Chapters 16 and 23, a rigorous theoretical description of flow with a concentrated suspension of particles is not available. Unfortunately, such a description is necessary for a quantitatively accurate understanding of the flow of blood in blood vessels.

The blood vessels through which the blood flows has dimensions that are small enough so that the effects of the particulate nature of blood should not be ignored. As noted earlier, blood consists of a suspension of red blood cells, white blood cells, and platelets in plasma. Strictly speaking, the describing equations and calculation presented earlier for both flow and pressure drop are valid only under restricted conditions. The equations are not strictly valid if[8]:

1. The particle is not "very" small.
2. The particle is not a smooth rigid sphere.
3. The particle is located "near" the surrounding walls containing the fluid.
4. The particle is located "near" one or more other particles.
5. The motion of the fluid and particle is multidimensional.
6. Brownian motion effect is significant.

Each of the above topics is treated briefly below.[8] Despite the above limitations, it should be noted that these effects are rarely included in any traditional engineering analysis of a fluid-particle system. It is more common to use an empirical constant or factor that would account for all of these various effects.

1. At very low values of the Reynolds number, when particles approach sizes comparable to the mean free path of the fluid molecules, the medium can no longer be regarded as continuous since particles can move between the molecules at a faster rate than predicted by the aerodynamic theories, which leads to standard drag coefficients. To allow for this "slip," Cunningham introduced a multiplying correction factor to Stokes' law; details are available in Chapter 23.

2. For particles having shapes other than spherical, it is necessary to specify the size and geometric form of the body and its orientation with respect to the direction of flow of the blood. One major dimension is chosen as the characteristic length and other important dimensions are given as ratios to the chosen one. Such rations are called shape factors. Nonspherical bodies generally tend to orient in a preferred direction during motion; actually, orientation is another effect that needs to be considered. For example, at high Reynolds numbers, a disk always settles horizontally with its flat face perpendicular to its motion; a streamlined shape, on the other hand, falls nose down into its position of least resistance. At low Reynolds numbers, a particle such as a disk or ellipsoid with three perpendicular symmetry planes can settle in any position.

3. In most engineering applications, the particles are negligibly small when compared to the dimensions of the conduit; therefore, wall effects can usually be neglected. However, in blood flow wall effects can be more pronounced. Theoretical considerations or experimental work has established factors for modifying the describing equation to account for wall effects under different sets of circumstances.

4. In most biomedical applications, it is almost inevitable that large numbers of particles will be involved. It is also very likely that the particles will influence one another. Therefore, equations for the fluid resistance to the motion of single particles have to be modified to account for such interactions between particles. Particle interactions can become appreciable even at very low concentrations. Even a particle-volume concentration (the ratio of particle volume to total volume) of 0.2% will increase the fluid resistance to particle motion by about 1%. In general, for volume concentrations below 1%, the effect of particle interactions may be neglected.

5. Previous discussions on particle motion were limited to the unidimensional case, that is, the parallel movement of a particle relative to the fluid. However, this is often not the general case. This situation is defined as multidimensional flow. Equations must then be developed to describe each of the velocity components of the particle. The main complication arises if more than one relative velocity component exists.

6. As a result of bombardment by the molecules of the fluid medium, suspended particles will be subjected to a random motion known as Brownian movement. This effect becomes significant only when the particles are very small and their mass approaches that of the fluid molecules. Einstein[9] showed that Brownian movement, in general, becomes significant only for particles less than about 0.05 μm.

Illustrative Example 33.13   A 10-year biomedical research study generated data for the average number of heart attacks per 100 individuals in the same age group (H) and gender as a function of the number of quarts of beer that have been consumed per month (q). See Table 33.4.

Table 33.4   Data for Illustrative Example 33.13

H	q
3	10
3.6	30
4.44	100
5.19	250

Using the data in Table 33.4, estimate the coefficients a and b in the heart attack equation below:

$$H = a(q)^b$$

Solution   Linearize the equation

$$\ln(H) = \ln(a) + (b)\ln(q)$$

Change variables to Y and X.
   Set $\ln(H)$ equal to Y and $\ln(q)$ equal to X

$$Y = A + BX$$

where

$$A = \ln(a)$$
$$B = b$$

Regress the above data (4 data points) using the method of least squares. The method of least squares requires that the sum of the errors squared between the data and model is minimized

$$\ln(3) = A + B\ln(10)$$
$$\ln(3.6) = A + B\ln(30)$$
$$\ln(4.44) = A + B\ln(100)$$
$$\ln(5.19) = A + B\ln(250)$$

Generate the linear equation coefficients A and B. These may be obtained through a longhand calculation. However, they are more often obtained with the aid of computer software:

$$A = 0.704$$
$$B = 0.171$$

Obtain constants a and b by taking the inverse natural logarithm of A to obtain a

$$a = 2.02$$
$$b = 0.171$$

The heart attack equation is therefore

$$H = 2.02q^{0.171}$$

Information on data regression using the method of least squares is available in the literature.[10]

The reader should note that if more heart attack–quarts of beer data becomes available, thus increasing the number of points, then the calculated line may not be the best representation of all the data; the least squares solution should then be recomputed using all the data. In addition, the assumed model, for example, linear, may not be the "best" model.

33.7 BIOMEDICAL ENGINEERING OPPORTUNITIES

Key activities in this profession include[2,3]:

1. Development of improved species of plants and animals for food production.
2. Invention of new medical diagnostic tests for diseases.
3. Production of synthetic vaccines from clone cells.
4. Bioenvironmental engineering to protect human, animal, and plant life from toxicants and pollutants.
5. Study of protein-surface interactions.
6. Modeling of flow dynamics.
7. Modeling of mass transfer through membranes.
8. Modeling of the growth of kinetics of yeast and hybridoma cells.
9. Research in immobilized enzyme technology.
10. Development of therapeutic proteins and monoclonal antibodies.
11. Development of artificial hearts.

New applications that have emerged over the last half-century include[2,3]:

1. Application of engineering system analysis (physiologic modeling, simulation, and control of biological problems).
2. Detection, measurement, and monitoring of the physiologic signals (i.e., biosensors and biomedical instrumentation).
3. Diagnostic interpretation via signal-processing techniques of bioelectric data.

4. Therapeutic and rehabilitation procedures and devices (rehabilitation engineering).

5. Devices for replacement or augmentation of bodily functions (artificial organs).

6. Computer analysis of patient-related data and clinical decision-making (i.e., medical informatics and artificial intelligence).

7. Medical imaging, i.e., the graphical display of anatomic detail or physiologic functions.

8. The creation of new biologic products (i.e., biotechnology and tissue engineering).

Job-related activities of biomedical engineers include[2,3]:

1. Research into new materials for implanted artificial organs.

2. Development of new diagnostic instruments for blood analysis.

3. Writing software for analysis of medical research data.

4. Analysis of medical device hazards for safety and efficacy.

5. Development of new diagnostic imaging systems.

6. Design of telemetry systems for patient monitoring.

7. Design of biomedical sensors.

8. Development of expert systems for diagnosis and treatment of diseases.

9. Design of closed-loop control systems for drug administration.

10. Modeling of the physiologic systems of the human body.

11. Design of instrumentation for sports medicine.

12. Development of new dental materials.

13. Design of communication aids for individuals with disabilities.

14. Study of pulmonary fluid dynamics.

15. Study of biomechanics of the human body.

16. Development of material to be used as replacement for human skin.

17. Applications of nanotechnology to many of the above activates.

Obviously, these three lists are not intended to be all-inclusive. Many other applications are evolving that use the talents and skills of the biomedical engineer. This is a field where there is continual change and creation of new areas due to the rapid advancement in technology.

In terms of job opportunities, biomedical engineers are employed in universities, in industry, in hospitals, in research facilities of educational and medical institutions, in teaching, and in government and regulatory agencies. They can often serve a coordinating or interfacing function, using their background in both the engineering and medical fields to combine sound knowledge of engineering and the sciences.

Regulatory issues are a constant cause for concern and, perhaps, justifiably so. To satisfy safety regulations, most biomedical systems must have documented analysis of risk to show that they were designed, built, tested, delivered, managed, and used according to a planned and approved process. The two key regulatory agencies in the US are the Food and Drug Administration (FDA) and the Consumer Product Safety Commission.

In conclusion, biomedical engineering is now an important vital interdisciplinary field. The ultimate role of the biomedical engineer and the profession is to serve society. The great potential, challenge, and promise in this relatively new endeavor offers both technological and humanitarian benefits. The possibilities appear to be unlimited.

REFERENCES

1. J. Enderle, S. Blanchard, and J. Bronzino, "Introduction to Biomedical Engineering," 2nd edition, Elsevier/Academic Press, New York, 2000.
2. J. Bronzino (ed.), "Biomedical Engineering Fundamentals," 3rd edition, CRC/Taylor & Francis, Boca Raton, FL, 2000.
3. S. Vogel, "Life in Moving Fluids," 2nd edition, Princeton University Press, Princeton, NJ, 1994.
4. L. Theodore, personal notes, 2008.
5. D. Green and R. Perry (eds), "Perry's Chemical Engineers' Handbook," 8th edition, McGrawHill, New York, 2008.
6. K. Konstantopoulos, class notes.
7. M. LaBarbara, "Principles of Design of Fluid Transport Systems in Zoology," *Science*, 249, 992–1000, 1990.
8. L. Theodore, "Air Pollution Control Equipment," John Wiley & Sons, Hoboken, NJ, 2009.
9. A. Einstein, *Ann. Physics*, 19, 289, 1906.
10. S. Shafer and L. Theodore, "Probability and Statistics Application for Environmental Science," CRC Press/Taylor & Francis, Boca Raton, FL, 2007.

NOTE: Additional problems are available for all readers at www.wiley.com. Follow links for this title.

OPEN-ENDED PROBLEMS

34.1 INTRODUCTION

The educational literature provides frequent references to individuals, particularly engineers and scientists, that have different learning styles, and in order to successfully draw on these different styles, a variety of approaches can be employed. One such approach involves the use of open-ended problems.

The term "open-ended problem" has come to mean different things to different people. It basically describes an approach to the solution of a problem and/or situation for which there is usually not a unique solution. Three literature sources[1–3] provide sample problems that can be used when this educational tool is employed.

The authors of this book have applied this somewhat unique approach and has included numerous open-ended problems in several course offerings at Manhattan College. Student comments for the graduate course "Accident and Emergency Management" were recently tabulated. Student responses to the question "What aspects of this course were most beneficial to you?" are listed below:

1. "The open-ended questions gave engineers a creative license. We don't come across many of these opportunities."
2. "Open-ended questions allowed for candid discussions and viewpoints that the class may not have been otherwise exposed to."
3. "The open-ended questions gave us an opportunity to apply what we were learning in class with subjects we have already learned and gave us a better understanding of the course."

Fluid Flow for the Practicing Chemical Engineer. By J. Patrick Abulencia and Louis Theodore
Copyright © 2009 John Wiley & Sons, Inc.

4. "Much of the knowledge that was learned in this course is applicable to everyday situations and our professional lives."

5. "Open-ended problems made me sit down and research the problem to come up with ways to solve them."

6. "I thought the open-ended problems were inventive and made me think about problems in a better way."

7. "I felt that the open-ended problems were challenging. I, like most engineers, am more comfortable with quantitative problems than qualitative."

In effect, the approach requires teachers to ask questions, to not always accept things at face value, and to select a methodology that provides the most effective and efficient solution. Those who conquer this topic have taken the first step toward someday residing in an executive suite.

The remainder of this chapter addresses a host of topics involved with open-ended problems. The following sections are entitled: Developing Student's Power of Critical Thinking; Creativity; Brainstorming; Inquiring Minds; and Angels on a Pin. The chapter concludes with an applications section that contains eight open-ended Illustrative Examples primarily in the fluid-flow field.

34.2 DEVELOPING STUDENTS' POWER OF CRITICAL THINKING[4]

It has often been noted that we are living in the middle of an information revolution. For more than a decade, that revolution has had an effect on teaching and learning. Teachers are hard-pressed to keep up with the advances in their fields. Often their attempts to keep the students informed are limited by the difficulty of making new material available.

The basic need of both teacher and student is to have useful information readily accessible. Then comes the problem of how to use this information properly. The objectives of both teaching and studying such information are: to assure comprehension of the material and to integrate it with the basic tenets of the field it represents; and, to use comprehension of the material as a vehicle for critical thinking, reasoning, and effective argument.

Information is valueless unless it is put to use; otherwise it becomes mere data. To use information most effectively, it should be taken as an instrument for understanding. The process of this utilization works on a number of incremental levels. Information can be: absorbed; comprehended; discussed; argued in reasoned fashion; written about; and integrated with similar and contrasting information.

The development of critical and analytical thinking is the key to the understanding and use of information. It is what allows the student to discuss and argue points of opinion and points of fact. It is the basis for the student's formation of independent ideas. Once formed, these ideas can be written about and integrated with both similar and contrasting information.

34.3 CREATIVITY

Engineers bring mathematics and science to bear on practical problems, molding materials and harnessing technology for human benefit. Creativity is often a key component in this synthesis; it is the spark motivating efforts to devise solutions to novel problems, design new products, and improve existing practices. In the competitive marketplace, it is a crucial asset in the bid to win the race to build better machines, decrease product delivery times, and anticipate the needs of future generations.[5]

One of the keys to the success of an engineer or a scientist is to generate fresh approaches, processes and products, i.e., they need to be creative. Gibney[5] has detailed how some schools and institutions are attempting to use certain methods that essentially share the same objective: open students' minds to their own creative potential.

Gibney[5] provides information on "The Art of Problem Definition" developed by Rensselaer Polytechnic Institute. To stress critical thinking, they teach a seven step methodology for creative problem development. These steps are provided below:

1. Define the problem.
2. State objective.
3. Establish functions.
4. Develop specifications.
5. Generate multiple alternatives.
6. Evaluate alternatives.
7. Build.

In addition, Gibney[5] identified the phases of the creative process set forth by psychologists. They essentially break the process down into five basic stages:

1. Immersion.
2. Incubation.
3. Insight.
4. Evaluation.
5. Elaboration.

Psychologists have ultimately described the creative process as recursive. At any one of these stages, a person can double back, revise ideas, or gain new knowledge that reshapes his or her understanding. For this reason, being creative requires patience, discipline, and hard work.

Finally, Della Femina[6] recently outlined five "secrets" regarding the creative process:

1. Creativity is ageless.
2. You don't have to be Einstein.

3. Creativity is not an eight hour job.
4. Failure is the mother of all creativity.
5. Dead men don't create.

The reader is left with a thought from Theodore[7]: Creativity usually experiences a quick and quiet death in rooms that house conference tables.

34.4 BRAINSTORMING

Panitz[8] has demonstrated how brainstorming strategies can help engineering students generate an outpouring of ideas. Brainstorming guidelines include:

1. Carefully define the problem upfront.
2. Allow individuals to consider the problem before the group tackles it.
3. Create a comfortable environment.
4. Record all suggestions.
5. Appoint a group member to serve as a facilitator.
6. Keep brainstorming groups small.

A checklist for change was also provided, as detailed below:

1. Adapt.
2. Modify.
3. Magnify.
4. Minify.
5. Put to other uses.
6. Substitute.
7. Rearrange.
8. Reverse.
9. Combine.

34.5 INQUIRING MINDS

In an exceptional and well-written article by Lih[9] entitled *Inquiring Minds*, Lih commented on inquiring minds by saying "You can't transfer knowledge without them." His thoughts (which have been edited) on the inquiring or questioning process follow:

1. Inquiry is an attitude—a very important one when it comes to learning. It has a great deal to do with curiosity, dissatisfaction with the status quo, a desire to dig deeper, and having doubts about what one has been told.

2. Questioning often leads to believing—there is a saying that has been attributed to Confucius: "Tell me, I forget. Show me, I remember. Involve me, I understand." It might also be fair to add: "Answer me, I believe."

3. Effective inquiry requires determination to get to the bottom of things.

4. Effective inquiry requires wisdom and judgment. This is especially true for a long-range intellectual pursuit that is at the forefront of knowledge.

5. Inquiry is the key to successful life-long learning. If one masters the art of questioning, independent learning is a breeze.

6. Questioning is good for the questionee as well. It can help clarify issues, uncover holes in an argument, correct factual and/or conceptual errors, and eventually lead to a more thoughtful outcome.

7. Teachers and leaders should model the importance of inquiry. The teacher/leader must allow and encourage questions and demonstrate a personal thirst for knowledge.

Ultimately, the degree to which one succeeds (or fails) is based in part on one's state of mind or attitude. As President Lincoln once said: "Most people are about as happy as they make their minds to be." William James once wrote: "The greatest discovery of my generation is that human beings can alter their lives by altering their attitude of mind." So, no matter what one does, it is in the hand of that individual to make it a meaningful, pleasurable, and positive experience. This experience will ultimately bring success.

34.6 ANGELS ON A PIN[10]

There is a tale that appeared in print many years ago (there is some uncertainty regarding the source) that dissected the value of an open-ended approach to a particular problem. That story is presented below.

Some time ago I received a call from a colleague who asked if I would be the referee on the grading of an examination question. He was about to give a student a zero for his answer to a physics question, while the student claimed he should receive a perfect score and would if the system were not set up against the student: The instructor and the student agreed to submit this to an impartial arbiter, and I was selected.

I went to my colleague's office and read the examination question: "Show how it is possible to determine the height of a tall building with the aid of a barometer."

The student had answered: "Take a barometer to the top of the building, attach a long rope to it, lower the barometer to the street and then bring it up, measuring the length of the rope. The length of the rope is the height of the building."

I pointed out that the student really had a strong case for full credit since he had answered the question completely and correctly. On the other hand, if full credit was given, it could well contribute to a high grade for the student in his physics course. A high grade is supposed to certify competence in physics, but the answer did not confirm this. I suggested that the student have another try at answering the question I was not surprised that my colleague agreed, but I was surprised that the student did.

I gave the student six minutes to answer the question with the warning that the answer should show some knowledge of physics. At the end of five minutes, he had not written anything. I asked if he wished to give up, but he said no. He had many answers to this problem; he was just thinking of the best one. I excused myself for interrupting him and asked him to please go on. In the next minute, he dashed off his answer which read:

"Take the barometer to the top of the building and lean over the edge of the roof. Drop the barometer, timing its fall with a stopwatch. Then using the formula $S = 1/2\ at^2$, calculate the height of the building.

At this point, I asked my colleague if he would give up. He conceded and I gave the student almost full credit.

In leaving my colleague's office, I recalled that the student had said he had many other answers to the problem, so I asked him what they were. "Oh yes," said the student. "There are a great many ways of getting the height of a tall building with a barometer. For example, you could take the barometer out on a sunny day and measure the height of the barometer and the length of its shadow, and the length of the shadow of the building and by the use of a simple proportion, determine the height of the building."

"Fine," I asked. "And the others?"

"Yes," said the student. "There is a very basic measurement method that you will like. In this method you take the barometer and begin to walk up the stairs. As you climb the stairs, you mark off the length of the barometer along the wall. You then count the number of marks, and this will give you the height of the building in barometer units. A very direct method."

"Of course, if you want a more sophisticated method, you can tie the barometer to the end of a string, swing it as a pendulum, and determine the value of 'g' at the street level and at the top of the building. From the difference of the two values of 'g', the height of the building can be calculated."

Finally, he concluded, there are many other ways of solving the problem. "Probably the best," he said, "is to take the barometer to the basement and knock on the superintendent's door. When the superintendent answers, you speak to him as follows: "Mr. Superintendent, here I have a fine barometer. If you tell me the height of this building, I will give you this barometer."

At this point I asked the student if he really did know the conventional answer to this question. He admitted that he did, said that he was fed up with high school and college instructors trying to teach him how to think, using the "scientific method," and to explore the deep inner logic of the subject in a pedantic way, as is often done in the new mathematics, rather than teaching him the structure of the subject. With this in mind, he decided to revive scholasticism as an academic lark to challenge the Sputnik-panicked classrooms of America.

34.7 APPLICATIONS

Several of the open-ended Illustrative Examples have been drawn from the literature[1–2] and class notes from Theodore,[3] keying primarily on fluid flow issues.

Illustrative Example 34.1 You are asked to think of as many ways as possible to measure the viscosity of a fluid (consider both a gas and a liquid).

Solution Several types of equipment are available; they fall into five general categories:

1. Rotational type. This type measures the torque resulting from the rotation of a spindle inside a sample chamber through which the sample flows continuously.
2. Float or piston type. The float type measures the position of a specially shaped float inside a tapered tube through which the fluid flows at a constant rate. This equipment is similar to the rotameters used for flow measurement.
3. Time the discharge through a restriction, for example, an orifice or nozzle.
4. Time the fall of a ball (or obstacle) or rise of a bubble.
5. Capillary type. This type measures the pressure drop resulting from constant flow of the fluid through a capillary tube of specified diameter and length.

Because viscosity depends on temperature, the viscosity measurement must be thermostated with a heater or cooler.

Illustrative Example 34.2 Devise any method that involves the use of a "keftethe" (Greek version of an Italian meatball) to obtain the viscosity of a fluid.

Solution The simplest approach is to assume the "keftethe" can be physically modeled as a sphere. The sphere's terminal settling velocity in the fluid can be measured. By applying one of the particle dynamics equations developed in Chapter 23, the viscosity can be calculated if all physical property data is available. For example, if Stokes' law applies,

$$v = \frac{g\rho_p d_p^{\,2}}{18\mu}$$

can be rearranged and solved for μ

$$\mu = \frac{g\rho_p d_p^{\,2}}{18v}$$

Illustrative Example 34.3 A proposal plans to add fine particulates to a natural gas pipeline to reduce pressure drop. Comment on the value of the proposal.

Solution This a particularly interesting proposal. There is evidence that indicates that there is a reduction in pressure drop when fine particulates are added to a flowing fluid. Several research projects were initiated in this area in the 1980s and the data gathered verified the above statement. The proposal, however, has questionable merit because of the ultimate fate/disposal of the particulates. In effect, by "solving" one problem, another is being created. Specifically, the particles might not be combustible; if combustible, they might form harmful end products.

Illustrative Example 34.4   An undergraduate environmental engineering student has suggested that the kinetic energy of a moving gas stream normally discharged from a plant be recovered as part of an energy conservation measure. Comment on the idea.

Solution   On the surface, this appears to be a project worthy of consideration. However, the energy content of a gas stream, even at extremely high velocities, possesses an insignificant (relatively speaking) amount of energy. For example, if 10,000 acfm (60°F, 1 atm) of a gas stream (that may be considered air) is discharged from a stack at 50 ft/s, its kinetic energy (KE) is:

$$\text{KE} = \frac{1}{2}mv^2 = \frac{10,000}{379}\frac{(29)(50)^2}{32.2(60)}$$
$$= 1200\,\text{ft} \cdot \text{lb}_f/\text{s} = 2.0\,\text{Btu/s}$$

This would not appear to be a cost-effective option for recovering energy.

Illustrative Example 34.5   List factors that need to be considered in selecting the pipe diameter for an underground Alaskan crude oil pipeline.[11]

Solution

1. *Temperature of the pipe and crude oil.* The temperature at which the crude oil must be kept at for the length of the run is important to selecting the material for the pipe and the diameter. The temperature also effects the tension on the pipe in expansion and contraction at different temperatures.
2. *Pressure.* The pressure throughout the length of the pipe is an important factor that diameter could greatly effect. Smaller diameters allow for lower pressures and larger diameters need larger operating pressures. The discharge pressure is the pressure at the end of the pipeline that is important to the design diameter and the length of the pipe.
3. *Velocity.* The desired velocity at which the crude oil will flow effects the pipe diameter and length since a smaller diameter is needed for higher velocities.
4. *Number of bends.* Bending in pipes greatly affects pressure throughout the pipeline and should be considered when choosing the pipe diameter. On a pipeline, there could be many bends that will effect discharge pressure and velocity.
5. *Surface finish.* The finish of the pipe surface effects the movement of the crude oil through the pipeline. Since crude oil is very viscous the surface finish could effect the velocity and pressure throughout the system. A smooth surface would decrease friction and allow a better flow.

6. *Total length of run.* Since the Alaskan pipeline will be very long, this is a key consideration in choosing a diameter that could hold up to harsh conditions. Throughout the length of the run there could be many bends and lifting of the pipeline that could damage the pipe and cause a leak. Therefore, the thickness of the pipe is very important.

7. *Pitching of pipe.* Pitching refers to whether the pipe is lifted above ground at different areas. This results in straining of the pipe and velocity changes. Therefore, a diameter that can withstand pitching is also a factor.

8. *Climate.* Permafrost is defined as any rock or soil material that has remained below 32°F for more than two years. Warm permafrost remains just below 32°F. Therefore, any additional heat will cause thawing, which in turn could lower the stability of the soil.

9. *Slope.* The pipeline should (if possible) gradually slope downward to reduce the chance of particles settling out of the crude oil, which could eventually clog the pipe.

10. *Earthquake protection.* Due to certain faults that the pipeline will cross, there must be an earthquake monitoring and protection system in place for the pipeline.

11. *Properties of oil.* Oil pumped out from different sites has different physical as well as chemical properties. For example, if the oil contains a lot of impurities, the pipe is more likely to get corroded and hence proper coating will be required to prevent corrosion. Zinc coating is commonly used for Alaskan crude oil pipelines. Another physical property is the viscosity of oil that is being pumped. In addition, the denser the oil, the greater will be the pressure and thus the pipe size will need to be larger.

Illustrative Example 34.6   A ventilation system in a laboratory is no longer capable of pulling the required air flow to properly prevent toxic fumes from escaping work hoods. Rather than purchase a new fan to increase the flow rate, suggest other solutions to the problem.

Solution

1. Reduce the generation of the fumes.
2. Reduce the pressure drop across the valves, piping, etc., of the system.
3. Increase the opening (if possible) of the hood to reduce the pressure drop.
4. Provide another exhaust.

Illustrative Example 34.7   A cooling water pump is no longer capable of delivering the required flow rate to a highly exothermic reactor. Rather than purchase a new pump, you have been asked to list and/or describe what steps can be taken to resolve the problem.

Solution   The obvious option is to replace the pump. Since that is not a viable option, other options can include:

1. Carefully check the pump, including the clogging of screens and/or intakes, impellers.
2. Increase pipe size(s), that is, the diameter.
3. Decrease pipe length.
4. Eliminate unimportant valves, expansion and contraction joints, etc., in order to reduce the pressure drop.
5. Decrease the viscosity of the water by increasing its temperature.
6. Use a different cooling medium altogether or add an "additive" to the cooling water to decrease the viscosity and/or increase the heat capacity of the cooling medium.
7. Create an endothermic side reaction in the system.
8. Use synthetic lubricants to reduce friction.

Illustrative Example 34.8[12]   During the heat of the space race in the 1960s, the U.S. National Aeronautics and Space Administration (NASA) decided it needed a ball point pen to write in the zero gravity confines of its space capsules. Prepare a solution to the problem.

Solution   Since the gravity is zero, a pressure force is needed to force the ink to flow out. A simple device to solve the problem is provided in Fig. 34.1.

When the screw moves, ink will be transferred to the edge of the ball point.

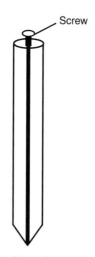

Figure 34.1   Zero gravity pen.

After considerable research and development, the "Astronaut Pen" was developed at a cost of $1 million. The pen worked and also enjoyed some modest success as a novelty item back here on Earth. The Soviet Union, faced with the same problem, used a pencil.

REFERENCES

1. A. Flynn and L. Theodore, "An Air Pollution Control Equipment Design Course for Chemical and Environmental Engineering Students Using an Open-Ended Problem Approach", ASEE Meeting, Rowan University, NJ, 2001.
2. A. Flynn, J. Reynolds, and L. Theodore, "Courses for Chemical and Environmental Engineering Students Using an Open-Ended Problem Approach", AWMA Meeting, San Diego, CA, 2003.
3. L. Theodore, class notes, 1999–2003.
4. Manhattan College Center for Teaching, "Developing Students' Power of Critical Thinking", Bronx, NY, January 1989.
5. K. Gibney, "Awakening Creativity", *ASEE Promo*, March 1988.
6. J. Della Femina, "Jerry's Rules", *Modern Maturity*, March–April 2000.
7. L. Theodore, personal notes, 1998.
8. B. Panitz, "Brain Storms", *ASEE Promo*, March 1998.
9. M. Lih, "Inquiring Minds", *ASEE Promo*, December 1998.
10. "Angels on a Pin", author, source, and date unknown.
11. D. Posillico, homework solution submitted to L. Theodore, 2002.
12. *New York Times*, November 11, 1993.

NOTE: Additional problems are available for all readers at www.wiley.com. Follow links for this title.

APPENDIX

A. TABLES

B. FIGURES

A. TABLES

Table A.1 Common engineering conversion factors

Length

1 ft = 12 in. = 0.3048 m, 1 yard = 3 ft

1 mi = 5280 ft = 1609.344 m

1 nautical mile (nmi) = 6076 ft

Mass

1 slug = 32.174 lb = 14.594 kg

1 lb = 0.4536 kg = 7000 grains

Acceleration and Area

1 ft/s^2 = 0.3048 m/s^2

1 ft^2 = 0.092903 m^2

Mass Flow and Mass Flux

1 slug/s = 14.594 kg/s, 1 lb/s = 0.4536 kg/s

1 kg/m$^2 \cdot$s = 0.2046 lb/ft$^2 \cdot$s = 0.00636 slug/ft$^2 \cdot$s

Pressure

1 lb$_f$/ft^2 = 47.88 Pa, 1 torr = 1 mm Hg

1 psi = 144 psf, 1 bar = 10^5 Pa

1 atm = 2116.2 psf = 14.696 psi = 101,325 Pa = 29.9 in Hg = 33.9 ft H$_2$O

Power

1 hp = 550 (ft$\cdot$lb$_f$)/s = 745.7 W

1 (ft$\cdot$lb$_f$)/s = 1.3558 W

1 Watt = 3.4123 Btu/h = 0.00134 hp

Specific Weight

1 lb$_f$/ft^3 = 157.09 N/m^3

Viscosity

1 slug/(ft$\cdot$s) = 47.88 kg/(m$\cdot$s) = 478.8 poise (p)

1 p = 1 g/(cm$\cdot$s) = 0.1 kg/(m$\cdot$s)= 0.002088 slug/(ft$\cdot$s)

Temperature Scale Readings

°F = (9/5)°C + 32

°C = (5/9)(°F − 32)

°R = °F + 459.69

K = °C + 273.16

°R = (1.8)K

Specific Heat or Gas Constant*

1 (ft$\cdot$lb$_f$)/(slug$\cdot$°R) = 0.16723 (N$\cdot$m)/(kg$\cdot$K)

1 Btu/(lb$\cdot$°R) = 4186.8 J/(kg$\cdot$K)

Volume

1 ft^3 = 0.028317 m^3 = 7.481 gal, 1 bbl = 42 U.S. gal

1 U.S. gal = 231 in^3 = 3.7853 L = 4 qt = 0.833 Imp. gal

1 L = 0.001 m^3 = 0.035315 ft^3 = 0.2642 U.S. gal

Density

1 slug/ft^3 = 515.38 kg/m^3, 1 g/cm^3 = 1000 kg/m^3

1 lb/ft^3 = 16.0185 kg/m^3, 1 lb/in^3 = 27.68 g/cm^3

Velocity

1 ft/s = 0.3048 m/s, 1 knot = 1 nmi/h = 1.6878 ft/s

1 mi/h = 1.4666666 ft/s (fps) = 0.44704 m/s

Volume Flow

1 gal/min = 0.002228 ft^3/s = 0.06309 L/s

1 million gal/day = 1.5472 ft^3/s = 0.04381 m^3/s

Force and Surface Tension

1 lb$_f$ = 4.448222 N = 16 oz, 1 dyne = 1 g$\cdot$cm/s^2 = 10^{-5} N

1 kg$_f$ = 2.2046 lb$_f$ = 9.80665 N

1 U.S. (short) ton = 2000 lb$_f$, 1 N = 0.2248 lb$_f$

1 N/m = 0.0685 lb$_f$/ft

Energy and Specific Energy

1 ft$\cdot$lb$_f$ = 1.35582 J, 1 hp$\cdot$h = 2544.5 Btu

1 Btu = 252 cal = 1055.056 J = 778.17 ft$\cdot$lb$_f$

1 cal = 4.1855 J, 1 ft$\cdot$lb$_f$/lb$_m$ = 2.9890 J/kg

Heat Flux

1 W/m^2 = 0.3171 Btu/(h$\cdot$ft^2)

Kinematic Viscosity

1 ft^2/h = 2.506 × 10^{-5} m^2/s, 1 ft^2/s = 0.092903 m^2/s

1 stoke (st) = 1 cm^2/s = 0.0001 m^2/s = 0.001076 ft^2/s

Thermal Conductivity*

1 cal/(s$\cdot$cm$\cdot$°C) = 242 Btu/(h$\cdot$ft$\cdot$°R)

1 Btu/(h$\cdot$ft$\cdot$°R) = 1.7307 W/(m$\cdot$K)

---

*Note that the intervals in absolute (Kelvin) and °C are equal. Also, 1°R = 1°F.

Latent heat: 1 J/kg = 4.2995 × 10^{-4} Btu/lb = 10.76 lb$_f$ $\cdot$ ft/slug = 0.3345 lb$_f$ $\cdot$ ft/lb, 1 Btu/lb = 2325.9 J/kg.

Heat transfer coefficient: 1 Btu/(h$\cdot$ft$^2 \cdot$°F) = 5.6782 W/(m$^2 \cdot$°C).

Heat generation rate: 1 W/m^3 = 0.09665 Btu/(h$\cdot$ft^3).

Heat transfer per unit length: 1 W/m = 1.0403 Btu/(h$\cdot$ft).

Mass transfer coefficient: 1 m/s = 11.811 ft/h, 1 lbmol/(h$\cdot$ft^2) = 0.0013562 kgmol/(s$\cdot$m^2).

Liquid	Density, kg/m^3	Dynamic Viscosity, μ, $kg/m \cdot s$ ($\times 10^4$)	Kinematic Viscosity, ν, m^2/s ($\times 10^6$)	Surface Tension, N/m ($\times 10^2$)	Vapor Pressure, kPa	Sound Velocity, m/s
Acetone	785	3.16	0.403	2.31	27.6	1174
Ammonia	608	2.20	0.362	2.13	910.0	
Benzene	881	6.51	0.739	2.88	10.1	1298
Carbon disulfide	1272					
Carbon tetrachloride	1590	9.67	0.608	2.70	1.20	924
Castor oil	970	9000	927.8			1474
Crude oil	856	72	8.4	3.0		
Engine oil (unused)	888	7994	900.2			
Ethanol (or ethyl alcohol)	789	11	1.4	2.28	5.7	1144
Ethylene glycol	1117	214	19.16	3.27		1644
Freon-12	1330	2.63	0.198	1.58		
Fuel oil, heavy	908	1324	145.9			
Fuel oil, medium	854	32.7	3.82			
Gasoline	680	2.92	0.429	2.16	55.1	
Glycerin	1260	14,900	1183	6.33	0.14	1909
Kerosene	804	1.92	0.239	2.8	3.11	1320
Mercury	13,550	15.6	0.115	48.4	1.1×10^{-6}	1450
Methanol	791	5.98	0.756	2.25	13.4	1103
Milk (skimmed)	1041	14	1.34			
Milk (whole)	1030	21.2	2.06			
Olive oil	919	840	91.4			
Pentane	624					
Soybean oil	919	400	43.5	3.6		
SAE 10 oil	917	1040	113.4	3.5		
SAE 30 oil	917	2900	316.2			
Seawater	1025	10.7	1.04	7.28	2.34	1535
Turpentine	862	14.9	1.73			
Water	998	10.0	1.06	7.28	2.34	1498

Example: At 20°C, the properties liquid methanol are: density = 791 kg/m^3 (or SG = 0.791), dynamic viscosity = 0.000598 $kg/m \cdot s$ (or 0.598 cP), kinematic viscosity = 0.756 × 10⁻⁶ m^2/s (0.756 cP = 8.14 × 10⁻⁶ ft^2/s), surface tension = 0.0225 N/m (0.00154 lb_f/ft), vapor (or saturation) pressure = 13,400 Pa (1.943 psi).

Table A.3 Properties of selected gases at 1 atm and 20°C (68°F)

Gas	Molecular Weight	Density, kg/m³	Viscosity Dynamic, μ, kg/m·s ($\times 10^5$)	Viscosity Kinematic, ν, m²/s ($\times 10^6$)	Ratio of Specific Heats, k	T_{crit}, K	P_{crit}, atm
Acetylene	26	1.09	0.97	8.3	1.30	309.5	61.6
Air (dry)	28.96	1.20	1.80	15.0	1.40	133	37
Ammonia	17.03	0.74	1.01	13.6	1.31	405	111.3
Argon	39.944	1.66	2.24	13.5	1.67		
Butane	58.1	2.49			1.11	425.2	37.5
Carbon dioxide	44.01	1.83	1.48	8.09	1.30	304	72.9
Carbon monoxide	28.01	1.16	1.82	15.7	1.40	133	34.5
Chlorine	70.91	2.95	1.03	3.49	1.34	417	76.1
Ethane	30.07	1.25	0.85	6.8	1.19	305	48.2
Ethylene	28	1.17	0.97	8.3	1.22	283.1	50.5
Helium	4.003	0.166	1.97	118.7	1.66	5.26	2.26
Hydrogen	2.016	0.0838	0.905	108.0	1.41	33	12.8
Hydrogen chloride	36.5	1.53	1.34	8.76	1.41	324.6	81.5
Hydrogen sulfide	34.1	1.43	1.24	8.67	1.30	373.6	88.9
Methane	16.04	0.667	1.34	20.1	1.32	190	45.8
Methyl chloride	50.5	2.15			1.20	416.1	65.8
Natural gas	19.5	0.804			1.27		
Nitrogen	28.02	1.16	1.76	15.2	1.40	126	33.5
Nitrogen oxide (NO)	30.01	1.23	1.90	15.4	1.40	179	65.0
Nitrous oxide (N₂O)	44.02	1.83	1.45	7.92	1.31	309	71.7
Oxygen	32.0	1.36	2.00	14.7	1.40	154	49.7
Propane	44.1	1.88			1.15	369.9	42.0
Sulfur dioxide	64	2.66	1.38	5.2	1.29	430	77.8
Water vapor	18.02	0.749	1.02	13.6	1.33	647	218.3

Example: At 20°C, the properties argon gas are: molecular weight = 39.944, density = 1.66 kg/m³ (0.00322 slug/ft³ = 0.104 lb/ft³), dynamic viscosity = 0.0000224 kg/m·s (0.0224 cP = 4.68 × 10⁻⁷ slug/ft·s = 1.51 × 10⁻⁵ lb/ft·s), kinematic viscosity = 13.5 × 10⁻⁶ m²/s (13.5 cSt = 1.45 × 10⁻⁴ ft²/s = 0.523 ft²/h), specific heat ratio = 1.67.

Table A.4 Properties of water at 1 atm (critical point 374°C, 22.09 MPa)

Temperature		Density, ρ		Absolute (Dynamic) Viscosity, μ		Kinematic Viscosity, ν		Surface Tension, σ, N/m	Vapor Pressure, p′, kPa
°C	°F	kg/m³	slug/ft³	kg/m·s ($\times 10^3$)	slug/ft·s ($\times 10^5$)	m²/s ($\times 10^6$)	ft²/s ($\times 10^5$)		
0	32	1000	1.940	1.788	0.373	1.788	1.925	0.0756	0.611
5	41	1000	1.940	1.518	0.317	1.519	1.635	0.0749	0.87
10	50	1000	1.940	1.307	0.273	1.307	1.407	0.0742	1.227
15	59	999	1.938	1.139	0.238	1.139	1.226	0.0735	1.70
20	68	998	1.937	1.003	0.209	1.005	1.082	0.0728	2.337
25	77	997	1.934	0.890	0.186	0.893	0.961	0.0720	3.17
30	86	996	1.932	0.799	0.167	0.802	0.864	0.0712	4.242
40	104	992	1.925	0.657	0.137	0.662	0.713	0.0696	7.375
50	122	988	1.917	0.548	0.114	0.555	0.597	0.0679	12.34
60	140	983	1.908	0.467	0.975	0.475	0.511	0.0662	19.92
70	158	978	1.897	0.405	0.846	0.414	0.446	0.0644	31.16
80	176	972	1.886	0.355	0.741	0.365	0.393	0.0626	47.35
90	194	965	1.873	0.316	0.660	0.327	0.352	0.0608	70.11
100	212	958	1.859	0.283	0.591	0.295	0.318	0.0589	101.33

Example: At 50°C (122°F) $\rho = 988$ kg/m³ (1.917 slug/ft³), $\mu = 0.548 \times 10^{-3}$ kg/m·s (0.114 $\times 10^{-5}$ slug/ft·s), $\nu = 0.555 \times 10^{-6}$ m²/s (0.597 $\times 10^{-5}$ ft²/s), $\sigma = 0.0679$ N/m (0.00465 lb$_f$/ft), vapor pressure = 12,340 Pa (1.79 psi).

557

Table A.5 Dimensions, capacities, and weights of standard steel pipes

Nominal Pipe Size, in	Outside Diameter, in	Schedule Number	Wall Thickness, in	Inside Diameter (ID), in	Cross-sectional Area of Metal, in^2	Inside Sectional Area, ft^2	Pipe Weight, lb/ft
$\frac{1}{8}$	0.405	40	0.068	0.269	0.072	0.00040	0.24
		80	0.095	0.215	0.093	0.00025	0.31
$\frac{1}{4}$	0.540	40	0.088	0.364	0.125	0.00072	0.42
		80	0.119	0.302	0.157	0.00050	0.54
$\frac{3}{8}$	0.675	40	0.091	0.493	0.167	0.00133	0.57
		80	0.126	0.423	0.217	0.00098	0.74
$\frac{1}{2}$	0.840	40	0.109	0.622	0.250	0.00211	0.85
		80	0.147	0.546	0.320	0.00163	1.09
$\frac{3}{4}$	1.050	40	0.113	0.824	0.333	0.00371	1.13
		80	0.154	0.742	0.433	0.00300	1.47
1	1.315	40	0.133	1.049	0.494	0.00600	1.68
		80	0.179	0.957	0.639	0.00499	2.17
$1\frac{1}{4}$	1.660	40	0.140	1.380	0.668	0.01040	2.27
		80	0.191	1.278	0.881	0.00891	3.00
$1\frac{1}{2}$	1.900	40	0.145	1.610	0.800	0.01414	2.72
		80	0.200	1.500	1.069	0.01225	3.63

2	2.375	40	0.154	2.067	1.075	0.02330	3.65
		80	0.218	1.939	1.477	0.02050	5.02
$2\frac{1}{2}$	2.875	40	0.203	2.469	1.704	0.03322	5.79
		80	0.276	2.323	2.254	0.02942	7.66
3	3.500	40	0.216	3.068	2.228	0.05130	7.58
		80	0.300	2.900	3.016	0.04587	10.25
$3\frac{1}{2}$	4.000	40	0.226	3.548	2.680	0.06870	9.11
		80	0.318	3.364	3.678	0.06170	12.51
4	4.500	40	0.237	4.026	3.17	0.08840	10.79
		80	0.337	3.826	4.41	0.07986	14.98
5	5.563	40	0.258	5.047	4.30	0.1390	14.62
		80	0.375	4.813	6.11	0.1263	20.78
6	6.625	40	0.280	6.065	5.58	0.2006	18.97
		80	0.432	5.761	8.40	0.1810	28.57
8	8.625	40	0.322	7.981	8.396	0.3474	28.55
		80	0.500	7.625	12.76	0.3171	43.39
10	10.75	40	0.365	10.020	11.91	0.5475	40.48
		80	0.594	9.562	18.95	0.4987	64.40
12	12.75	40	0.406	11.938	15.74	0.7773	53.36
		80	0.688	11.374	26.07	0.7056	88.57

Table A.6   Dimensions of heat exchanger tubes

Tube OD, in	B.W.G. Gauge	Thickness, in	Tube Inside Diameter (ID), in	Flow Area, in^2	Surface Area, Per Foot of Length, ft	
					External	Internal
$\frac{1}{4}$	22	0.028	0.194	0.0295	0.0655	0.0508
$\frac{1}{4}$	24	0.022	0.206	0.0333	0.0655	0.0539
$\frac{1}{2}$	18	0.049	0.402	0.1269	0.1309	0.1052
$\frac{1}{2}$	20	0.035	0.430	0.1452	0.1309	0.1126
$\frac{1}{2}$	22	0.028	0.444	0.1548	0.1309	0.1162
$\frac{3}{4}$	10	0.134	0.482	0.1825	0.1963	0.1262
$\frac{3}{4}$	14	0.083	0.584	0.2679	0.1963	0.1529
$\frac{3}{4}$	16	0.065	0.620	0.3019	0.1963	0.1623
$\frac{3}{4}$	18	0.049	0.652	0.3339	0.1963	0.1707
1	8	0.165	0.670	0.3526	0.2618	0.1754
1	14	0.083	0.834	0.5463	0.2618	0.2183
1	16	0.065	0.870	0.5945	0.2618	0.2278
1	18	0.049	0.902	0.6390	0.2618	0.2361
$1\frac{1}{4}$	8	0.165	0.920	0.6648	0.3272	0.2409
$1\frac{1}{4}$	14	0.083	1.084	0.9229	0.3272	0.2838
$1\frac{1}{4}$	16	0.065	1.120	0.9852	0.3272	0.2932
$1\frac{1}{4}$	18	0.049	1.152	1.042	0.3272	0.3016
2	11	0.120	1.760	2.433	0.5236	0.4608
2	12	0.109	1.782	2.494	0.5236	0.4665
2	13	0.095	1.810	2.573	0.5236	0.4739
2	14	0.083	1.834	2.642	0.5236	0.4801

(1 in. = 25.4 mm; 1 in.^{2} = 645.16 mm^2; 1 ft = 0.3048 m; 1 ft^2 = 0.0929 m^2).

Table A.7 Properties of saturated liquids

Temperature T, °C	Density ρ, kg/m^3	Heat Capacity c_p, kJ/kg·°C	Kinematic Viscosity ν, m^2/s	Thermal Conductivity k, W/m·°C	Diffusivity α, m^2/s	Prandtl Number Pr	Thermal Expansion Coefficient β, K^{-1}
			Ammonia, NH$_3$				
−50	703.69	4.463	0.435×10^{-6}	0.547	1.742×10^{-7}	2.60	
−40	691.68	4.467	0.406	0.547	1.775	2.28	
−30	679.34	4.476	0.387	0.549	1.801	2.15	
−20	666.69	4.509	0.381	0.547	1.819	2.09	
−10	653.55	4.564	0.378	0.543	1.825	2.07	
0	640.10	4.635	0.373	0.540	1.819	2.05	
10	626.16	4.714	0.368	0.531	1.801	2.04	
20	611.75	4.798	0.359	0.521	1.775	2.02	2.45×10^{-3}
30	596.37	4.890	0.349	0.507	1.742	2.01	
40	580.99	4.999	0.340	0.493	1.701	2.00	
50	564.33	5.116	0.330	0.476	1.654	1.99	
			Carbon Dioxide, CO$_2$				
−50	1156.34	1.84	0.119×10^{-6}	0.0855	0.4021×10^{-7}	2.96	
−40	1117.77	1.88	0.118	0.1011	0.4810	2.46	
−30	1076.76	1.97	0.117	0.1116	0.5272	2.22	
−20	1032.39	2.05	0.115	0.1151	0.5445	2.12	
−10	983.38	2.18	0.113	0.1099	0.5133	2.20	
0	926.99	2.47	0.108	0.1045	0.4578	2.38	
10	860.03	3.14	0.101	0.0971	0.3608	2.80	
20	772.57	5.0	0.091	0.0872	0.2219	4.10	14.00×10^{-3}
30	597.81	36.4	0.080	0.0703	0.0279	28.7	

(*Continued*)

Table A.7 Continued

Temperature T, °C	Density ρ, kg/m³	Heat Capacity c_p, kJ/kg·°C	Kinematic Viscosity ν, m²/s	Thermal Conductivity k, W/m·°C	Diffusivity α, m²/s	Prandtl Number Pr	Thermal Expansion Coefficient β, K⁻¹
				Sulfur Dioxide, SO_2			
−50	1560.84	1.3595	0.484×10^{-6}	0.242	1.141×10^{-7}	4.24	
−40	1536.81	1.3607	0.424	0.235	1.130	3.74	
−30	1520.64	1.3616	0.371	0.230	1.117	3.31	
−20	1488.60	1.3624	0.324	0.225	1.107	2.93	
−10	1463.61	1.3628	0.288	0.218	1.097	2.62	
0	1438.46	1.3636	0.257	0.211	1.081	2.38	
10	1412.51	1.3645	0.232	0.204	1.066	2.18	
20	1386.40	1.3653	0.210	0.199	1.050	2.00	1.95×10^{-3}
30	1359.33	1.3662	0.190	0.192	1.035	1.83	
40	1329.22	1.3674	0.173	0.185	1.019	1.70	
50	1299.10	1.3683	0.162	0.177	0.999	1.61	
			Dichlorodifluoromethane (Freon), CCl_2F_2				
−50	1546.75	0.8750	0.310×10^{-6}	0.067	0.501×10^{-7}	6.2	
−40	1518.71	0.8847	0.279	0.069	0.514	5.4	
−30	1489.56	0.8956	0.253	0.069	0.526	4.8	
−20	1460.57	0.9073	0.235	0.071	0.539	4.4	
−10	1429.49	0.9203	0.221	0.073	0.550	4.0	
0	1397.45	0.9345	0.214×10^{-6}	0.073	0.557×10^{-7}	3.8	2.63×10^{-3}
10	1364.30	0.9496	0.203	0.073	0.560	3.6	
20	1330.18	0.9659	0.198	0.073	0.560	3.5	
30	1295.10	0.9835	0.194	0.071	0.560	3.5	
40	1257.13	1.0019	0.191	0.069	0.555	3.5	
50	1215.96	1.0216	0.190	0.067	0.545	3.5	

Glycerin, $C_3H_5(OH)_3$

Temp.							
0	1276.03	2.261	0.00831	0.282	0.983×10^{-7}	84.7×10^{3}	
10	1270.11	2.319	0.00300	0.284	0.965	31.0	
20	1264.02	2.386	0.00118	0.286	0.947	12.5	0.50×10^{-3}
30	1258.09	2.445	0.00050	0.286	0.929	5.38	
40	1252.01	2.512	0.00022	0.286	0.914	2.45	
50	1244.96	2.583	0.00015	0.287	0.893	1.63	

Ethylene Glycol, $C_2H_4(OH)_2$

Temp.							
0	1130.75	2.294	57.53×10^{-6}	0.242	0.934×10^{-7}	615	
20	1116.65	2.382	19.18	0.249	0.939	204	0.65×10^{-3}
40	1101.43	2.474	8.69	0.256	0.939	93	
60	1087.66	2.562	4.75	0.260	0.932	51	
80	1077.56	2.650	2.98	0.261	0.921	32.4	
100	1058.50	2.742	2.03	0.263	0.908	22.4	

Engine Oil (unused)

Temp.							
0	899.12	1.796	0.00428	0.147	0.911×10^{-7}	47,100	
20	888.23	1.880	0.00090	0.145	0.872	10,400	
40	876.05	1.964	0.00024	0.144	0.834	2870	
60	864.04	2.047	0.839×10^{-4}	0.140	0.800	1050	0.70×10^{-3}
80	852.02	2.131	0.375	0.138	0.769	490	
100	840.01	2.219	0.203	0.137	0.738	276	
120	828.96	2.307	0.124	0.135	0.710	175	
140	816.94	2.395	0.080	0.133	0.686	116	
160	805.89	2.483	0.056	0.132	0.663	84	

(*Continued*)

Table A.7 *Continued*

Temperature T, °C	Density ρ, kg/m³	Heat Capacity c_p, kJ/ kg·°C	Kinematic Viscosity ν, m²/s	Thermal Conductivity k, W/m·°C	Diffusivity α, m²/s	Prandtl Number Pr	Thermal Expansion Coefficient β, K⁻¹
				Mercury, Hg			
0	13,628.22	0.1403	0.124×10^{-6}	8.20	42.99×10^7	0.0288	1.82×10^{-4}
20	13,759.04	0.1394	0.114	8.69	46.06	0.0249	
50	13,505.84	0.1386	0.104	9.40	50.22	0.0207	
100	13,384.58	0.1373	0.0928	10.51	57.16	0.0162	
150	13,264.68	0.1365	0.0853	11.49	63.54	0.0134	
200	13,144.94	0.1570	0.0802	12.34	69.08	0.0116	
250	13,025.60	0.1357	0.0765	13.07	74.06	0.0103	
315.5	12,857	0.134	0.0673	81.5	0.0083		

Table A.8 Properties of gases at atmospheric pressure[a]

Temperature T, K	Density ρ, kg/m^3	Heat Capacity c_p, kJ/kg·°C	Dynamic Viscosity μ, kg/m·s	Kinematic Viscosity v, m^2/s	Thermal Conductivity k, W/m·°C	Diffusivity α, m^2/s	Prandtl Number Pr
			Helium				
144	0.3379	5.200	125.5×10^{-7}	37.11×10^{-6}	0.0928	0.5275×10^{-4}	0.70
200	0.2435	5.200	156.6	64.38	0.1177	0.9288	0.694
255	0.1906	5.200	181.7	95.50	0.1357	0.1375	0.70
366	0.13280	5.200	230.5	173.6	0.1691	2.449	0.71
477	0.10204	5.200	275.0	269.3	0.197	3.716	0.72
589	0.08282	5.200	311.3	375.8	0.225	5.215	0.72
700	0.07032	5.200	347.5	494.2	0.251	6.661	0.72
800	0.06023	5.200	381.7	634.1	0.275	8.774	0.72
			Hydrogen				
150	0.16371	12.602	5.595×10^{-6}	34.18×10^{-6}	0.0981	0.475×10^{-4}	0.718
200	0.12270	13.540	6.813	55.53	0.1282	0.772	0.719
250	0.09819	14.059	7.919	80.64	0.1561	1.130	0.713
300	0.08185	14.314	8.963	109.5	0.182	1.554	0.706
350	0.07016	14.436	9.954	141.9	0.206	2.031	0.697
400	0.06135	14.491	10.864	177.1	0.228	2.568	0.690
450	0.05462	14.499	11.779	215.6	0.251	3.164	0.682
500	0.04918	14.507	12.636	257.0	0.272	3.817	0.675
550	0.04469	14.532	13.475	301.6	0.292	4.516	0.668
600	0.04085	14.537	14.285	349.7	0.315	5.306	0.664
700	0.03492	14.574	15.89	455.1	0.351	6.903	0.659
800	0.03060	14.675	17.40	569	0.384	8.563	0.664
900	0.02723	14.821	18.78	690	0.412	10.217	0.676

(*Continued*)

Table A.8 *Continued*

Temperature T, K	Density ρ, kg/m^3	Heat Capacity c_p, kJ/kg·°C	Dynamic Viscosity μ, kg/m·s	Kinematic Viscosity ν, m^2/s	Thermal Conductivity k, W/m·°C	Diffusivity α, m^2/s	Prandtl Number Pr
Oxygen							
150	2.6190	0.9178	11.490×10^{-6}	4.387×10^{-6}	0.01367	0.05688×10^{-4}	0.773
200	1.9559	0.9131	14.850	7.593	0.01824	0.10214	0.745
250	1.5618	0.9157	17.87	11.45	0.02259	0.15794	0.725
300	1.3007	0.9203	20.63	15.86	0.02676	0.22353	0.709
350	1.1133	0.9291	23.16	20.80	0.03070	0.2968	0.702
400	0.9755	0.9420	25.54	26.18	0.03461	0.3768	0.695
450	0.8682	0.9567	27.7	31.99	0.03828	0.4609	0.694
500	0.7801	0.9722	29.91	38.34	0.4173	0.5502	0.697
550	0.7096	0.9881	31.97	45.05	0.04517	0.641	0.700
Nitrogen							
200	1.7108	1.0429	12.947×10^{-6}	7.568×10^{-6}	0.01824	0.10224×10^{-4}	0.747
300	1.1421	1.0408	17.84	15.63	0.02620	0.22044	0.713
400	0.8538	1.0459	21.98	25.74	0.03335	0.3734	0.691
500	0.6824	1.0555	25.70	37.66	0.03984	0.5530	0.684
600	0.5687	1.0756	29.11	51.19	0.04580	0.7486	0.686
700	0.4934	1.0969	32.13	65.13	0.05123	0.9466	0.691
800	0.4277	1.1225	34.84	81.46	0.05609	1.1685	0.700
900	0.3796	1.1464	37.49	91.06	0.06070	1.3946	0.711
1000	0.3412	1.1677	40.00	117.2	0.06475	1.6250	0.724
1100	0.3108	1.1857	42.28	136.0	0.06850	1.8591	0.736
1200	0.2851	1.2037	44.50	156.1	0.07184	2.0932	0.748
Carbon Dioxide							
220	2.4733	0.783	11.105×10^{-6}	4.490×10^{-6}	0.010805	0.05920×10^{-5}	0.818
250	2.1657	0.804	12.590	5.813	0.012884	0.07401	0.793

T	ρ	c_p	μ	ν	k	α	Pr
300	1.7973	0.871	14.958	8.321	0.016572	0.10588	0.770
350	1.5362	0.900	17.205	11.19	0.02047	0.14808	0.755
400	1.3424	0.942	19.32	14.39	0.02461	0.19463	0.738
500	1.0732	1.013	23.26	21.67	0.03352	0.3084	0.702
550	0.9739	1.047	25.08	25.74	0.03821	0.3750	0.685
600	0.8938	1.076	26.83	30.02	0.04311	0.4483	0.668

Ammonia

T	ρ	c_p	μ	ν	k	α	Pr
273	0.7929	2.177	9.353×10^{-6}	1.18×10^{-5}	0.0220	0.1308×10^{-4}	0.90
323	0.6487	2.177	11.035	1.70	0.0270	0.1920	0.88
373	0.5590	2.236	12.886	2.30	0.0327	0.2619	0.87
423	0.4934	2.315	14.672	2.87	0.0391	0.3432	0.87
473	0.4405	2.395	16.49	3.74	0.0467	0.4421	0.84

Water Vapor

T	ρ	c_p	μ	ν	k	α	Pr
380	0.5863	2.060	12.71×10^{-6}	2.16×10^{-5}	0.0246	0.2036×10^{-4}	1.060
400	0.5542	2.014	13.44	2.42	0.0261	0.2338	1.040
450	0.4902	1.980	15.25	3.11	0.0299	0.307	1.010
500	0.4405	1.985	17.04	3.86	0.0339	0.387	0.996
550	0.4005	1.997	18.84	4.70	0.0379	0.475	0.991
600	0.3652	2.026	20.67	5.66	0.0422	0.573	0.986
650	0.3380	2.056	22.47	6.64	0.0464	0.666	0.995
700	0.3140	2.085	24.26	7.72	0.0505	0.772	1.000
750	0.2931	2.119	26.04	8.88	0.0549	0.883	1.005
800	0.2739	2.152	27.86	10.20	0.0592	1.001	1.010
850	0.2579	2.186	29.69	11.52	0.0637	1.130	1.019

^{a}Values of dynamic viscosity μ, thermal conductivity k, specific heat c_p, and Prandtl number Pr, are not strongly pressure-dependent for He, H_2, O_2, and N_2 and may be used over a fairly wide range of pressures.

Table A.9 Properties of air at 1 atm

Temperature, °C	Density, ρ, kg/m³	Dynamic Viscosity, μ, kg/m·s ($\times 10^5$)	Kinematic Viscosity, ν, m²/s ($\times 10^5$)	Capacity Heat, c_p, J/kg·K	Thermal Conductivity, k, W/m·K ($\times 10^2$)	Thermal Expansion Coefficient, β, K (10^3)	Prandtl Number, Pr
−40	1.52	1.51	0.98		2.0		0.715
−20	1.40	1.61	1.15	1004.8	2.21		0.713
0	1.29	1.71	1.32	1004.8	2.42	3.65	
10	1.248	1.76	1.41	1004.8	2.49	3.53	
20	1.205	1.81	1.50	1004.8			
30	1.165	1.86	1.60	1004.8			
40	1.128	1.90	1.68	1004.8	2.7		
50	1.09	1.95	1.79	1007.0	2.8		
60	1.060	2.00	1.87	1009.0			
80	1.000	2.09	2.09	1009.0			
100	0.946	2.17	2.30	1009.0	3.12		
150	0.835	2.38	2.85	1017.0	3.53		
200	0.746	2.57	3.45	1025.8	3.88		0.686
250	0.675	2.75	4.07	1034.1	4.24		0.680
300	0.616	2.93	4.76				
400	0.525	3.25	6.19				
500	0.457	3.55	7.77		5.73		0.709

Example: At 50°C, the air properties are: density = 1.09 kg/m³ (0.00211 slug/ft³ = 0.679 lb/ft³), dynamic viscosity = 0.0000195 kg/m · s (4.073 × 10⁻⁷ slug/ft · s = 1.31 × 10⁻⁵ lb/ft · s), thermal conductivity, k = 0.028 W/m · K, coefficient of thermal expansion, $\beta = 1/T = 1/(273 + 50) = 0.0031$ K⁻¹. The Prandtl number, Pr = $c_p \mu / k = 0.7$.

Table A.10   Properties of water (saturated liquid)

Temp. °F	Temp. °C	Heat Capacity c_p, kJ/ kg·K	Density ρ, kg/m³	Viscosity μ, kg/m·s	Thermal Conductivity k, W/m·°C	Prandtl Number Pr	$\dfrac{g\beta\rho^2 c_p}{\mu k}$ 1/m³·°C
32	0	4.225	999.8	1.79×10^{-3}	0.566	13.25	1.91×10^{9}
40	4.44	4.208	999.8	1.55×10^{-3}	0.575	11.35	1.91×10^{9}
50	10	4.195	999.2	1.31×10^{-3}	0.585	9.40	6.34×10^{9}
60	15.56	4.186	998.6	1.12×10^{-3}	0.595	7.88	1.08×10^{10}
70	21.11	4.179	997.4	9.8×10^{-4}	0.604	6.78	1.46×10^{10}
80	26.67	4.179	995.8	8.6×10^{-4}	0.614	5.85	1.91×10^{10}
90	32.22	4.174	994.9	7.65×10^{-4}	0.623	5.12	2.48×10^{10}
100	37.78	4.174	993.0	6.82×10^{-4}	0.630	4.53	3.3×10^{10}
110	43.33	4.174	990.6	6.16×10^{-4}	0.637	4.04	4.19×10^{10}
120	48.89	4.174	988.8	5.62×10^{-4}	0.644	3.64	4.89×10^{10}
130	54.44	4.179	985.7	5.13×10^{-4}	0.649	3.30	5.66×10^{10}
140	60	4.179	983.3	4.71×10^{-4}	0.654	3.01	6.48×10^{10}
150	65.55	4.183	980.3	4.3×10^{-4}	0.659	2.73	7.62×10^{10}
160	71.11	4.186	977.3	4.01×10^{-4}	0.665	2.53	8.84×10^{10}
170	76.67	4.191	973.7	3.72×10^{-4}	0.668	2.33	9.85×10^{10}
180	82.22	4.195	970.2	3.47×10^{-4}	0.673	2.16	1.09×10^{11}
190	87.78	4.199	966.7	3.27×10^{-4}	0.675	2.03	
200	93.33	4.204	963.2	3.06×10^{-4}	0.678	1.90	
220	104.4	4.216	955.1	2.67×10^{-4}	0.684	1.66	
240	115.6	4.229	946.7	2.44×10^{-4}	0.685	1.51	
260	126.7	4.250	937.2	2.19×10^{-4}	0.685	1.36	
280	137.8	4.271	928.1	1.98×10^{-4}	0.685	1.24	
300	148.9	4.296	918.0	1.86×10^{-4}	0.684	1.17	
350	176.7	4.371	890.4	1.57×10^{-4}	0.677	1.02	
400	204.4	4.467	859.4	1.36×10^{-4}	0.655	1.00	
450	232.2	4.585	825.7	1.20×10^{-4}	0.646	0.85	
500	260	4.731	785.2	1.07×10^{-4}	0.616	0.83	
550	287.7	5.024	735.5	9.51×10^{-5}			
600	315.6	5.703	678.7	8.68×10^{-5}			

Note: $Gr_x Pr = $ Rayleigh number, $Ra_x = \left(\dfrac{g\beta\rho^2 c_p}{\mu k} \right) L^3 \Delta T.$

B. FIGURES

No	Liquid	X	Y
1	Acetaldehyde	15.2	4.8
2	Acetic acid, 100%	12.1	14.2
3	Acetic acid, 70%	9.5	17.0
4	Acetic anhydride	12.7	12.8
5	Acetone, 100%	14.5	7.2
6	Acetone, 35%	7.9	15.0
7	Allyl alcohol	12.6	14.3
8	Ammonia, 100%	10.1	2.0
9	Ammonia, 26%	11.8	13.9
10	Amyl acetate	7.5	12.5
11	Amyl alcohol	8.1	18.4
12	Aniline	12.3	18.7
13	Anisole	12.5	13.5
14	Arsenic trichloride	13.9	14.5
15	Benzene	12.5	10.9
16	Bimethyl oxalate	12.3	15.8
17	Biphenyl	12.0	18.3
18	Brine, CaCl$_2$, 25%	6.6	15.9
19	Brine, NaCl, 25%	10.2	16.6
20	Bromine	14.2	13.2
21	Bromotoluene	20.0	15.9
22	Butyl acetate	12.3	11.0
23	Butyl alcohol	8.6	17.2
24	Butyric acid	12.1	15.3
25	Carbon dioxide	11.6	0.3
26	Carbon disulfide	16.1	7.5
27	Carbon tetrachloride	12.3	13.1
28	Chlorobenzene	12.3	12.4
29	Chloroform	14.4	10.2
30	Chlorosulfonic acid	11.2	18.1
31	o-Chlorotoluene	13.0	13.3
32	m-Chlorotoluene	13.3	12.5
33	p-Chlorotoluene	13.3	12.5
34	m-Cresol	2.5	20.8
35	Cyclohexanol	2.9	24.3
36	Dibromoethane	12.7	15.8
37	Dichloroethane	13.2	12.2
38	Dichloromethane	14.6	8.9
39	Diethyl oxalate	11.0	16.4
40	Dipropyl oxalate	10.3	17.7
41	Ethyl acetate	13.7	9.1
42	Ethyl alcohol, 100%	10.5	13.8
43	Ethyl alcohol, 95%	9.8	14.3
44	Ethyl alcohol, 40%	6.5	16.6
45	Ethyl benzene	13.2	11.5
46	Ethyl bromide	14.5	8.1
47	Ethyl chloride	14.8	6.0
48	Ethyl ether	14.5	5.3
49	Ethyl formate	14.2	8.4
50	Ethyl iodide	14.7	10.3
51	Ethyl glycol	6.0	23.6
52	Formic acid	10.7	15.8
53	Freon-11	14.4	9.0
54	Freon-1	16.8	5.6
55	Freon-21	15.7	7.5

No	Liquid	X	Y
56	Freon-22	17.2	4.7
57	Freon-113	12.5	11.4
58	Glycerol, 100%	2.0	30.0
59	Glycerol, 50%	6.9	19.6
60	Heptane	14.1	8.4
61	Hexane	14.7	7.0
62	Hydrochloric acid, 31.5%	13.0	16.6
63	Isobutyl alcohol	7.1	18.0
64	Isobutyric acid	12.2	14.4
65	Isopropyl alcohol	8.2	16.0
66	Kerosene	10.2	16.9
67	Linseed oil, raw	7.5	27.2
68	Mercury	18.4	16.4
69	Methanol, 100%	12.4	10.5
70	Methanol, 90%	12.3	11.8
71	Methanol, 40%	7.8	15.5
72	Methyl acetate	14.2	8.2
73	Methyl chloride	15.0	3.8
74	Methyl ethyl ketone	13.9	8.6
75	Naphtalene	7.9	18.1
76	Nitric acid, 95%	12.8	13.8
77	Nitric acid, 60%	10.8	17.0
78	Nitrobenzene	10.6	16.2
79	Nitrotoluene	11.0	17.0
80	Octane	13.7	10.0
81	Octyl alcohol	6.6	21.1
82	Pentachloroethane	10.9	17.3
83	Pentane	14.9	5.2
84	Phenol	6.9	20.8
85	Phosphorus tribromide	13.8	16.7
86	Phosphorus trichloride	16.2	10.9
87	Propionic acid	12.8	13.8
88	Propyl alcohol	9.1	16.5
89	Propyl bromide	14.5	9.6
90	Propyl chloride	14.4	7.5
91	Propyl iodide	14.1	11.6
92	Sodium	16.4	13.9
93	Sodium hydroxide, 50%	3.2	25.8
94	Stannic chloride	13.5	12.8
95	Sulfur dioxide	15.2	7.1
96	Sulfuric acid, 110%	7.0	27.4
97	Sulfuric acid, 98%	7.0	24.8
98	Sulfuric acid, 60%	10.2	21.3
99	Sulfuryl chloride	15.2	12.4
100	Tetrachloroethane	11.9	15.7
101	Tetrachloroethylene	14.2	12.7
102	Titanium tetrachloride	14.4	12.3
103	Toluene	13.7	10.4
104	Trichloroethylene	14.8	10.5
105	Turpentine	11.5	14.9
106	Vinyl acetate	14.0	8.8
107	Water	10.2	13.0
108	o-Xylene	13.5	12.1
109	m-Xylene	13.9	10.6
110	p-Xylene	13.9	10.9

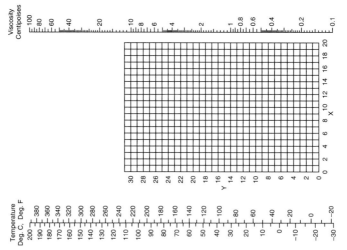

Figure B.1 Absolute viscosity of liquids.

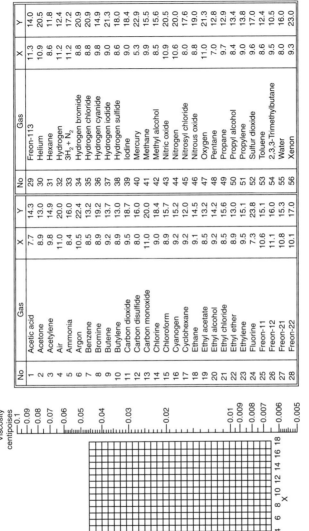

No	Gas	X	Y	No	Gas	X	Y
1	Acetic acid	7.7	14.3	29	Freon-113	11.3	14.0
2	Acetone	8.9	13.0	30	Helium	10.9	20.5
3	Acetylene	9.8	14.9	31	Hexane	8.6	11.8
4	Air	11.0	20.0	32	Hydrogen	11.2	12.4
5	Ammonia	8.4	16.0	33	$3H_2 + N_2$	11.2	17.2
6	Argon	10.5	22.4	34	Hydrogen bromide	8.8	20.9
7	Benzene	8.5	13.2	35	Hydrogen chloride	8.8	20.9
8	Bromine	8.9	19.2	36	Hydrogen cyanide	9.8	14.9
9	Butene	9.2	13.7	37	Hydrogen iodide	9.0	21.3
10	Butylene	8.9	13.0	38	Hydrogen sulfide	8.6	18.0
11	Carbon dioxide	9.5	18.7	39	Iodine	9.0	18.4
12	Carbon disulfide	8.0	16.0	40	Mercury	5.3	22.9
13	Carbon monoxide	11.0	20.0	41	Methane	9.9	15.5
14	Chlorine	9.0	18.4	42	Methyl alcohol	8.5	15.6
15	Chloroform	8.9	15.7	43	Nitric oxide	10.9	20.5
16	Cyanogen	9.2	15.2	44	Nitrogen	10.6	20.0
17	Cyclohexane	9.2	12.0	45	Nitrosyl chloride	8.0	17.6
18	Ethane	9.1	14.5	46	Nitrous oxide	8.8	19.0
19	Ethyl acetate	8.5	13.2	47	Oxygen	11.0	21.3
20	Ethyl alcohol	9.2	14.2	48	Pentane	7.0	12.8
21	Ethyl chloride	8.5	15.6	49	Propane	9.7	12.9
22	Ethyl ether	8.9	13.0	50	Propyl alcohol	8.4	13.4
23	Ethylene	9.5	15.1	51	Propylene	9.0	13.8
24	Fluorine	7.3	23.8	52	Sulfur dioxide	9.6	17.0
25	Freon-11	10.6	15.1	53	Toluene	8.6	12.4
26	Freon-12	11.1	16.0	54	2,3,3-Trimethylbutane	9.5	10.5
27	Freon-21	10.8	15.3	55	Water	8.0	16.0
28	Freon-22	10.1	17.0	56	Xenon	9.3	23.0

Figure B.2 Absolute viscosity of gases and vapors at 1 atm.

INDEX

Fluid Flow for the Practicing Chemical Engineer. By J. Patrick Abulencia and Louis Theodore
Copyright © 2009 John Wiley & Sons, Inc.

573